ADVANCES IN
Applied Microbiology
VOLUME 34

ADVANCES IN

Applied Microbiology

Edited by SAUL L. NEIDLEMAN

Emeryville, California

VOLUME 34

Academic Press, Inc.
Harcourt Brace Jovanovich, Publishers
San Diego   New York   Berkeley   Boston
London   Sydney   Tokyo   Toronto

COPYRIGHT © 1989 BY ACADEMIC PRESS, INC.
ALL RIGHTS RESERVED.
NO PART OF THIS PUBLICATION MAY BE REPRODUCED OR
TRANSMITTED IN ANY FORM OR BY ANY MEANS, ELECTRONIC
OR MECHANICAL, INCLUDING PHOTOCOPY, RECORDING, OR
ANY INFORMATION STORAGE AND RETRIEVAL SYSTEM, WITHOUT
PERMISSION IN WRITING FROM THE PUBLISHER.

ACADEMIC PRESS, INC.
San Diego, California 92101

United Kingdom Edition published by
ACADEMIC PRESS LIMITED
24-28 Oval Road, London NW1 7DX

LIBRARY OF CONGRESS CATALOG CARD NUMBER: 59-13823

ISBN   0-12-002634-1   (alk. paper)

PRINTED IN THE UNITED STATES OF AMERICA
89 90 91 92   9 8 7 6 5 4 3 2 1

CONTENTS

PREFACE vii

What's in a Name?—Microbial Secondary Metabolism

J. W. BENNETT AND RONALD BENTLEY

I.	Historical Overview	1
II.	Primary and Secondary Metabolites	4
III.	How Shall a Metabolite Be Named?	11
IV.	Nomenclature	16
V.	Epilogue	23
	References	24

Microbial Production of Gibberellins: State of the Art

P. K. R. KUMAR AND B. K. LONSANE

I.	Introduction	30
II.	Historical Highlights	31
III.	Chemistry of Gibberellins	34
IV.	Mode of Action	39
V.	Commercial and Potential Uses	45
VI.	Routes for Production	45
VII.	Microorganisms Producing Gibberellins	49
VIII.	Biosynthesis Pathways	51
IX.	Liquid Surface Fermentation	57
X.	Submerged Fermentation	59
XI.	Use of Immobilized Whole Cells	98
XII.	Solid-State Fermentation	100
XIII.	Analytical Methods	111
XIV.	Economic Considerations	116
XV.	Epilogue	121
	References	123

Microbial Dehydrogenations of Monosaccharides

MILOŠ KULHÁNEK

I.	Introduction	141
II.	Dehydrogenases and Mechanisms of Dehydrogenations	142

III.	Special Nature of Microbial Dehydrogenations of Monosaccharides	147
IV.	Relationships between Structure and Dehydrogenation of Monosaccharides	150
V.	Microorganisms and Fermentation Technique	155
VI.	Preparative and Industrial Applications	160
	References	174

Antitumor and Antiviral Substances from Fungi

Shung-Chang Jong and Richard Donovick

I.	Introduction	183
II.	Historical Background	184
III.	Screening of Fungi	185
IV.	Antibiotics Produced by Fungi	186
V.	Antitumor–Antiviral Substances Produced by Fungi	202
	References	244

Biotechnology—The Golden Age

V. S. Malik

I.	Introduction	263
II.	The Industrial Organism	264
III.	The Technology	269
IV.	Microbial Degradation of Toxic Pollutants	275
V.	Enzymes: The Catalysts of the Future	276
VI.	The Energy	277
VII.	Engineering Tomorrow's Antibiotics	280
VIII.	Crop Improvement	281
IX.	Human Proteins of Therapeutic Value	289
X.	Vaccines for the Future	292
XI.	Hybridomas, Monoclonal Antibodies, and Diagnostic Kits	294
XII.	Inherited Diseases	296
XIII.	Embryo Transfer and Animal Husbandry	297
XIV.	Future Prospects	300
	References	301

Index 307

PREFACE

The badge of office conferred upon the editor of *Advances in Applied Microbiology* has passed from Dr. Allen Laskin to me. The change is not a result of an untimely death, treachery, or any malignant circumstance; it is a simple matter of time, change, and other pressures which are making new demands on Dr. Laskin, while leaving me eager to undertake the new challenge. In all candor, I enjoy writing more than editing, but the task before me is suitably compelling. During his 8 years at the helm, Dr. Laskin has set a high standard for this series for which we thank him.

I intend to make succeeding volumes of this series as cosmopolitan as possible—applied microbiology knows few boundaries. We can write about drugs and soy sauce, agriculture and malaria, athlete's foot and evolution. Any of these topics may find their way into future volumes.

I encourage you, the reader of this publication, to help me fill future volumes with your writing, your science, and your thoughts. While I shall seek out specific authors for specific papers, it would be equally rewarding to receive for consideration proposals for manuscripts from the readers of these volumes. My thanks to all contributors past and present who have helped shape *Advances* and to those future contributors whose work will carry it forward.

SAUL L. NEIDLEMAN

What's in a Name?—Microbial Secondary Metabolism

J. W. Bennett* and Ronald Bentley†

*Department of Biology
Tulane University
New Orleans, Louisiana 70118

†Department of Biological Sciences
University of Pittsburgh
Pittsburgh, Pennsylvania 15260

I. Historical Overview
   A. Introduction
   B. The Origins
II. Primary and Secondary Metabolites
   A. Microbial Metabolites
   B. More Jargon
   C. The Wrong Term in the Right Place at the Right Time?
III. How Shall a Metabolite Be Named?
   A. Overproduction
   B. Growth Phase
   C. Function
IV. Nomenclature
   A. Alternatives
   B. Some Semantics
   C. Semantics and Secondary Metabolism
   D. The Definitions
V. Epilogue
   References

I. Historical Overview

A. Introduction

Of the prodigious number of carbon compounds, many are materials produced by living organisms. They are termed "natural products" and, significantly, are responsible for the naming of the discipline of *organic* chemistry. Natural products constitute a very large and diverse group of compounds that are hard to characterize. One subgroup of the whole, with which this chapter is concerned, comprises what modern biologists call "secondary metabolites." We address the following questions: What are secondary metabolites? Where did the term "secondary metabolite" originate? How is "secondary metabolite" used today in the biological sciences? Moreover, there is a final and highly provocative question: When a metabolite is called "secondary," does the act of so naming it influence our thinking about it?

B. THE ORIGINS

Prior to the origins of modern science, materials now termed "secondary metabolites" (or containing such compounds) were known indirectly. Dyes (e.g., indigo), perfumes (e.g., myrrh), spices (e.g., cinnamon), medicinals (e.g., digitalis), poisons (e.g., hemlock), pharmacological agents (e.g., opium), and cosmetics (e.g., henna), were widely exploited. Indeed, modern drug companies still use ethnobotanical studies to discover new bioactive agents.

Most of the materials that have been used for centuries were obtained from plants. However, certain macrofungi were known for their toxic, hallucinogenic, or putative aphrodisiacal effects, and a few microbial products have also left a record. For example, many strains of the bacterium *Serratia marcescens* produce the red pigment, prodigiosin. This common bacterium frequently contaminated human food; through the ages, such growth was mistaken for fresh blood, causing fear in the superstitious, the proclamation of holy miracles by the Church, and the precipitation of anti-Semitic riots during the Middle Ages (Gaughran, 1969). Another historical example was provided by the filamentous fungus *Claviceps purpurea*. In parasitizing rye, mycelial infiltration of the grains produced a macroscopic sclerotium that contained ergot alkaloids. Ingestion of this "ergot of rye" caused "St. Anthony's fire," a terrible gangrenous disease with associated convulsions and hallucinations, documented in art and history from recurring epidemics throughout Medieval Europe. The last major outbreak of St. Anthony's fire occurred in France as recently as 1951 (Fuller, 1968).

Modern organic chemistry began with the isolation and characterization of certain compounds from human body fluids and from plant and animal tissues. Between 1769 and 1787, the Swedish scientist Carl Wilhelm Scheele characterized, *inter alia*, citric acid from lemons, malic acid from apples, and gallic acid from a mold fermentation of gall "nuts." Pharmacologically active botanicals also attracted early chemists. Thus, morphine was first isolated in 1817. Within a few years, other alkaloids, such as strychnine and quinine, as well as the terpenes borneol and camphor were characterized (Geissman and Crout, 1969). Liebig's classic text, *Handbuch der Organische Chemie* (1843), listed approximately 2000 substances derived from natural sources (Haslam, 1985).

In the second half of the nineteenth century, the number of known metabolites increased almost exponentially. To organic chemists, these strange metabolites were simply "natural products" providing never-ending challenges of structure determination and total synthesis. For biologists, the various alkaloids, resins, and other unique botanical products were more enigmatic. The Darwinian paradigm required that organismal attributes impart adaptive value. Since most of these strange natural products served no discernible role in active metabolism, the early plant physiologists categorized them as waste or storage products (Sachs, 1882). The first seriously scientific use of the "primary"/"secondary" classification for metabolites in modern times is apparently found in the following sentence from a lecture by Albrecht Kossel (1891): "Ich schlage vor, diese wesentlichen Bestandtheile der Zelle als primäre zu bezeichnen, hingegen diejenigen, welche nicht in jeder entwicklungsfähigen Zelle gefunden werden als secundäre" ("I suggest to call the essential components of the cell as primary ones, however those which cannot be found in every growing cell we shall call secondary ones").

The genesis of such usage may ultimately date to the Greek "protobiochemists." Some 2000 years ago, Galen expressed the belief that the organism was a combination of primary and secondary qualities (Florkin, 1972). Be this as it may, Kossel's distinction had no immediate influence and was not found in the encyclopedic *Physiology of Plants* (Pfeffer, 1900). Pfeffer, in fact, classified plant metabolites as (1) building or formative substances, (2) plastic or trophic substances stored as reserve foods before being drawn into metabolism, or (3) aplastic or atrophic substances taking no further part in metabolism.

The term "secondary" for plant products such as alkaloids was revived by Czapek (1925), who wrote: " . . . dass es sich bei der Bildung solcher Stoffe um Prozesse handelt, die nicht jedem Zellplasma eigen sind, sondern mehr sekundären Charakter haben." This sentence has been translated as "the production of such substances (i.e., alkaloids) involves processes not inherent to every cell plasma, but more of a secondary character" (Mothes, 1980) and could be rendered more colloquially as "their synthesis follows those processes which are not primary for every cell, but rather have more secondary characteristics." A book chapter entitled "Sekundäre Pflanzenstoffe" included materials such as fats and phosphatides as well as alkaloids and terpenes (Kostyschew, 1926). A somewhat later German text, *Biochemie und Physiologie der Sekundären Pflanzenstoffe* (Paech, 1950), contributed to the broad application of the term "secondary." Not long after the appearance of this work, the influential American text, *Principles of Plant Physiology* (Bonner and Galston, 1952), spoke of the "highways and byways in metabolism," with the clear implication that secondary metabolism constituted a byway.

II. Primary and Secondary Metabolites

A. MICROBIAL METABOLITES

It would be fascinating to trace the influence of fungal metabolites on the development of organic chemistry. There exists a long history in the manufacture of gallic acid by using fungi to hydrolyze plant tannins; the first citation of tannin fermentation in a topical bibliography is to a Scheele paper dated 1787 (Thom and Raper, 1945). The carbohydrate mannitol was isolated from higher fungi as early as 1813 (Birkinshaw *et al.*, 1931) and oxalate crystals were recognized in various fungi even before this acid was described as a metabolite of *Aspergillus niger* by Wehmer in 1891 (Foster, 1949). Wehmer also demonstrated citric acid formation by fungi 2 years later, and a plant for the commercial manufacture of this metabolite by fungi was actually operated at Thann (Alsace) from 1893 to 1903 (Miall, 1975). More complex fungal metabolites, such as kojic acid, mycophenolic acid, and penicillic acid, were discovered early in the twentieth century.

Systematic studies of fungal metabolites, which were obtained by growing pure cultures on chemically defined media, were begun in 1922 by Harold Raistrick and colleagues in the seemingly unlikely environment of the Ardeer factory of Nobel's Explosives Co. Ltd. at Stevenston in Scotland (now Imperial Chemical Industries, PCL). They were continued after 1929 at the London School of Hygiene and Tropical Medicine. Some 200 mold metabolites were isolated in a state of purity (Raistrick, 1949) and more than 120 publications reported this research, the last in 1965 (Divekar *et al.*, 1965).

In several cases, the structural study of such natural products forced chemists to expand their conceptual horizons. One example is Raistrick's discovery of the *Penicillium stipitatum* metabolite, $C_8H_6O_5$, named stipitatic acid (Birkinshaw *et al.*, 1942) and the similar materials, puberulic and puberulonic acids, from other Penicillia (Birkinshaw and Raistrick, 1932). Although Raistrick could not assign structures to these materials, Dewar (1945) suggested that they contained the previously unknown, seven-membered, aromatic ring structure named tropolone. The subsequent developments concerning these and related compounds have been prodigious. In the same way, the discovery of the four-membered β-lactam ring in penicillin presented organic chemists with provocative and formidable challenges. The biosynthetic virtuosity of these humble fungi induced one of Raistrick's colleagues to quip that "if ever there was a field in which a pentavalent carbon atom could turn up, it would be mould metabolism" (Raphael, 1948).

Classical microbiologists were somewhat slower to appreciate the importance of mold metabolism. The late nineteenth century and early twentieth century constituted a golden age for medical bacteriology. During these years, fungi were, in the main, viewed as nothing more than contaminants of pure cultures of medically important bacteria. Then came the discovery of penicillin.

At about the time that Raistrick moved to London, Fleming, working in the same city, discovered the antibacterial action of culture filtrates of *Penicillium notatum* (Fleming, 1929) and gave the name penicillin to his "mould juice." Of his own volition, Raistrick carried out work on penicillin, but was able neither to obtain a pure preparation nor to determine its structure (Clutterbuck et al., 1932). It was 10 years after Fleming's discovery that Florey and colleagues at Oxford took up the challenge and conducted the first experiments to obtain significant amounts of penicillin and to prove its clinical efficacy (Bickel, 1972). The development of penicillin from a rare elixir of mold juice to an effective, inexpensive antibiotic took place during World War II. The exciting story of those events has been told many times (Hare, 1970; Macfarlane, 1979, 1984; Hobby, 1985).

The lesson of penicillin was not lost on microbiologists, and massive screening programs were instituted to seek new wonder drugs from microbial sources. Actinomycetes provided an even richer lode of bioactive natural products than did molds, and for the first time the number of pharmaceutical agents of microbial origin eclipsed that of botanicals. Many microbiologists and chemists found employment in pharmaceutical companies after World War II. The Golden Age of Medical Bacteriology had been supplanted by the Golden Age of Antimicrobial Agents, and a new type of chemical engineer, the fermentation technologist, had come into being. What the pioneer Raistrick had called "a region of biosynthesis" (Raistrick, 1949) was now of wide general interest. Moreover, the huge screening programs had uncovered many more metabolites lacking detectable antibiotic activity than they had found new antibiotics.

A new generation of biologists had to face the issue of function. Although many of these diverse natural products were indisputably pharmacologically active, many others were not. For all of them, the evolutionary advantage to the *producing organism* remained obscure.

An influential hypothesis was developed by Jackson Foster in his ground-breaking book, *Chemical Activities of Fungi* (Foster, 1949). He argued that fungi produced large amounts of some metabolic products, other than structural cell components, because of a deranged or pathological condition. When artificially high concentrations of carbohydrates were made available in a culture medium, the normal enzyme systems became overloaded and a metabolic shunt to "overflow pathways" occurred. Foster applied the terms "shunt metabolite" and "shunt metabolism" to situations in which either a "normal" metabolite (e.g., citrate) was produced in excessive amounts or a novel material (e.g., kojate) was produced. Foster was likely influenced by earlier workers. For instance, more than two decades before his text appeared, Derx (1925) had summarized the relationship with "pathology" as follows: "In a similar way an accumulation of oxalic acid in cultures of *A. niger* is possible only when there is a lack of assimilable nitrogen, or when growth takes place at lower temperatures. Under favourable circumstances there is very little formation of oxalic acid. In consequence, I am of opinion that a wider meaning may be taken from the words of Duclaux regarding formation of oxalic acid: 'C'est un produit de souffrance' " ("It is a product of suffering").

In a 1944 review entitled "Biochemistry of Fungi," Tatum attempted to rationalize the metabolic processes by which typical mold products are formed. It was a daunting task, since so little was then known—even so fundamental a unit as acetyl-CoA was not identified until 1951, the tricarboxylic acid cycle was not completely accepted until 1948, and roles for shikimate and mevalonate were yet to come. Tatum wrote as follows: " . . . it seems reasonable to conclude that the various mold products—acids, pigments, and antibiotic substances—may represent normal metabolic products or cell constituents, or perhaps represent closely related substances formed from these. The formation of a pigmented product from p-aminobenzoic acid by bacteria may illustrate such a *secondary* reaction" (Tatum, 1944, emphasis added). He then presented a figure illustrating some possible biosynthetic relationships. In this figure, sugar gives rise to primary intermediates which could yield simple heterocyclic compounds such as kojic acid or be converted to secondary intermediates, in which the glucopyranose structure has undergone various modifications and/or polymerizations to form metabolites such as fumigatin or helminthosporin (Tatum, 1944, p. 693). The "paper chemistry" leaves much to the imagination, and later research revealed that the aromatic compounds Tatum discussed were actually biosynthesized by pathways that were not envisaged in 1944.

It is noteworthy, however, that Tatum did speak of secondary reactions and of primary and secondary intermediates. Tatum's main contribution, of course, was that for which he later shared the Nobel prize—namely, the insight that "the inherent capacities for the formation of these various substances are genetically controlled." Foster, in his classic book *Chemical Activities of Fungi* (1949), generally accepted Tatum's thesis. He also made the important point that the "very numerous different principal metabolic activities of molds can be looked upon as manifestations of a few main types of metabolic activity. The great majority of them can be considered merely as extensions of the preceding ones so that gradually a series is built up, with comparatively simple examples on one end compounding successively to extreme complexity on the other" (Foster, 1949, p. 172). Foster suggested that "products of primary condensation" could participate in further "secondary condensations." Thus, complex pigments, and materials such as the anthraquinones, might have been formed from simple unicyclic rings (the "primary condensation rings").

While Tatum (1944) and Foster (1949) must be credited with the hypothesis that strange natural products were derived from primary intermediates of normal metabolism, it was Bu'Lock (1961) who first explicitly applied the term "secondary metabolite" in microbiology. He noted that . . . "just because the antibiotics are primarily defined, detected, examined and ultimately marketed as agents acting upon other organisms, the fact that they are themselves organic products has become relatively overshadowed." In a very elegant and much quoted paragraph, he continued:

> Now ever since Perkin, failing to make quinine, founded the dyestuffs industry, the organic chemists have found the study of "natural products" an inexhaustible source of exercises, which can be performed out of pure curiosity even when paid for in the hope of a more commercial reward. As a result, the organic chemist's view of nature is unbalanced, even lunatic, but still in some ways more exciting than that of the biochemist. While the enzymologist's garden is a dream of uniformity, a green meadow where the cycles of Calvin and Krebs tick round in disciplined order, the organic chemist walks in an untidy jungle of uncouthly named extractives, rainbow displays of pigments, where in every bush there lurks the mangled shape of some alkaloid, the exotic perfume of some new terpene, or some shocking and explosive polyacetylene. To such a visionary, both the diatretynes are equal prizes, to be set together as "natural products." We shall do the same, but since to a more sober eye both are in a sense less "natural" than, say, glycine or adenosine triphosphate (ATP), we may prefer the term "secondary metabolites." (Bu'Lock, 1961, p. 294)

Bu'Lock then defined his terms: "Given the generally acceptable view that there are basic patterns of general metabolism, on which the variety of organic systems imposes relatively minor modifications, we can define secondary metabolites as having, by contrast, a restricted distribution (which is almost species specific) and no obvious function in general metabolism" (Bu'Lock, 1961, p. 294). Bu'Lock provided no citations of earlier uses of the word "secondary" in connection with metabolism, but was aware of its use by plant physiologists (Bu'Lock, 1965; also personal communication to R. Bentley, July 6, 1982). It will be seen that Bu'Lock contrasted "general" with "secondary"; however, the substitution of "primary" for "general" has followed naturally, although he himself did not emphasize the "primary"/"secondary" usage (see later).

It should also be noted that just prior to Bu'Lock's paper, Gaden (1959), writing from the perspective of a chemical engineer, had created a taxonomy for all microbial products based on fermentation processes. In this classification, Class I metabolites accumulated as a result of primary energy metabolism (e.g., ethanol and lactate). Class II metabolites arose indirectly from reactions of energy metabolism or by side reactions from direct metabolic processes (e.g., citrate and itaconate). Class III metabolites were not concerned with energy metabolism but were independently elaborated or accumulated by cells (e.g., penicillin and other antibiotics). Gaden's classification was particularly important for the development of fermentation kinetics as a key aspect of the required reorientation of chemical engineering aspects of the new "biotechnology." Gaden and others (Doskočil et al., 1958) distinguished between "primary mycelium" formed early in a fermentation and "secondary mycelium" associated with the phase of rapid antibiotic synthesis; they seemed close to using the word "secondary" in connection with Class III metabolites.

B. More Jargon

The production of secondary metabolites is frequently associated with the cessation of growth, or at least with its slowing-down. In higher plants, this is displayed in organ-specific, tissue-specific, and cell-specific patterns of metabolite biosynthesis. In microbial submerged batch fermentations, the onset of secondary metabolism is usually roughly correlated with the end of the growth phase ("balanced growth") and the beginning of the stationary phase (Borrow et al., 1961). In a study of 6-methylsalicylate production in *Penicillium urticae*, Bu'Lock et al. (1965) confirmed the tendency of batch fermentations to display such distinct phases of growth and secondary metabolism. When applied to a filamentous organism such as *P. urticae*, terms such as "multiplication" and "logarithmic growth" did not seem appropriate. Bu'Lock was again responsible for suggesting alternatives:

> . . . we prefer the general concepts of a *trophophase* (Greek, nutrient) based on Borrow's concept of balanced nutrient uptake, and an *idiophase* (Greek, peculiar) in which are displayed metabolic idiosyncracies, either quantitative (e.g., accelerated lipid synthesis) or qualitative (e.g., antibiotic production). The concept of primary and secondary metabolic processes, as now generally understood, is clearly related to this distinction. (Bu'Lock et al., 1965, p. 774)

Since molds and actinomycetes, the major antibiotic-producing groups being grown in most industrial fermentations, are all filamentous, the jargon filled a terminological vacuum, and soon found wide acceptance. Subsequently, new terms were spawned: "idiolite" (peculiar metabolite) as a synonym for "secondary metabolite" (Walker, 1974) and "idiotroph" for "mutants which grow in minimal medium but fail to produce an idiolite unless supplemented with a precursor of that secondary metabolite" (Nagaoka and Demain, 1975, p. 628). In later writings, the notion of growth phase is often included in the definition of secondary metabolism, with or without allusion to idiophase. For example, in a major review on trace metal influences, this succinct statement appears: "Secondary metabolites are defined as natural products that have a restricted distribution, possess no obvious function in cell growth, and are synthesized by cells that have stopped dividing" (Weinberg, 1970, p. 1). Unfortunately, the seemingly-precise word "growth" is used by many microbiologists in an exceedingly ambiguous and imprecise manner (Bu'Lock, 1975).

C. The Wrong Term in the Right Place at the Right Time?

Despite this lengthy history, and the rather general contemporary acceptance of primary and secondary metabolism, there are difficulties. Following are some of the major objections.

1. Secondary is an unfortunate choice of words. It is defined in the *Oxford English Dictionary* as "Belonging to the second class in respect of dignity or importance: entitled to consideration only in the second place. Also, and usually, in less precise sense: Not in the first class; not chief or principal; of minor importance, subordinate." These meanings have "led many workers to assume that such compounds are not involved in the dynamic pathways of metabolism, but merely exist as a

means of ridding the cell of excess metabolites or excretory products derived from primary pathways, which otherwise would be deleterious to their functioning." (Swain, 1974).

In a preface to a volume entitled *Secondary Plant Products*, Bell and Charlwood (1980) complain, "It is a pity indeed that the term 'secondary' should ever have been applied to these compounds, as the word gives the unfortunate impression that they are all relatively unimportant." The derogatory connotation of "secondary" may also have practical consequences. In this connection, the frankly blunt statement of Floss (1979) cannot be improved—"Secondary metabolites does not sound well when we apply for grants for our scientific work."

2. The term "secondary metabolism" has been used in at least two different senses in biochemistry. Thus, the pentose monophosphate pathway and the glucuronate–ascorbate pathway have been said to constitute "part of the secondary metabolism of glucose" (Lehninger, 1982).

"Secondary" and "specific" have also been used in connection with comparative biochemistry with reference to vitamin requirements, cytochromes, uricolytic enzymes, and polysaccharides such as cellulose and chitin. In this context, Baldwin (a particularly influential teacher of British biochemists) wrote that "there exists a common and fundamental metabolic ground-plan to which living organisms of every kind must probably conform. But here again, in addition to the fundamental features, we find secondary and specific adaptational features . . ." (Baldwin, 1949).

3. Other complications arise because the "primary" metabolic pathways are not necessarily exactly the same in all species, and, in a multicellular organism, not necessarily the same in all organs. It is unfortunate that the "unity of biochemistry" has been emphasized to the extent that biochemical diversity is often ignored. One authority has even gone so far as to remark that "to the classical biochemist, all organisms must be the same; if necessary, you put them in a Waring Blendor until they are" (J. D. Bu'Lock, letter to J. W. Bennett, Aug. 11, 1987). To counteract this tendency, a few specific exceptions to the "unity of biochemistry" may be noted: (1) The genetic code in mitochondria is slightly different from the "universal" genetic code; (2) prokaryotes possess many unusual metabolic capabilities not found in eukaryotes, e.g., the ability to fix nitrogen and to use reduced inorganic compounds as electron sources; (3) Two pathways exist for lysine biosynthesis; in fungi, 2-oxoglutarate is the precursor, whereas in most other organisms, lysine derives from aspartate and pyruvate via diaminopimelate; (4) the incorporation of CO_2 during photosynthesis may

follow the C_3 or C_4 pathway; and (5) the intermediate, δ-aminolevulinate, required for porphyrin biosynthesis, is obtained in bacteria and animals from a reaction between succinyl-CoA and glycine; in plants and cyanobacteria this intermediate probably arises by reduction of the 1-carboxyl group of 2-oxoglutarate.

Many other examples could be quoted. The point is simply that the "unity of biochemistry" is merely a convenient generalization. The term "primary metabolism" perhaps unfortunately reinforces the notion that there is one central family of "housekeeping" functions, and the many differences in the "highways" of cellular metabolism, as a result, tend to be overlooked.

III. How Shall a Metabolite Be Named?

From the beginning of the use of the primary/secondary terminology, it was recognized that making the distinction between primary and secondary metabolism was not always easy (Kossel, 1891). Also at an early date, it was pointed out that caution was necessary in drawing conclusions (Czapek, 1921). Not so generally recognized is the extent to which certain underlying assumptions about the nature of the dichotomy affect our classifications.

A. Overproduction

Many workers assume that primary products of metabolism do not accumulate to substantial levels in cells, since, generally, the steady-state concentrations of such materials are low. When significant accumulations do occur, the phenomenon may be termed "overproduction" (Krumphanzl et al., 1982; Hütter, 1986) and the metabolite that accumulates, even an intermediate in a bona fide primary metabolic pathway, is often reclassified as a secondary metabolite. The most famous example is citrate, which has been characterized as a "metabolic swinger."

This philosophy probably reflects Jackson Foster's lingering influence. His original definition of shunt metabolites included "normal metabolites" such as citrate, when they were produced in "excessive amounts." Similarly, Bu'Lock's circumscription of secondary metabolites included "some special cases of substances which are extraordinary not in their structure but in the amount in which they are sometimes formed, such as the riboflavin accumulated by Eremothecium ashbyii" (Bu'Lock, 1961). Haslam (1986) has written persuasively that the accumulation (i.e., overproduction) of citrate, malate, shikimate, riboflavin, and others should be viewed as examples of defective regulatory patterns of primary metabolism. Industrial microbiologists have, of course, manipulated both environmental and genetic parameters in order to increase yields for commercial fermentations, thus accentuating the "overproduction." There is a tendency to characterize all overproducing strains as "sick." Thus Vaněk et al. (1981b) speak of the "pathophysiology" of production strains, while Neijssel and Tempest (1979) say they are "functional cripples and suffer from genetic diseases." Or, as Duclaux phrased it more than half a century earlier, the metabolites are "produits de souffrance" (quoted in Derx, 1925).

It is important to keep in mind that both Foster's writings and Bu'Lock's early papers on overproduction and secondary metabolism predated the advances in microbial genetics that led to the elucidation of the *lac* operon and eventually of other molecular models of induction and repression. At the same time, they drew attention to the need to provide proper explanations of precisely these regulatory phenomena. In light of the new insights concerning the regulation of microbial metabolism, there is no longer any need to use "overproduction" as a distinguishing criterion for secondary metabolism. Materials such as citrate, riboflavin, vitamin B_{12}, glutamate, and lysine can be permanently and unequivocally removed from the ranks of secondary metabolism.

Two other attributes variously imputed to secondary metabolism, and often included in formal definitions, similarly affect the classification of microbial products. The first concerns fermentation kinetics and is relatively easy to address; the second concerns evolutionary function and is more intractable.

B. Growth Phase

The onset of microbial secondary metabolism is often associated with the "cessation" of growth. The trophophase/idiophase distinction is a useful way of describing this relationship in many industrial fermentations. However, alteration of the nutritional environment and other growth parameters frequently changes the dynamics of a fermentation, and numerous examples of antibiotic synthesis during the growth phase have been described. Some practitioners, based on such growth-related biosynthesis, have implied or concluded that compounds synthesized during the growth phase are not secondary metabolites. The literature on this topic has been reviewed by Calam (1979) and Aharonowitz and Demain (1980). The latter authors write:

> ... the definition of "secondary metabolite" should not include the stage of the growth cycle in which the compound is produced. ... Metabolites are ... not "secondary" because they are produced after growth; they are "secondary" because they are not *essential* for growth. Metabolism is not "secondary" because it occurs after growth; it is "secondary" because it is not *necessary* for growth. (Aharonowitz and Demain, 1980, p. 9)

We concur with their judgment. Although a definition of secondary metabolism might be embellished by noting a frequent correlation between growth arrest and metabolite production, the stage of the growth cycle is not critical to distinguishing between primary and secondary metabolism. Further discussion is provided by Bu'Lock (1975) and Campbell (1983).

C. Function

The matter of function is more problematic. A subordinate role for secondary metabolism has been underscored by the perception that these processes are unworthy of study since many of the metabolites have no (apparent) selective advantage for the producing organism. Indeed, those who have argued that the diversity of secondary plant products is accidental have been described as the "waste-product lobby" (Swain, 1976).

The apparent lack of function has intrigued—and baffled—all theorists in the field (see, *inter alia*, Bu'Lock, 1961; Weinberg, 1971; Demain, 1974; Haavik, 1979; Lancini and Parenti, 1982; Bennett and Ciegler, 1983; Campbell, 1984). Speculations about the function of microbial secondary metabolites range from the view that they are evolutionary neutral laboratory artifacts (Woodruff, 1966) or "players in the evolution game" (Zähner *et al.*, 1982) to an all-purpose role of "maintaining mechanisms essential to cell multiplication when cell multiplication is no longer possible" (Bu'Lock, 1961) to an unshaken conviction that individual metabolites serve individual selective advantages, albeit usually unknown ones (Demain, 1974; Haavik, 1979; Bennett, 1983).

Paradoxically, the debate has persisted in chemistry and microbiology with little recognition of parallel advances in ecological biochemistry (sometimes termed "chemical ecology") that render many of the speculations moot. Plant physiologists have accumulated an extremely convincing body of literature demonstrating that plant secondary metabolites confer numerous selective advantages in an ecological sense, frequently mediating interspecific interactions such as defense and competition (Dethier, 1954; Fraenkel, 1959; Ehrlich and Raven, 1964; Whittaker, 1970; Gilbert and Raven, 1975; Siegler and Price, 1976; Harborne, 1977, 1978; Swain, 1977; Rosenthal and Janzen, 1979; Bell, 1980; Torssell, 1983; Haslam, 1985). The special name "allelopathic substance" or "allelochemical," is applied to plant secondary metabolites that inhibit the germination, growth, or occurrence of other plants (for a recent review, see Harborne, 1986). Recently, a soil-dwelling Basidiomycete fungus, *Laetisaria arvalis*, was found to produce an allelochemical inducing rapid hyphal lysis in some phytopathogenic fungi such as *Pythium ultimum* (Bowers et al., 1986). The allelochemical, named "laetisaric acid," was identified as 8-hydroxy-(9Z,12Z)-octadecadienoic acid. In addition, functions have been identified or suggested for a number of other microbial metabolites as sex hormones, metal chelators, repellants, protective agents, and the like (Janzen, 1977; Calam, 1979; Haavik, 1979; Thomas, 1979; Zähner, 1979; Bennett and Christensen, 1983; Campbell, 1984).

With reference to plant secondary metabolites, Whittaker (1970) wrote: "The evolution of these substances is probably not, in fact, comprehensible except in an ecological context, including organisms other than the plants producing them" and Vickery and Vickery (1981) state unequivocally that "The once popular definition of a [plant] secondary metabolite—that it does not play an indispensable role in the plant and is not ubiquitous—is no longer applicable." A recent text on natural product chemistry provides somewhat of an exception in writings by chemists: "We could on good grounds define secondary metabolites as non-nutritive chemicals controlling the biology of other species in the environment or, in other words, secondary metabolites play a prominent role in the coexistence and coevolution of species . . . " (Torssell, 1983, p. 6).

Despite these many examples to refute the accusation that secondary metabolites are without function, and despite the many converts to the belief that seondary metabolites do have a *raison d'être*, the issue of function continues to confound analyses of the subject. Weinberg (1971) wrote, "Perhaps we can tentatively conclude that the majority of cells of plant and animal tissue have, during evolution, divested themselves of the need to form large quantities of completely functionless secondary metabolites." In this connection, a major review by Haslam (1986) contains a table illustrating the "ambiguities of definition of secondary metabolism" when metabolites are classified on the basis of function. Examples included polyamines such as putrescine and spermidine, which "have the aura" of essential metabolites yet are ubiquitous in nature without having a clearly defined role in cellular function, and conversely, the bicyclic sesquiterpene diol, sirenin, and the gallic acid glucoside, turgorin, that, as Haslam phrased it, have "the look of secondary metabolites" yet mediate clearly defined biological functions. Haslam concluded, "In terms of the earlier definitions, there are clear problems in the classification of such natural products"; but he did not go so far as to reject "clearly defined function" as an appropriate criterion.

Even more forthrightly, a recent book review stated, "By definition, secondary metabolites offer no apparent selective advantage to the producer cell. . . . When the natural function of a metabolite is understood, the metabolite might then be classified as a primary metabolite" (James, 1987).

In all of this, it has been overlooked that Bu'Lock (1961) referred to ". . . no *obvious* function" (emphasis added) and understood the dilemma ("No one can spend long in consideration of the so-called natural products before asking, or being asked, the question 'What are they supposed to do?', and to evade that question here would be unfair since we have already raised it, in defining secondary metabolites as having 'no *obvious* function!' . . . "). Much earlier, Raistrick, never a member of the waste-product lobby, had written in a review, "There has been a tendency, occasionally expressed to the reviewer, to regard these mould metabolic products as microbiological curiosities, or merely as by-products (whatever that may mean) of no particular interest. That this view is surely a mistaken one appears to be indicated by the very large amounts in which they are often formed" (Raistrick, 1940, p. 571).

Lack of function and lack of obvious function are two different things. Similarly, lack of selective function (something advantageous in evolution) and lack of essential function (something necessary for cellular life) are also quite different. Here, an analogy from zoology may be of heuristic value. Many differentiated animal cells produce compounds that have been termed "luxury metabolites":

> Luxury molecules have been defined as those molecules, which though of great utility to the intact organism, are not essential to the viability of the cells synthesizing them. Myosin or fibrinogen, for example, are not essential for the survival of muscle or liver cells. Other luxury molecules would be hemoglobin, insulin, thryoxine, and chondroitin sulfate, as well as their associated mRNA's and repressors and derepressors. In conntrast, there are those molecules produced by most cells which are essential to the viability of the cell synthesizing them. This category includes molecules such as the cyto-

chromes, hexokinase, the enzymes synthesizing amino acids, the ubiquitous sRNA's [sic] and rRNA's, cholesterol, glycogen, etc. (Holtzer et al., 1969, p. 19)

The analogies to primary and secondary metabolism in microorganisms are obvious. A metabolite need not be essential to the viability of the cell to be of utility in the survival of the organism, the colony, or the species.

We suggest that a definition of secondary metabolism should allude to the known ecological functions served by some of the secondary metabolites. The notion of "lack of function" should, in any case, be removed from the definition. Whether or not a metabolite has a known function in the economy of the producing organism should not be a critical distinguishing feature of secondary metabolism.

What should be emphasized is the fact that secondary metabolism is a manifestation of cellular differentiation. Secondary metabolism is genotypically specific: ". . . it is here that the evolved individuality of microbial species is manifested at the molecular level" (Bu'Lock, 1980). Like many aspects of cellular differentiation, the expression of secondary metabolism is dependent on many nongenetic factors. The genotype is not always expressed in the phenotype, and expressivity varies with numerous environmental parameters. Optimal growth conditions are frequently not optimal conditions for secondary biosynthesis; hence, secondary metabolism can be viewed as an aspect of the differentiation which limited growth implies (Luckner, 1972; Bu'Lock, 1975; Bennett, 1983).

IV. Nomenclature

A. Alternatives

In the early development of a science, naming and classifying are important. Physical scientists have often accused biologists of being mere labelers and classifiers, just as today biologists often accuse social scientists of the same. Nor is the matter of giving elaborate names to poorly understood phenomena limited to scientists. It is traditional to hide ignorance behind terminology: the schoolmaster of *The Deserted Village* used "words of learned length, and thund'ring sound" in debate with the parson and thereby "amazed the gazing rustics rang'd around" (Goldsmith, 1770).

We have seen that there are problems with the use of "primary"/ "secondary." Nature's biosynthetic diversity has elicited many names and phrases in the scientific literature, so that there are alternative possibilities. Early on, Raistrick distinguished between "general biochemical reactions, common to many species" on the one hand, and "certain highly specific products produced in some cases by a single species" on the other (Raistrick and Rintoul, 1931). Raistrick's colleague, Birkinshaw (1937) spoke of the "general and specific biochemical characterization of moulds." We have already mentioned Bonner and Galston's (1952) "highways and byways in metabolism" and Walker's (1974) "idiolites."

The following represents a selection of some of the more colorful phrases that have been applied. Many of them carry judgments about function: "functionless anomalies, flotsam on the metabolic beach" (Geissman and Crout, 1969); "an extraordinary bestiary of organic compounds" (Weinberg, 1970); "considered to be detritus: useless but structurally interesting compounds" (Mann, 1978); "the organic compounds of primary metabolism are the stations on the main lines of [a] railway, the compounds of secondary metabolism the termini of branch lines (Herbert, 1981); "biology's Caliban—a thing of darkness" (Campbell, 1984); "biochemistry's Cinderella" (Dutton, 1986); "products made on the playground of metabolism" and the "wood-shavings [by-products] of metabolism" (Schlegel, 1986).

In order to classify secondary metabolism, others have used nononsense categories of chemical reaction types such as oxidation, reduction, and hydroxylation (Gale, 1951; Wallen et al., 1959; Bentley and Campbell, 1968; Campbell, 1984), while still others have based their taxonomies on pathways of biosynthetic origin such as acetate-derived, amino acid-derived, and the like (Turner, 1971; Turner and Aldridge, 1983). These classifications have much to recommend them but do not solve the problem of naming the whole class of metabolites.

Of the many terms and phrases, perhaps Raistrick's "general"/"specific" has the most merit. "Specific" or "special" does not carry the negative connotations of "secondary," nor does "general" imply the preeminence of "primary." There is also historical precedent, since as early as 1925, Derx had noted that certain natural products were referred to as "specialities." The use of "general"/"specific" has, in fact, been recommended by several authors (Janzen, 1978; Zähner, 1979; Rose, 1979; Martin and Demain, 1980; Campbell, 1984; Whiting, 1985). It is also important to recollect that Bu'Lock (1961) referred mostly to general and secondary metabolism. In his 1961 paper there is only the following sentence, considering relationships between morphological and biochemical features, which links primary and secondary: ". . . any attempt to reconstruct the ancestry of microorganisms should consider both primary and secondary metabolic features as well as morphological taxonomy." For the most part, Bu'Lock contrasted general and secondary biosynthesis, referred to substrates which are common to both general and secondary metabolism, and described materials as "prime precursors" (e.g., acetate and mevalonate), and as "common" (e.g., ribose) or "unusual" (e.g., desosamine). More recently, Bu'Lock (1975) has linked together ". . . the so-called 'special' or 'secondary metabolites.' "

B. Some Semantics

Before stipulating definitions for primary/general and secondary/special metabolism and metabolites, one last diversion is necessary.

A definition is a statement of the meaning of a word or group of words; it is an attempt at a precise qualification. Most scientists believe that the formulation of technical language ("coining words," "creating scientific jargon") is a straightforward proposition, and try to be careful in setting forth precise definitions that will circumscribe the meaning of things and phenomena being described. Nevertheless, most scientists, at one time or another, have found themselves involved in controversies that concern differences in the definition of terms, and discover that the meaning of a given scientific term may not be as straightforward as they had assumed. Semantics, the branch of linguistics that studies the theory of meaning, can shed light on scientific nomenclatural disputes. Indeed, semanticists are well aware of the importance of their discipline to science:

> One sphere of activity where precise definition is imperative is that of *science*. All scientists are linguists to some extent. They are responsible for devising a consistent terminology, a skeleton language to talk about their subject-matter. . . . Botany gave perhaps the first example of a rich and articulate nomenclature enabling research to master the complexities of nature; the author of this system of classification, Linnaeus, is credited with the saying: "if you do not know the names, your knowledge of things will be lost." Yet terminologies have often been allowed to grow up uncontrolled, with the result that all varieties of multiple meaning have freely developed with them: . . . the same term can denote different things, and the same things [can] be denoted by different terms. . . . Any major compilation of scientific terminologies should in future be based on the principles evolved by semantics. (Ullmann, 1951, p. 107)

Although one might hope that no discipline would be more scrupulously consistent in its terminology than semantics, this is woefully not the case (Ullmann, 1957). In their classic book, *The Meaning of Meaning*, Ogden and Richards (1923) compiled 16 main definitions and

22 subdefinitions of "meaning." Since then, semanticists have employed numerous overlapping and sometimes conflicting terminologies in the language they use to talk about language, and the semantic literature is filled with a formidable proliferation of verbal ambiguities. Even the more clearly circumscribed philological aspects of "meaning" suffer from a fundamental ambiguity; some scholars treat it as a relational term, whereas others identify it with the traditional dictionary usage (Ullmann, 1957).

Recognizing that semanticists use "meaning" in many different and subtle ways (so much so that there are those who believe the term has become meaningless) we here introduce some elementary semantic vocabulary in the school of a lexicographer which will be useful in our analysis of secondary metabolism. The lexicographic model assumes a difference between things (extension, denotation, reference; in German, *Sinn*) and the concepts suggested in the mind by these things ("meaning," intension, connotation, signification; in German *Bedeutung*). The words "extension" and "intension" are borrowed from logic; "denotation" and "connotation" are from the field of literary criticism (Ullmann, 1957; Hayakawa, 1941). Semanticists go on to distinguish two kinds of intensional meanings or connotations, the informative and the affective. Informative connotations are the "impersonal meanings," while affective connotations are the aura of personal feelings that a term arouses.

Ideally, scientists try to devise and use terms that are largely or entirely informative in connotation, and which have stipulated definitions with precise and unambiguous "meaning." Nevertheless, scientific terminology may inherit or acquire affective content. The clinical entity caused by *Mycobacterium leprae* can be called "leprosy" or "Hansen's disease." Many modern physicians prefer the latter because it moderates the stigmatic connotations of the disease that exist when it is called leprosy. Contemporary examples of scientific terms that acquired emotional content include the methodological advances that are grouped together under the rubric of "genetic engineering" and the retroviral disease named "acquired immunodeficiency syndrome."

C. Semantics and Secondary Metabolism

Within this elementary semantic framework, we see that "secondary metabolites" denotes (refers to) certain chemical compounds. The natural products themselves (e.g., penicillin, prodigiosin, or aflatoxin) are the extensions (denotations) of "secondary metabolite." Penicillin denotes a β-lactam with strong antibacterial properties; prodigiosin, a tripyrrole with intense red pigmentation; aflatoxin, a bisdihydrofuranocoumarin with toxic effects in vertebrates; and so on.

The intensions (connotations) of "secondary metabolite" (what most natural scientists simply call the "definition" or the "meaning") are the ideas, concepts, and feelings suggested when the term is used. The connotations of "secondary metabolite" include a number of informative attributes: low molecular weight, nonubiquity, nonessentiality for cell survival, and genotypic specificity. We have argued that some of the informative connotations that have been applied to "secondary metabolite" are spurious—overproduction, relationship with growth phase of production, and evolutionary function fall into this latter category. Finally, "secondary metabolite" also has affective connotations. As we have pointed out earlier, the customary meaning of the word "secondary" is "of lesser importance." Indeed, the original appellation by German plant physiologists reflected the (presumable) lack of cellular function displayed by these compounds, hence their "secondary" nature.

When one accepts that linguistic abstractions carry both informative and emotional content, it becomes clear that the word "secondary" affixed to "metabolite" can influence scientific thinking. On one level, it affects the ways individual metabolites are included among or excluded from the ranks of secondary metabolism. If the word "secondary" is construed to be synonymous with "lack of function," then any metabolite that is shown to have a useful function in cellular metabolism will no longer be "secondary." But if, as we have argued, known evolutionary function is irrelevant to the circumscription of "secondary metabolite," then we must stipulate a definition utilizing more appropriate and accurate attributes.

This still leaves the matter of emotional content. There are those who automatically assume that metabolites which are described as "secondary" are of lesser importance than the sugars, amino acids, lipids, nucleotides, and other compounds of general metabolism. For that reason, "secondary" will always remain an unfortunate choice of adjectives. Although it is clear that secondary is an unfortunate word to apply to a process of considerable metabolic significance, "change is not made without inconvenience, even from worse to better" (Johnson, 1755). Changes in well-entrenched scientific nomenclature have sometimes been achieved (few if any biochemists today refer to DPN, itself a change from coenzyme I), but one has only to contemplate the number of books with the word "secondary" in their titles (see Table I) to realize that a proposal to replace "secondary" with a more realistic name is not very likely to succeed. Indeed, at a recent international

TABLE I

A Chronological Listing of Selected Book Titles to Illustrate the
Well-Entrenched Usage of "Secondary Metabolism" and Related Terms[a]

Title	Author(s) or editor(s)	Date
Lehrbuch der Pflanzenphysiologie, Vol. 1 (Biochemie und Physiologie der Sekundären Pflanzenstoffe)	Paech	1950
Encylopedia of Plant Physiology, Vol. X (The Metabolism of Secondary Plant Products)	Rühland[b]	1958
The Biosynthesis of Natural Products: An Introduction to Secondary Metabolism	Bu'Lock	1965
Organic Chemistry of Secondary Plant Metabolism	Geissman and Crout	1969
Secondary Metabolism in Plants and Animals	Luckner	1972
Recent Advances in Phytochemistry, Vol. 8 (Metabolism and Regulation of Secondary Plant Products)	Runeckles and Conn[b]	1974
Secondary Metabolism and Coevolution	Luckner et al.[b]	1976
Secondary Metabolism and Cell Differentiation	Luckner et al.	1977
Antibiotics and Other Secondary Metabolites	Hütter et al.[b]	1978
Secondary Metabolism[c]	Mann	1978
Regulation of Secondary Product and Plant Hormone Metabolism	Luckner and Schreiber[b]	1979
Economic Microbiology, Vol. 3 (Secondary Products of Metabolism)	Rose[b]	1979
Herbivores: Their Interaction with Secondary Plant Metabolites	Rosenthal and Janzen[b]	1979
Encyclopedia of Plant Physiology, New Series, Vol. 8 (Secondary Plant Products)	Bell and Charlwood[b]	1980
The Biochemistry of Plants, Vol. 7 (Secondary Plant Products)	Conn[b]	1981
The Biosynthesis of Mycotoxins: A Study in Secondary Metabolism	Steyn[b]	1980
The Biosynthesis of Secondary Metabolites	Herbert	1981
Secondary Plant Metabolism	Vickery and Vickery	1981
Secondary Metabolism and Differentiation in Fungi	Bennett and Ciegler	1983
Natural Product Chemistry: A Mechanistic and Biosynthetic Approach to Secondary Metabolism	Torssell	1983
Secondary Metabolism in Microorganisms, Plants, and Animals	Luckner	1984
Metabolites and Metabolism: A Commentary on Secondary Metabolism	Haslam	1985
Regulation of Secondary Metabolite Formation	Kleinkauf et al.[b]	1985

[a] Full citations are provided in the reference list. Were this compilation to include individual articles and book chapters containing the terms "secondary metabolite" and "secondary metabolism," it would be impossibly long.
[b] Editors.
[c] A second edition was published in 1987.

workshop, Luckner (1985) declared emphatically and unequivocally that use of the term "secondary metabolism" is so widespread "that it is useless to discuss whether you can replace it or not."

Nevertheless, we suggest that there would be many advantages (and little inconvenience) in supplementing the use of "secondary" with "special" or "specific" whenever possible, and in returning to the original (Raistrick and Rintoul, 1931; Bu'Lock, 1961) use of "general" rather than "primary."

D. THE DEFINITIONS

> We need not restrict lexicography by requiring that a definition be a *perfect* rendition of a meaning. (Weinrich, 1980)

General metabolite (hence, *general metabolism*). A metabolic intermediate or product, found in most living systems, essential to growth and life, and biosynthesized by a limited number of biochemical pathways.

Secondary metabolite (hence, *secondary metabolism*). A metabolic intermediate or product, found as a differentiation product in restricted taxonomic groups, not essential to growth and life of the producing organism, and biosynthesized from one or more general metabolites by a wider variety of pathways than is available in general metabolism.

While we believe these definitions are self-explanatory and will stand by themselves, the following elaborations expand on the basic concepts which lead us to propose them.

General metabolites are involved in functions such as energy transduction, or the production of materials necessary for cell structure, maintenance, and reproduction. Many general metabolites are ubiquitous in living systems. General metabolism is characteristic of undifferentiated cells and provides the broad metabolic foundation for the concept of the unity of biochemistry. Examples of general metabolites are the essential amino acids, sugars such as D-ribose, and nucleotides such as adenosine triphosphate (ATP). The enzymes which interconvert them would be an important component of general metabolism. A widely used synonym for "general metabolite" (or "metabolism") is "primary metabolite" (or "metabolism").

Secondary metabolites are often materials of low molecular weight and are frequently accumulated in large quantities, after the cessation of growth, in "families" of related compounds (congeners). Most secondary metabolites are produced by bacteria, fungi, or plants, but a few are produced by animals; enormous intergeneric, interspecific, and intraspecific variation in secondary metabolism is the rule. The best known secondary metabolites exhibit pharmacological activity as drugs, toxins, etc., but the function of secondary metabolites in the producing organism is not always obvious. Nevertheless, numerous secondary metabolites do impart known selective advantages upon their producers, primarily of an ecological nature. Secondary metabolism is characteristic of differentiated cells and provides the raw material for most of the studies in natural products chemistry. Examples of secondary metabolites include alkaloids, polyketides, terpenes, nonribosomal oligopeptides, and the products of numerous other "unusual" pathways. Suggested synonyms for "secondary metabolite" (or "metabolism") are "specific" and "special metabolite" (or "metabolism").

Tables displaying the major differences betweeen general (primary) and secondary (specific) metabolites are given by Thomas (1979), Campbell (1984), Bennett (1985), and Hütter (1986).

V. Epilogue

"Language is the means by which existentially unique processes are equated, ordered, and understood" (Nagel, 1945, p. 617). Scientific language, by stipulating meaning and refining definitions as new information becomes available, seeks to minimize ambiguity and to maximize clarity.

It has been over a quarter of a century since Bu'Lock (1961) introduced the term secondary metabolite to microbiology. During that time, there have been great advances in analytical chemistry, ecological biochemistry, and molecular biology which have expanded our understanding of secondary metabolism and secondary metabolites. The definitions proposed here are intended to clarify contemporary thinking by accommodating these new findings. They should not be used to force a Procrustian dichotomy upon natural products, but, rather, should be applied with flexibility as useful abstractions to help find order in diversity.

We recognize that it is far more difficult to establish a canonical definition than a canonical DNA sequence. Nevertheless, we hope that by regularizing usage of "general metabolism" and "secondary metabolism," some of the misleading and incorrect applications that persist in the literature may be remedied.

General metabolism represents "ancient, successful, and general solutions to global biological problems" (Campbell, 1984), while secondary metabolism represents the splendid, idiosyncratic diversity of nature, endowing different species with specific solutions to biological problems. Much of the morphological diversity that is the basis of our great taxonomic systems is readily discerned with the naked eye. With the exception of pigments, the chemical diversity expressed in secondary metabolism is less obvious and therefore, perhaps, more mysterious. Yet it, too, represents the beguiling and beautiful manifestations of genotypic differentiation that species have perfected during their evolutionary history. There is little that is "secondary" about secondary metabolism.

> Little do ye know your own blessedness; for to travel hopefully is a better thing than to arrive, and the true success is to labour. (Stevenson, 1881)

Acknowledgments

Our thanks to F. Gottlieb for assistance with German translations, and to I. M. Campbell for many useful discussions. In addition, insights and comments from J. D. Bu'Lock, J. V. Catano, A. L. Demain, and J. D. Harborne are gratefully acknowledged.

References

Aharonowitz, Y., and Demain, A. L. (1980). *Biotechnol. Bioeng.* **22,** 5.
Baldwin, E. (1949). "An Introduction to Comparative Biochemistry," pp. 148–149. Cambridge Univ. Press, London and New York.
Bell, E. A. (1980). In "Encyclopedia of Plant Physiology" (E. A. Bell and B. V. Charlwood, eds.), New Series, Vol. 8, pp. 11–21. Springer-Verlag, Berlin and New York.
Bell, E. A., and Charlwood, B. V., eds. (1980). "Encyclopedia of Plant Physiology," New Series, Vol. 8, (Secondary Plant Products). Springer-Verlag, Berlin and New York.
Bennett, J. W. (1983). In "Secondary Metabolism and Differentiation in Fungi" (J. W. Bennett and A. Ciegler, eds.), pp. 1–32. Dekker, New York.
Bennett, J. W. (1985). In "Fungal Protoplasts" (J. F. Peberdy and L. Ferenczy, eds.), pp. 189–203. Dekker, New York.
Bennett, J. W., and Christensen, S. B. (1983). *Adv. Appl. Microbiol.* **29,** 53.
Bennett, J. W., and Ciegler A., eds. (1983). "Secondary Metabolism and Differentiation in Fungi." Dekker, New York.
Bentley, R., and Campbell, I. M. (1968). *Compr. Biochem.* **20,** 415–489.
Bickel, L. (1972). "Rise Up To Life." Angus & Robertson, London.
Birkinshaw, J. H. (1937). *Biol. Rev. Cambridge Philos. Soc.* **12,** 357.
Birkinshaw, J. H., Charles, J. H. V., Hetherington, A. C., and Raistrick, H. (1931). *Trans. R. Soc. London, Ser. B* **220,** 153.
Birkinshaw, J. H., and Raistrick, H. (1932). *Biochem. J.* **26,** 441.
Birkinshaw, J. H., Chambers, A. R., and Raistrick, H. (1942). *Biochem. J.* **36,** 242.
Bonner, J., and Galston, A. W. (1952). "Principles of Plant Physiology," 1st ed. Freeman, San Francisco, California.
Borrow, A., Jefferys, E. G., Kessell, R. H. J., Lloyd, E. C., Lloyd, P. D., and Nixon, I. S. (1961). *Can. J. Microbiol.* **7,** 227.

Bowers, W. S., Hoch, H. C., Evans, P. H., and Katayama, M. (1986). *Science* **232**, 105.
Bu'Lock, J. D. (1961). *Adv. Appl. Microbiol.* **3**, 293.
Bu'Lock, J. D. (1965). "The Biosynthesis of Natural Products: An Introduction to Secondary Metabolism." McGraw-Hill, London.
Bu'Lock, J. D. (1975). *In* "The Filamentous Fungi" (J. E. Smith and D. R. Berry, eds.), Vol. 1, pp. 34–58. Edward Arnold, London.
Bu'Lock, J. D. (1980). *In* "The Biosynthesis of Mycotoxins: A Study in Secondary Metabolism" (P. S. Steyn, ed.), pp. 1–16. Academic Press, New York.
Bu'Lock, J. D., Hamilton, D., Hulme, M. A., Powell, A. J., Smalley, H. M., Shepherd, D., and Smith, G. N. (1965). *Can. J. Microbiol.* **11**, 765.
Calam, C. T. (1979). *Folia Microbiol.* (Prague) **24**, 276.
Campbell, I. M. (1983). *J. Nat. Prod.* **46**, 60.
Campbell, I. M. (1984). *Adv. Microb. Physiol.* **25**, 1.
Clutterbuck, P. W., Lovell, R., and Raistrick, H. (1932). *Biochem. J.* **26**, 1907.
Conn, E. E., ed. (1981). "The Biochemistry of Plants," Vol. 7. Academic Press, New York.
Czapek, F. (1925). "Biochemie der Pflanzen," 3rd ed., Vol. 3, p. 220. Fischer, Jena. (See also 1921 edition).
Demain, A. L. (1974). *Ann. N. Y. Acad. Sci.* **235**, 601.
Derx, H. G. (1925). *K. Akad. Wet. Amsterdam, Proc. Sect. Sci.* **28**, 96.
Dethier, V. C. (1954). *Evolution(Lawrence, Kans.)* **8**, 33.
Dewar, M. J. S. (1945). *Nature (London)* **155**, 50.
Divekar, P. V., Raistrick, H., Dobson, T. A., and Vining, L. C. (1965). *Can. J. Chem.* **43**, 1835. (Although this is the last paper of the famous "Studies in the Biochemistry of Micro-organisms" series, a posthumous Raistrick paper was published much later - see Raistrick and Rice, 1971.)
Doskočil, J., Sikyta, B., Kašparová, J., Doskočilová, D., and Zajicek, J. (1958). *J. Gen. Microbiol.* **18**, 302.
Dutton, M. F. (1986). *Biochem. S. Afr.* **2**, 21.
Ehrlich, P. R., and Raven, P. H. (1964). *Evolution (Lawrence, Kans.)* **18**, 568.
Fleming, A. (1929). *Br. J. Exp. Pathol.* **10**, 226.
Florkin, M. (1972). *Compr. Biochem.* **30**, 44.
Floss, H. G. (1979). Verbal remark reported by Vaněk et al. (1981a).
Foster, J. W. (1949). "Chemical Activities of Fungi." Academic Press, New York.
Fraenkel, G. S. (1959). *Science* **129**, 1466.
Fuller, J. G. (1968). "The Day of St. Anthony's Fire." Macmillan, New York.
Gaden, E. L. (1959). *J. Biochem. Microbiol. Technol. Eng.* **1**, 413.
Gale, E. F. (1951). "The Chemical Activities of Bacteria," 3rd ed., pp. 1–9. University Tutorial Press, London.
Gaughran, E. R. L. (1969). *Trans. N. Y. Acad. Sci.* [2] **31**, 3.
Geissman, T. A., and Crout, D. H. G. (1969). "Organic Chemistry of Secondary Plant Metabolism," pp. 2–19. Freeman, Cooper, & Co., San Francisco, California.
Gilbert, L. E., and Raven, P. H., eds. (1975). "Coevolution of Animals and Plants." Univ. of Texas Press, Austin.
Goldsmith, O. (1770). "The Deserted Village." See "Collected Works of Oliver Goldsmith" (A. Friedman, ed.), pp. 273–304. Oxford Univ. Press, London and New York, 1966.
Haavik, H. I. (1979). *Folia Microbiol.* (Prague) **24**, 365.
Harborne, J. D. (1977). "Introduction to Ecological Biochemistry." Academic Press, New York.
Harborne, J. D., ed. (1978). "Biochemical Aspects of Plant and Animal Coevolution." Academic Press, New York.

Harborne, J. D. (1986). *Nat. Prod. Rep.* **3**, 323.
Hare, R. (1970). "The Birth of Penicillin." Allen & Unwin, London.
Haslam, E. (1985). "Metabolites and Metabolism: A Commentary on Secondary Metabolism." Oxford Univ. Press (Clarendon), London and New York.
Haslam, E. (1986). *Nat. Prod. Rep.* **3**, 217.
Hayakawa, S. I. (1941). "Language in Action." Harcourt, Brace, New York.
Herbert, R. B. (1981). "The Biosynthesis of Secondary Metabolites." Chapman & Hall, New York.
Hobby, G. L. (1985). "Penicillin: Meeting the Challenge." Yale Univ. Press, New Haven, Connecticut.
Holtzer, H. Bischoff, R., and Chacko, S. (1969). In "Cellular Recognition" (R. T. Smith and R. A. Good, eds.), pp. 19–25. Appleton-Century-Crofts, New York.
Hütter, R. (1986). In "Biotechnology: A Comprehensive Treatise" (H. Pape and H.-J. Rehm, eds.), Vol. 4, pp. 4–17. Verlagsges., Weinheim.
Hütter, R., Leisinger, T., Nuesch, J., and Wehrli, W., eds. (1978). "Antibiotics and Other Secondary Metabolites." Academic Press, London.
James, J. B. (1987). *Soc. Ind. Microbiol. News* **37** (1), 32.
Janzen, D. H. (1977). *Am. Nat.* **111**, 691.
Janzen, D. H. (1978). In "Biochemical Aspects of Plant and Animal Coevolution" (J. B. Harborne, ed.), pp. 163–206. Academic Press, New York.
Johnson, S. (1755). "A Dictionary of the English Language." W. Strahan, London; see also the 1967 edition, AMS Press, New York. (The quotation occurs in the Preface and is attributed to R. Hooker, 1554[?]–1600.)
Kleinkauf, H., von Döhren, H., Dornauer, H., and Neseman, G., eds. (1985). "Regulation of Secondary Metabolite Formation," Workshop Conf. Hoechst, Vol. 16. Verlagsges., Weinheim.
Kossel, A. (1891). *Arch. Anat. Physiol. (Leipzig)* **15**, 181 (In this publication, Albrecht Kossel is referred to as Hr. Kossel and the initial is not given. The spelling used in the quotation is that of the original publication.)
Kostytschew, S. (1926). "Lehrbuch der Pflanzenphysiologie," Vol. 1, pp. 390–451. Springer, Berlin.
Krumphanzl, V., Sikyta, B., and Vaněk, Z., eds. (1982). "Overproduction of Microbial Products." Academic Press, New York.
Lancini, G., and Parenti, F. (1982). "Antibiotics: An Integrated View." Springer-Verlag, New York.
Lehninger, A. L. (1982). "Principles of Biochemistry," pp. 456–459. Worth, New York.
Luckner, M. (1972). "Secondary Metabolism in Plants and Animals." Academic Press, New York.
Luckner, M. (1984). "Secondary Metabolism in Microorganisms, Plants, and Animals." VEB Gustav Fischer Verlag.
Luckner, M. (1985). Quoted by H. von Döhren, in Kleinkauf et al. (1985, p. 370).
Luckner, M., and Schreiber, K., eds. (1979). "Regulation of Secondary Product and Plant Hormone Metabolism," Proc. 12th FEBS Meet., Vol. 55, Symp. S8. Pergamon, Oxford.
Luckner, M., Mothes, K., and Nover, L., eds. (1976). "Secondary Metabolism and Coevolution," Nova Acta Leopold., Suppl. No. 7. Dtsch. Akad. Naturf. Leopold., Halle (Saale).
Luckner, M., Nover, L., and Böhm, H. (1977). "Secondary Metabolism and Cell Differentiation." Springer-Verlag, Berlin and New York.
Macfarlane, G. (1979). "Howard Florey: The Making of a Great Scientist." Oxford Univ. Press, London and New York.

Macfarlane, G. (1984). "Alexander Fleming: The Man and the Myth." Harvard Univ. Press, Cambridge, Massachusetts.
Mann, J. (1978). "Secondary Metabolism," p. 6. Oxford Univ. Press (Clarendon), London and New York.
Martin, J. R., and Demain, A. L. (1980). *Microbiol. Rev.* **44**, 220.
Miall, L. M. (1975). In "The Filamentous Fungi" (J. E. Smith and D. R. Berry, eds.), Vol. 1, pp. 104–121. Edward Arnold, London.
Mothes, K. (1980). *Encycl. Plant Physiol., New Ser.* **8**, 1–10.
Nagaoka, K., and Demain, A. L. (1975). *J. Antibiot.* **28**, 627.
Nagel, E. (1945). *J. Philos.* **42**, 617.
Neijssel, O. M., and Tempest, D. W. (1979). *Symp. Soc. Gen. Microbiol.* **29**, 50–60.
Ogden, C. K., and Richards, I. A. (1923). "The Meaning of Meaning: A Study of the Influence of Language upon Thought and of the Science of Symbolism." Kegan Paul, Trench, Trubner & Co., London.
Paech, K. (1950). "Lehrbuch der Pflanzenphysiologie," Vol. 1, Part 2. Springer-Verlag, Berlin and New York.
Pfeffer, W. (1900). In "The Physiology of Plants" (A. J. Ewart, transl. and ed.), 2nd ed. Oxford Univ. Press (Clarendon), London and New York.
Raistrick, H. (1940). *Annu. Rev. Biochem.* **9**, 571.
Raistrick, H. (1949). *Proc. R. Soc. London, Ser. A* **199**, 141.
Raistrick, H., and Rice, F. A. H. (1971). *J. Chem. Soc. C* p. 3069.
Raistrick, H., and Rintoul, W. (1931). *Philos. Trans. R. Soc. London, Ser. B* **220**, 2.
Raphael, R. A. (1948). *R. Coll. Sci. J.* **18**, 42.
Rose, A. H., ed. (1979). "Secondary Products of Metabolism." Academic Press, New York.
Rosenthal, G. A., and Janzen, D. H., eds. (1979). "Herbivores: Their Interaction with Secondary Plant Metabolites." Academic Press, New York.
Rühland, W., ed. (1958). "Encyclopedia of Plant Physiology," Vol. 10. Springer-Verlag, Berlin and New York.
Runeckles, V. C., and Conn, E. E., eds. (1974). "Recent Advances in Phytochemistry," Vol. 8. Academic Press, New York.
Sachs, J. (1882). "Vorlesungen über Pflanzenphysiologie." Engelmann, Leipzig.
Schlegel, H. G. (1986). "General Microbiology," 6th ed., Engl. transl., p. 340. Cambridge Univ. Press, London and New York.
Siegler, D., and Price, P. W. (1976). *Am. Nat.* **110**, 101.
Stevenson, R. L. (1881). "El Dorado." An essay published in collected form as "Virginibus Puerisque and Other Papers (The Travels and Essays of Robert Louis Stevenson)", Vol. 13, pp. 106–109. Charles Scribner's, New York, 1901.
Steyn, P. S., ed. (1980). "The Biosynthesis of Mycotoxins: A Study in Secondary Metabolism." Academic Press, New York.
Swain, T. (1974). *Compr. Biochem.* **29**, Part A, 125–302.
Swain, T. (1976). In "Secondary Metabolism and Coevolution" (M. Luckner, K. Mothes, and L. Nover, eds.), Nova Acta Leopold., Suppl. No. 7. Dtsch. Akad. Naturf. Leopold., Halle (Saale).
Swain, T. (1977). *Annu. Rev. Plant Physiol.* **28**, 479.
Tatum, E. L. (1944). *Annu. Rev. Biochem.* **13**, 667.
Thom, C., and Raper, K. B. (1945). "A Manual of the Aspergilli," p. 294. Williams & Wilkins, Baltimore, Maryland.
Thomas, R. (1979). In "Comprehensive Organic Chemistry" (D. H. R. Barton and W. D. Ollis, eds.), Vol. 5, pp. 869–914. Pergamon, Oxford.
Torssell, K. B. G. (1983). "Natural Product Chemistry: A Mechanistic and Biosynthetic Approach to Secondary Metabolism." Wiley, Chichester.

Turner, W. B. (1971). "Fungal Metabolites." Academic Press, New York.
Turner, W. B., and Aldridge, D. C. (1983). "Fungal Metabolites II." Academic Press, New York.
Ullmann, S. (1951). "Words and their Use." Frederick Muller, London.
Ullmann, S. (1957). "The Principles of Semantics," 2nd ed. Philosophical Library, New York.
Vaněk, Z., Cudlin, J., and Krumphanzl, J. (1981a). *Folia Microbiol. (Prague)* Sect. Discuss., 1978–1980, p. 96.
Vaněk, Z., Cudlin, J., Blumauerova, M., Hostalek, Z., Podojil, M., Rehacek, A., and Krumphanzl, V. (1981b). "Physiology and Pathophysiology of the Production of Excessive Metabolites." Inst. Microbiol., Czech. Acad. Sci., Prague.
Vickery, M. L., and Vickery, B. (1981). "Secondary Plant Metabolism." University Park Press, Baltimore, Maryland.
Walker, J. B. (1974). *J. Biol. Chem.* **249,** 2397.
Wallen, L. L., Stodola, F. H., and Jackson, R. W. (1959). "Type Reactions in Fermentation Chemistry," Agric. Res. Serv., U.S. Dept. of Agriculture, Washington, D.C.
Weinberg, E. D. (1970). *Adv. Microb. Physiol.* **4,** 1.
Weinberg, E. D. (1971). *Perspect. Biol. Med.* **14,** 565.
Weinrich, U. (1980). "On Semantics." Univ. of Pennsylvania Press, Philadelphia.
Whiting, D. A. (1985). *Nat. Prod. Rep.* **2,** 389.
Whittaker, R. H. (1970). *In* "Chemical Ecology" (E. Sondheimer and J. B. Simeone, eds.), pp. 43–70. Academic Press, New York.
Woodruff, H. B. (1966). *Symp. Soc. Gen. Microbiol.* **7,** 168.
Zähner, H. (1979). *Folia Microbiol. (Prague)* **24,** 435.
Zähner, H., Drautz, H., and Weber, W. (1982). *In* "Bioactive Microbial Products: Search and Discovery" (J. D. Bu'Lock, L. J. Nisbet, and D. J. Winstanley, eds.), pp. 51–70. Academic Press, New York.

Microbial Production of Gibberellins: State of the Art

P. K. R. KUMAR AND B. K. LONSANE

Fermentation Technology and Bioengineering Discipline
Central Food Technological Research Institute
Mysore 570 013, India

I. Introduction
II. Historical Highlights
   A. 1895–1935
   B. 1936–1959
   C. 1960–1987
III. Chemistry of Gibberellins
   A. Gibberellins
   B. Gibberellic Acid
   C. Decomposition Products of GA_3
   D. Gibberellin Glycosides
   E. Esters of Gibberellins
IV. Mode of Action
   A. Interactions with Plant Tissues
   B. Interactions with Microorganisms, Algae, and Animals
   C. Mechanism of Action
   D. Structure–Activity Relationships
V. Commercial and Potential Uses
VI. Routes for Production
   A. Chemical Synthesis
   B. Extraction from Plants
   C. Microbial Fermentation
VII. Microorganisms Producing Gibberellins
   A. Fungi, Actinomycetes, Yeasts, and Bacteria
   B. Industrially Used Cultures
VIII. Biosynthesis Pathways
   A. Formation of Isopentenyl Pyrophosphate
   B. Formation of Terpenes and Terpenoids
   C. Formation of *ent*-Kaurene
   D. Formation of GA_{12}-Aldehyde
   E. Pathways beyond GA_{12}-Aldehyde
   F. Inhibitors of the Biosynthesis
IX. Liquid Surface Fermentation
X. Submerged Fermentation
   A. The Technique and Utility
   B. Physical Factors
   C. Nutritional Factors
   D. Growth Phases in Fermentor
   E. Regulation of GA_3 Production
   F. Kinetic Studies

G. Mathematical Models
H. Process Operation Strategies
I. Concomitant Products
J. Downstream Processing
XI. Use of Immobilized Whole Cells
XII. Solid-State Fermentation
   A. The Technique and Its Potential
   B. Physical Factors
   C. Nutritional Factors
   D. Large-Scale Trial
   E. Fed-Batch Process
   F. Growth Pattern and GA_3 Biosynthesis
   G. Concomitant Products
   H. Downstream Processing
XIII. Analytical Methods
   A. Bioassays
   B. Physicochemical Instrumentation Methods
XIV. Economic Considerations
   A. Liquid Surface Fermentation
   B. Submerged Batch Fermentation
   C. Continuous Submerged Fermentation
   D. GA_3 Production by Immobilized Whole Cells
   E. Solid-State Fermentation
   F. Comparative Economics
XV. Epilogue
   References

I. Introduction

Gibberellins (GAs), a large family of closely related diterpenoid acids biologically derived from tetracyclic diterpenoid hydrocarbon (Graebe and Ropers, 1978), represent an important group of potent plant growth hormones. Among these, gibberellic acid (GA_3) has received the greatest attention (Jefferys, 1970). GA_3 regulates the rate of growth and final length of internodes (Graebe and Ropers, 1978) and is extensively used for a variety of beneficial effects (Phinney, 1983). The importance of GA_3 is reflected in the publication of a number of books (Krishnamoorthy, 1975; Lenton, 1980; Crozier, 1983), chapters in books related to plant growth regulators, and specific reviews (Gould, 1961; Cross, 1968; Sembdner et al., 1972; Agnistikova et al., 1974; Graebe and Ropers, 1978; Hedden, 1979; Phinney, 1979; Crozier, 1981; MacMillan, 1980; Goodwin and Mercer, 1983; Mahadevan, 1984). These cover mainly aspects of the chemistry, biosynthetic pathways, mode of action, structure–activity relationship, and uses. GA_3 is a high-value industrially important biochemical selling at $1–3/g (United States currency) in the international market, depending on the purity and potency. Therefore, its use at present is limited to high-premium crops. The world's annual requirement of GA_3 in 1980 was 12–15 tons (Martin, 1983).

The industrial process currently used for fermentative production of GA_3 is based on submerged fermentation (SmF) techniques. In spite of the use of the best process technology, the yield of GA_3 is low. As early as 1979, it was stressed that the SmF process used today is approaching a saturation point beyond which cost reduction is impossible (Vass and Jefferys, 1979). Thus, it is essential to look beyond the conventional SmF technique and explore other avenues to achieve an economical process. This might become possible by attaining various goals, such as production of the metabolite in higher concentration, lower medium cost, lesser capital investment and plant operating cost, and reduced expenses on downstream processing through the application of process operating strategies such as continuous fermentation, fed-batch culture, and use of immobilized growing cells. The recent developments, pioneered at the Central Food Technological Research Institute (CFTRI), Mysore, India, in the extension of the solid-state fermentation technique to the production of GA_3 showed promise in attaining some of the above goals.

The information on the fermentative production of GAs is scattered in the literature except for a few reviews in earlier years (Stowe and Yamaki, 1957; Stowe et al., 1961; Wakagi, 1958; Grove, 1961, 1963; Hanson, 1969; Jefferys, 1970; Grigorov and Angelova, 1976; Vass and Jefferys, 1979) and a book dealing with early information up to 1957–1958 (Stodola, 1958). In fact, the research and development efforts beyond 1965 on the fermentative production of GAs were on a low-key basis except for a renewed interest in the current decade. A critical analysis of all fermentation aspects to understand various unit parameters, process operation strategies and the evaluation of the possible uses of these in the production of GA_3 at higher yield and lower cost has not been dealt with systematically in one place. The present review, therefore, is an effort to fulfill this need and to critically examine the existing as well as the potential techniques of fermentation and product recovery in the production of GAs. The salient features of the chemistry, mode of action, and uses of GAs are also given, as these are of interest to fermentation technologists.

II. Historical Highlights

The events leading to the discovery of GAs and their acceptance as plant growth hormones indicate the problems and breakthroughs that are usually faced in understanding any scientific process or in the development of any new product. The discovery of GAs in higher plants and microorganisms is classified as an intriguing example of serendipity in science (Phinney, 1983), as it has resulted from research carried out to control the Bakanae disease of rice plants. These historical accounts were reviewed extensively by Phinney (1983).

A. 1895–1935

A disease of rice plants, resulting in excessive elongation of the seedlings and the lack of fruit even in mature plants was given many different names, such as Bakanae, Ahoine, Yurei, Somen nae, Naganae, Sasanae, Yarinae, Yarikatsugi, Oyakata, Otokonae, and Onnanae, by Japanese farmers (Hori, 1903). The early work established the induction of Bakanae disease in healthy rice plants by infecting them with Bakanae fungus (Hori, 1898) and attributed its occurrence to the stimulus given by the fungal mycelium (Sawada, 1912). The ability of sterilized culture filtrates of the fungus to cause Bakanae disease, when injected into hollow stems of the young plants, and the development of the disease in the plants germinated from rice grains soaked in sterile extract, established that the disease is caused by a metabolite produced by the fungus (Kurosawa, 1926). The ability of the fungus to produce another metabolite capable of inhibiting the growth of rice plants and suppression of the production of this inhibitor when the fungus was grown at pH 3.0 were reported subsequently (Kurosawa, 1930, 1934).

Further developments in the isolation of the active principle and studies on the chemical as well as the biological properties were due to the interest taken by other workers (Ito and Kimura, 1931; Nisikado, 1931, 1932). Though the early attempts to crystallize the metabolite were not successful, these studies revealed a number of useful chemical properties, such as adsorption of the active principle on charcoal, ability to pass through a semipermeable membrane, and stability of the metabolite even at boiling water temperature. The taxonomic position of the fungus responsible for the disease was ill defined up to 1930. Subsequently, the imperfect stage of the fungus was named *Fusarium moniliforme* (Sheldon) while the perfect stage was named *Gibberella fujikuroi* (Sawada) Wollen-Weber (Nisikado, 1931, 1932). The major breakthrough in obtaining crystalline material from the culture filtrate was achieved in 1934 but the crystallized material, identified as picolinic acid, was inhibitory to plants (Yabuta *et al.*, 1934). By changing the cultural conditions during growth, a noncrystalline substance, named "gibberellin" (GA), was obtained (Yabuta, 1935).

This was the first time the word "gibberellin" was used for a substance with plant growth-stimulating properties.

B. 1936–1959

The stimulation of seedling elongation in a number of economically significant plants by the noncrystalline, but purified, samples of GA was reported by Yabuta and Hayashi (1936, 1938). Subsequently, the crystallization of the active factor into GA A and B (Yabuta and Sumiki, 1938) and their chemical as well as biological properties were reported (Yabuta et al., 1941; Yabuta and Sumiki, 1944). The work on GAs was confined to Japan until its initiation in 1950 in England by the Imperial Chemical Industry (ICI) (Brian et al., 1954; Cross, 1954; Curtis and Cross, 1954). The initial work at ICI was confined to evaluation of a large number of Fusarium strains, chemical characterization of GAs, large-scale production trials under the SmF technique using Hoover's washing machine, interactions of GAs with auxins, and the presence of GAs in higher plants (Borrow et al., 1961; West and Phinney, 1956). The field trials were also carried out in the early 1950s by ICI on the physiological activities of GAs. The promotion of grass growth by 0.6–10.8 CWt(hundredweight)/acre, though with a small reduction in protein content, was reported (Brian, 1959).

At about the same time, work was under progress in the United States on the Bakanae symptoms in field-grown wheat caused by Fusarium sp. Subsequently, the optimum fermentation parameters for the production of GAs were reported. Fermentation experiments up to the 300-gallon level were also attempted in stainless steel fermentors under SmF conditions at the United States Department of Agriculture (Stodola et al., 1955). However, the product was found to be different than that reported by Japanese scientists and hence was referred to as gibberellin X (Stodola et al., 1955). The effects of GAs in barley malting and in other agricultural practices were also investigated. Some important results included acceleration of malting of barley, greatly increased fruit size in grapes, overcoming of dormancy and frost damage, and other desirable benefits such as fruit setting, flowering and senescence (Griggs and Iwakiri, 1961; Wittwer and Bukovac, 1958; Donoho and Walker, 1957; Weaver, 1959; Galun, 1959; Brian et al., 1959a,b). Consequently, the manufacture of GAs at ICI's penicillin plant in Trafford Park, Manchester, England, was initiated in 1952 to meet the growing demand. In addition, licences to five United States companies for the manufacture of GAs were granted by ICI. All earlier efforts to produce GAs were based on the use of liquid surface fermentation

(Yabuta et al., 1940; Curtis and Cross, 1954), but the SmF technique was used from 1955 (Stodola et al., 1955).

C. 1960–1987

In the early 1960s, the effects of GAs on a large number of crops, guinea pigs, poultry, and microorganisms were investigated extensively (Schwartz, 1963; Warden and Schaible, 1958; Aleksandrov, 1964; Dahlström et al., 1961). These resulted in unconfirmed claims of increased growth rates and other beneficial effects (Mees and Elson, 1978). Around 1961–1963, a number of attempts were also made to chemically synthesize GAs and their intermediate compounds using various approaches (Money et al., 1961; Kos and Loewenthal, 1963). These efforts were continued and a number of publications appeared in more recent years (Corey et al., 1978a,b; Corey and Smith, 1979; Corey and Munroe, 1982). The constant, continued interest in GAs has also culminated in reports of a number of GAs from both plant and microbial sources. The attempts to improve the yields or economics of the microbial production of GAs include the extension of continuous fermentation (Holme and Zacharias, 1965) and the production of GAs using immobilized whole cells (Heinrich and Rehm, 1981; Kahlon and Malhotra, 1986). The potential of the solid-state fermentation (SSF) technique was also examined in the early 1980s at the CFTRI (Prapulla et al., 1983).

III. Chemistry of Gibberellins

Progress in the chemistry of GAs was rapid after 1960 mainly due to the development of accurate and improved purification and analytical techniques (Graebe and Ropers, 1978). It has resulted in chemical characterization of a total of 71 GAs, which are distinguished from each other by a subscript numeral (Phinney, 1983; Yamane et al., 1985). It was stressed by Crozier (1981) that new additions to the existing list of GAs will undoubtedly continue, as many more potential permutations of GA structure are possible. The trivial nomenclature GA_1–GA_{71} also allows new additions by simply allocating subsequent numbers as per the procedure proposed by MacMillan and Takahashi (1968). The structures or chemical names of GA_1–GA_{66}, GA_{67}–GA_{68}, and GA_{69}–GA_{71} are reported by Crozier (1983), the American Chemical Society (1982–1986), and Yamane et al. (1985), respectively. Among these, a total of 25 GAs are produced by microorganisms and the information on carbon atoms present in them, molecular weight, sources, and references are given in Table I.

TABLE I
GAs Produced by Microorganisms

GAs	Class[a]	Molecular weight	Source	Reference
GA_1	A	344	Gibberella fujikuroi	Stodola et al. (1955), Grove et al. (1958)
GA_2	A	346	G. fujikuroi	Takahashi et al. (1955), Grove (1961)
GA_3	A	346	G. fujikuroi	Cross (1954), Stodola et al. (1955)
			Neurospora crassa	Kawanabe et al. (1983)
			Fusarium moniliforme	Ganchev et al. (1984), Gohlwar et al. (1984)
GA_4	A	328	G. fujikuroi	Takahashi et al. (1957)
			Sphaceloma manihoticola	Wilhelm and Graebe (1979)
GA_7	A	330	G. fujikuroi	Cross et al. (1962a,b)
GA_9	A	316	G. fujikuroi	Cross et al. (1962a,b)
GA_{10}	A	330	G. fujikuroi	Hanson (1966)
GA_{11}	A	330	G. fujikuroi	Brown et al. (1967)
GA_{12}	A	318	G. fujikuroi	Cross and Norton (1965)
GA_{13}	B	378	G. fujikuroi	Galt (1965)
GA_{14}	A	334	G. fujikuroi	Galt (1968)
GA_{15}	B	330	G. fujikuroi	Hanson (1967)
GA_{16}	A	348	G. fujikuroi	Galt (1968), Bearder and MacMillan (1973)
GA_{24}	B	346	G. fujikuroi	Harrison and MacMillan (1971)
GA_{25}	B	362	G. fujikuroi	Harrison and MacMillan (1971)
GA_{36}	B	370	G. fujikuroi	Bearder and MacMillan (1972)
GA_{37}	B	346	G. fujikuroi	Bearder and MacMillan (1973)
GA_{40}	A	332	G. fujikuroi	Yamaguchi et al. (1975).
GA_{41}	B	396	G. fujikuroi	Bearder and MacMillan (1973)

(continued)

TABLE I (continued)

GAs	Class[a]	Molecular weight	Source	Reference
GA_{42}	A	352	G. fujikuroi	Bearder and MacMillan (1973)
GA_{47}	A	348	G. fujikuroi	Beeley and MacMillan (1976)
GA_{54}	A	348	G. fujikuroi	Murofushi et al. (1979)
GA_{55}	A	364	G. fujikuroi	Murofushi et al. (1979)
GA_{56}	A	364	G. fujikuroi	Murofushi et al. (1979)
GA_{57}	A	364	G. fujikuroi	Murofushi et al. (1980)

[a] A, C-19 GAs; B, C-20 GAs.

A. Gibberellins

All GAs have the *ent*-gibberellane skeleton and are divided into two groups based on the possession of either 19 or 20 carbon atoms (Crozier, 1981). However, the basic carbon skeleton of all gibberellins and the numbering of carbons are similar (Crozier, 1983) and are shown in Fig. 1. The C_{19} GAs have 19 carbons and contain a 19 → 10 γ-lactone bridge, except for a 19 → 2 linkage in the case of GA_{11} (Crozier, 1981). The C_{20} GAs contain 20 carbons and this additional carbon exists as a CH_3, CH_2OH, CHO, or COOH function (Hanson, 1968). The 20-CH_2OH group is reported to form a 19 → 20 δ-lactone bridge, while an equilibrium with a 19 → 20 δ-lactol ring exists in the case of the 20-CHO function (Hanson, 1968). Variations and other modification to the above configurations (Hanson, 1968) are given in Table II.

B. Gibberellic Acid

Gibberellic acid (GA_3), having an empirical formula $C_{19}H_{22}O_6$, is chemically characterized as a tetracarbocyclic dihydroxy-γ-lactonic acid containing two ethylene bonds and one free carboxylic acid group (Cross, 1954; Hanson, 1968). Its structure was obtained by studying the acidic decomposition products (Hanson, 1968), while the absolute stereochemistry was deduced on the basis of optical rotary dispersion (Cross et al., 1959a,b; Stork and Newman, 1959; Aldridge et al., 1963; Bourne et al., 1963), circular dichroism (Scott et al., 1964) and X-ray analysis (McCapra et al., 1966; Hartsuck and Lipscomb, 1963). It gives

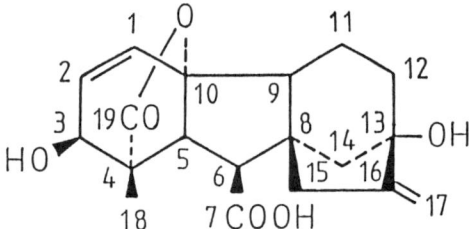

FIG. 1. Basic carbon skeleton of gibberellins and the numbering of carbon atoms.

an intense blue color and a strong blue fluorescence with cold sulfuric acid but does not reduce Fehling's solution or ammoniacal silver nitrate. It consumes two equivalents of an alkali on titration and did not form any color with ferric chloride (Hanson, 1968).

Amorphous or crystalline GA_3 is a colorless substance with a melting point at 233–235°C and optical rotation of +82° in ethyl alcohol (Sukh Dev and Misra, 1986). It is readily soluble in ethanol, methanol, acetone, dimethyl sulfoxide, and alkaline aqueous solution. Its solubility is very low in water, ethyl acetate, and ether, while it is soluble with difficulty in chloroform, carbon tetrachloride, and petroleum ether (Yabuta and Hayashi, 1939). The spectral data of GA_3 were compiled by Sukh Dev and Misra (1986). GA_3 is one of the least stable compounds among all GAs, including its aqueous solutions (Graebe and Ropers, 1978).

TABLE II

Variations and Additional Modifications to the Basic Configurations of C-20 GAs

Serial No.	Position	Group/modification
1	2, 3 and 1, 10	Epoxide
2	C-3 and C-12	Keto
3	C-1, C-2, C-12, and C-15	β-Hydroxylation
4	C-1, C-2, C-12, and C-16	α-Hydroxylation
5	18-Methyl group	Oxidation to carbonyl and carboxyl functions
6	1, 2 and 2, 3	Introduction of double bond
7	C-20	Oxidation state
8	3β- and 13α-Hydroxyl group	Presence/absence

C. Decomposition Products of GA_3

At room temperature, GA_3 in aqueous solution was reported to undergo decomposition to gibberellenic acid and iso-GA_3 to a significant extent (Cross et al., 1961; Pryce, 1973; Moffatt, 1960). A number of other decomposition products such as iso-GA_3 hydroxy acid, allogibberic acid, epi-allogibberic acid, and dehydroallogibberic acid were known to be formed upon autoclaving of aqueous solutions of GA_3 (Graebe and Ropers, 1978). The other decomposition products detected in fermentation broth are described in Section X,I. The structures of these decomposition products were reported by Graebe and Ropers (1978) and Cross et al. (1963). The biological activities of these decomposition products varied widely. The inhibition of flowering in Lemna perpusilla 6746 (Hodson and Hamner, 1971) and the presence of 10–25% residual activity (Graebe and Ropers, 1978) in autoclaved aqueous solutions of GA_3 are well known. Iso-GA_3, iso-GA_3 hydroxy acid, gibberellenic acid, and allogibberic acid are also reported to have some GA-like activity in a number of bioassays (Brian et al., 1967; Sembdner et al., 1964; Paleg et al., 1964). The biological activity of dehydrogibberellic acid is lower than that of GA_3 but is slightly higher than that of allogibberic acid at higher concentrations (Pryce, 1973). However, epi-allogibberic acid is reported to be biologically inactive in a number of bioassays (Brian et al., 1967).

D. Gibberellin Glycosides

Gibberellin glycosides have been isolated from immature seeds of Pharbitis nil (Yokota et al., 1969; 1971a,b) and also from other plants (Schreiber et al., 1970; Yamane et al., 1971). In these glucosides, glucose is always present in the pyranose form bonded by an anomeric atom to a hydroxyl group of the gibberellin (Russell, 1975). Glucose and free GAs or the nor-ketone can be liberated from GA glucosides by the action of glycosidases or by acid hydrolysis. The separation of these hydrolyzed products on thin-layer paper chromatography was reported (Yokota et al., 1971a,b). The R_f values of seven glucosides using three different systems were reported by Yokota et al. (1971a,b). The function of GA conjugates is still not clear. In fact, they were considered as storage or depot forms of the free GAs (Stoddart and Venis, 1980). It has also been proposed that the polar, water-soluble conjugates may act as transportable forms.

E. Esters of Gibberellins

Two types of gibberellin esters, i.e., acetyl or glucosyl esters, have been reported. The acetyl gibberellin is acidic in nature and is chemically more akin to gibberellin glucosides (Russell, 1975). The acetyl gibberellic acid was reported to be produced by G. fujikuroi and has been shown to possess a biological activity as high as gibberellic acid in bioassays (Schreiber et al., 1966). Several glucosyl esters (Hiraga et al., 1972) from mature seeds of Phaseolus vulgaris were also reported. Glucose and free gibberellins can be released by acid, alkali, or enzymatic hydrolysis (Russell, 1975). The ester linkage is attached to C-7 carboxylic group and was confirmed by nuclear magnetic resonance. Glucosyl esters of gibberellins A_1, A_4, A_{37}, and A_{38} have been isolated and characterized (Russell, 1975). These neutral substances have been reported to be the reserve forms of gibberellins (Hashimoto and Rappaport, 1966a,b).

IV. Mode of Action

A. Interactions with Plant Tissues

Extensive studies on interactions between GAs and plant tissues have indicated that GAs affect almost all of the plant organs and parts, although the most spectacular effects are on the stem elongation. The effects of GAs on plant tissues and organs are varied in nature.

Various reports are available on the effects of GAs on roots (Murashige, 1964; Brian et al., 1960; Paleg, 1965), shoots (Haissig, 1972; Brian and Hemming, 1955; Dostal, 1959; Maeda, 1960), leaves (Hayashi et al., 1956; Katsumi, 1970; Robbins, 1957), fruits and seeds (Lang, 1970; Luckwill et al., 1969; Jackson and Coombe, 1966), and germinating cereals and aleurone tissues (Laidman, 1983; Briggs, 1978). GAs were also reported to affect flowering (Zeevaart, 1969; Penner, 1960; Sironval, 1961) and senescence in leaves as well as fruits (Brian et al., 1959a; Coggins et al., 1960; Dostal and Leopold, 1967). The important results on the role of GAs in germinating cereals and aleurone tissues are given in Table III. These interactions of GAs with plant tissues are attributed to various phenomena, such as increased cell division and cell elongation as well as increased number of lateral roots (Burström, 1960; Butcher and Street, 1960), ability to counteract inhibition of stem growth by light (Kende and Lang, 1964), induction of cell expansion (Humphreys and Wheeler, 1960), stimulation of hydrolytic en-

TABLE III

THE ROLE OF GAs IN GERMINATING CEREALS AND ALEURONE TISSUES

Effect	Reference
Increased level of mRNA translatable for α-amylase	Ho and Varner (1974)
Induction of transcription of α-amylase-specific DNA sequences	Bernal-Lugo et al. (1981)
Increased level of α-amylase	Paleg (1960a,b), Yomo (1960a,b)
Two- to threefold increase in poly(A)	Berry and Sachar (1981)
Induction of other RNA species	Mann (1975)
Induction of fatty acid metabolism via β-oxidation and glyoxylate cycle	Newman and Briggs (1976)
de novo synthesis of phosphatases and diastase	Briggs (1973), Filner and Varner (1967), Hardie (1975)
Increased number of microbodies, i.e., glyoxysomes	Newman and Briggs (1976)
Initiation of carbohydrate and lipid metabolism	Buller et al. (1976)
Mobilization of mineral and protein reserve	Laidman (1983)
Controlling the release of hydrolysis products	Paleg (1960b)
A severalfold increased release of K^+, Mg^{2+}, Ca^{2+}, PO_4^{3-}, and amino acids	Laidman (1983)
Active turnover of phospholipid into triacyl glycerol	Laidman (1983)
Threefold stimulation of phospholipid metabolism	Koehler and Varner (1973)

zymes (Paleg, 1960a,b), regulation of DNA-dependent RNA synthesis (Fletcher and Osborne, 1965, 1966a), and control of cellular processes in germinating cereals and aleurone tissues (Laidman, 1983).

B. INTERACTIONS WITH MICROORGANISMS, ALGAE, AND ANIMALS

1. Microorganisms

Some reports are available in the literature on the varied effects of GAs on growth and metabolism of microorganisms (Table IV). Unfortunately, these results have not been followed up and, consequently, GAs remain unexplored in industrial fermentations.

TABLE IV
Effect of GAs on Microorganisms

Microorganism	Effect	Reference
Bacillus subtilis	Increased synthesis of α-amylase	Tang et al. (1973)
Candida spp.	Stimulation of growth	Santoro and Casida (1962)
Candida tropicalis	Reversion of toxic effect of trimethyl ammonium chloride and lauryl pyridinum chloride	Ohara et al. (1975)
Boletus spp. and a few other fungi	Inhibition of growth	Santoro and Casida (1962)
Fungal cultures	Formation of abnormal conidia	Roy (1964)
Rhizopus nigricans	Production of high amount of protein when cultured in molasses medium	Debska and Urbanek (1970)
Claviceps purpurea	Enhanced alkaloid content	Ostrovskii et al. (1961)
Fusarium solani	Inhibition of production of cellulase and polygalacturonase	Garg and Mehrotra (1977)

2. Algae

The presence of GA-like substances with biological activities was reported in Hypnea musciformis, Enteromorpha prolifera, Ecklonia sp., and other algae (Jennings and McComb, 1967; Henke and Schaller, 1965; Taylor and Wilkinson, 1977; Jennings, 1968). The extracts from these algae as well as pure GA_3 itself were shown to stimulate the growth of E. prolifera by these workers. The experimental data on Ecklonia radiata also indicated the role of endogenous GAs in regulation of the growth of algae (Jennings, 1971). The other effects of GAs on algae are presented in Table V and these indicate that literature reports are conflicting.

3. Animals

Various claims have been made for the beneficial pharmacological effects of GAs on animals, and these are reviewed extensively by Schwartz et al. (1983). Of these, the major effects are shown in Table VI.

TABLE V

Some Effects of GAs on Algae

Algae	Effect	Reference
Chlorella fusca	No effect on growth or sporulation	Lien et al. (1971)
Chlorella vulgaris, Chlorella ellipsoidea	No effect up to the 10^{-2} M level	Ciferri and Bertossi (1957), Tauriya et al. (1962)
Chlorella pyrenoidosa	Marked stimulation of growth and cell division	Kim and Greulach (1961), Bendana and Fried (1967)
Chlorella spp.	Inhibition at $>3 \times 10^{-6}$ M dose	Saono (1964)
Branch apices of Gracilaria verucosa	Stimulation of growth of slow-growing, but not of fast-growing, strains	Jennings (1971)
Excised branch of apices of Hypnea musciformis	No stimulation of growth by GA_3 or GA_7	Jennings (1971)
Gametophyte of Ecklonia radiata	Overcoming inhibition of growth induced by β-chloroethyl trimethyl ammonium chloride (CCC)	Jennings (1971)

TABLE VI

Pharmacological Effects of GAs on Animals

Animals	Effect	Reference
Mice, rats, rabbits, guinea pigs, dogs, chickens, etc.	Stimulation of body weight	Albertini et al. (1960), Kimura et al. (1959), Ratsimamanga and Boiteau (1964), Schwartz (1962), Warden and Schaible (1958)
Rats	Thyroid weight increase by 40% and testicles enlarged without increased spermatogenic activity	Gawienowski et al. (1977), Gawienowski and Chatterjee (1980)
Mice	Increased function of macrophages	Schwartz and Laginova (1966), Aleksandrowicz et al. (1975)
Guinea pigs	Stimulation of parenchymatous organs to make them less sensitive to tuberculosis infection	Schwartz (1963)
Guinea pigs	Protection against X-rays	Schwartz and Laginova (1966)

It is interesting to note that the increased body weight due to GAs in animals was not due to retention of water in tissues or the quality of the feed supplied to animals (Schwartz et al., 1983).

C. Mechanism of Action

Though the synthesis of GAs in plants was attributed to endoorganization evolved by the plants for speedy modification to cope with environmental fluctuations (Trewavas, 1979), the mechanism of action of GAs in promoting plant growth is not yet fully understood (Stoddart, 1983). However, it is well established that the mode of action of GAs is quite distinct from the animal hormones (Stoddart, 1983). In addition, the molecular basis for the action of GAs is also obscure (Trewavas, 1979). Interactions with receptor proteins (Stoddart, 1983), cellular membranes (Johnson and Kende, 1971; Evins and Varner, 1972; Jones, 1969a,b; Mann, 1975; Stoddart et al., 1978), and cell walls (Stoddart, 1983) were suggested as the possible mode of action of GAs in the plants.

D. Structure–Activity Relationships

These relationships are of vital importance in predicting the biological contribution of a particular part of the GA molecule and also for an insight into the necessity for certain molecular arrangements for biological activity and specificity in various bioassay systems (Brian et al., 1964; Reeve and Crozier, 1975; Graebe and Ropers, 1978; Agnistikova et al., 1974). These are also helpful in predicting, with some accuracy, biological activity and the most suitable bioassay system for any new GA (Hoad, 1983), although it is known that the activity of a particular GA depends on the type of bioassay used (Hill and Wimble, 1969; Bailiss and Hill, 1971). The structure–activity relationships were reviewed and discussed extensively by various workers (Graebe and Ropers, 1978; Crozier, 1981; Hoad, 1983; Hedden, 1979). The important features include (1) the GAs with 20 carbon atoms have lower activity as compared to the GAs with 19 carbon atoms; (2) among the GAs with 19 carbon atoms, those that were 3β-hydroxylated, $3\beta,\beta$-hydroxylated, or with a 1,2-unsaturated configuration showed higher activity and, among these, the latter showed highest activity; (3) much higher activity of C-20 GAs with δ-lactone or aldehyde group at C-20 position as compared with those with methyl or carboxylic groups at that position; (4) substantial or complete loss of biological activity due to the presence of a 2 β-hydroxyl group, depending on the bioassay employed (Sponsel et al., 1977); (5) 3β-hydroxylation of GA_4 is not a necessary prerequisite for expression of biological activity (Hoad et al., 1981); (6) 3α-OH GAs are less active than their 3β-OH counterparts (Brian et al., 1967; Murakami, 1970a,b; Agnistikova et al., 1974); (7) the dwarf rice bioassay gave a better response to C-20 GAs as compared to other GAs, although the response to these compounds by barley endosperm and lettuce hypocotyl assays was poor; (8) the cucumber hypocotyl assay responded to 13-hydroxylated GAs either poorly or negatively; and (9) the barley aleurone and dwarf pea assays are more dependent on 3β-hydroxylation.

The above generalizations indicate the involvement of the factors concerned with the shape of GAs and receptor molecules (Brian et al., 1967; Crozier et al., 1970). The application of the goodness of fit theory to these generalization also supports that the activity of a GA is based on the degree to which it fits a hypothetical receptor molecule or site. These generalizations also indicate the involvement of factors concerned with the metabolism of GAs by the plants. It is well known that the response of a bioassay is affected by the ability of plant tissue to metabolize the GA applied to it (Crozier, 1981). The enhancement of activity due to correct lipophilic–hydrophilic balance and subsequent effective transport also plays an important role (Hoad et al., 1981).

The above hypotheses on the structure–activity relationships of GAs are given credence by various experimental results, such as (1) a large number of GA glucosides from diverse plants exhibited large differences in their bioactivity (Sembdner et al., 1980); (2) GA glucosides showed reduced activity if the bioassays are carried out after adding inhibitors of β-glucosidase (Sembdner et al., 1972, 1973); (3) equal or lower activity of fluoro-substituted GAs as compared to the parent compounds (Stoddart, 1972; Hoal et al., 1983; Jones, 1976); (4) increased activity of hydroxylated GA_4 if the C-2 position was blocked to hydroxylation, but no such increase in case of GA_9 under similar conditions (Hoad et al., 1981); (5) a severalfold increase in the activity of GA_4 when modified to 2-2-diMe GA_4 (Hoad et al., 1981); (6) lower activity of 3 α-OH GA as compared to 3β-OH GA (Murakami, 1970a,b; Agnistikova et al., 1974); and (7) exhibition of bioactivity by the GA precursors, particularly ent-kaurene, ent-kaurenol, ent-kaurenoic acid, gradiflorenic acid, steviol, and ent-6α- and 7α-dihydroxy kaurenoic acids (Graebe and Ropers, 1978; Becker and Kempf, 1976; Murakami, 1972; Cross et al., 1970b).

V. Commercial and Potential Uses

Innumerable reports are available on the beneficial effects of GAs on plants, animals, and microorganisms. Among these, the current uses on a commercial scale are limited to those in agriculture, nurseries, viticulture, tea gardens, greenhouses, etc. GAs are used at ppm levels and their use results in a number of physiological effects, such as elimination of dormancy in seeds, acceleration of seed germinations, improvement in crop yields, marked stem elongation, promotion of fruit setting, induction of flowering in photoperiodically sensitive and cold-requiring plants, and the overcoming of dwarfism (Martin, 1983; Jones, 1983). It is of interest to note that the application of GAs to grapes resulted in an increased size of the berry as well as the cluster weight (Weaver, 1958). In addition, a mere application of 40 ppm of GAs at the time of defoliation overcame defoliation and reduction in berry weight, which were otherwise of the order of 50% and 38%, respectively at the time of harvest (Sidahmed and Kliewer, 1980). Another extensive use of GAs is in malting of barley for increasing the yield of malt and in reducing the steeping time (Paleg, 1960a,b; Yomo, 1960a,b; Gibbons, 1981; Briggs, 1978). GAs are also used in a variety of research projects and for pharmacological applications in animals (Schwartz et al., 1983). The well established and widely employed major commercial uses are presented in Table VII, while the major potential applications are listed in Table VIII.

VI. Routes for Production

A. Chemical Synthesis

A complete chemical synthesis of stereospecific GA_3 was reported by Corey et al. (1978a,b), using 2-allyloxyanisole as a starting material. A retrogressive synthesis was followed involving a number of steps, each of which varied in yield. By using another route, a tricyclic intermediate was obtained from the starting material, 4-benzyloxy-cyclohexanone (Corey and Smith, 1979). This route was reported to be similar to the previous method. The results of the simple synthesis of key intermediates were also reported by Corey and Munroe (1982). In addition, the total synthesis of GA_3 via a hydrofluorene route was reported by Hook et al. (1980). Some of the other reports available on suitable chemical synthesis of GAs include those on total synthesis of GA_4 (Cossey et al., 1980), partial synthesis of GA_{45} and GA_{63} (Dolan et

TABLE VII

Major Commercial Uses of GAs

Crop/commodity	Application technique	Beneficial effect	Reference
Grapes	Spraying, dipping	Thinning, size increase, harvest control	Weaver (1958)
Navel orange	Spraying	Harvest control	Coggins (1969, 1973)
Cotton	Spraying	Boll set	Walhood (1958)
Barley seeds	Soaking	Improved malting, reduction in steeping time	Paleg (1960a,b), Yomo (1960a,b), Gibbons (1981), Briggs (1978)
Potato seeds and other seeds	Soaking	Overcoming dormancy	Khan (1975)
Mango	Spraying	Delayed flowering	Kachru et al. (1972)
Artichoke	Spraying	Early maturation	De Angelis (1970)
Sugar cane	Spraying	Improved yield	Nickell (1980, 1982)
Cherries	Spraying	Increased fruit size, improved firmness, harvest control	Modlibowska and Wicken (1982)
Banana	Spraying	Increased fruit length	Lockhard (1975)
Lettuce	Spraying	Increased seed production	Harrington (1960)
Valencia orange	Spraying	Improved puffiness	Monselise et al. (1976)

TABLE VIII

IMPORTANT POTENTIAL USES OF GAs

Aspect	Beneficial effect	Reference
Long-day biennial plants	Flower acceleration	Wittwer and Bukovac (1958)
Dieffenbacteria maculata (Lodd)	More flowers per plant	Henry (1980)
Pea	Fruit set and fruit growth	Zucconi and Bukovac (1978)
Infection of potato leaves by Phytophthora infestans	Inhibition of sporangiophores of parasite	Weindhmayer (1964)
Cabbage black ringspot virus transmission to turnip plants by Myzus persicae	Reduced infection	Selman and Kandiah (1971)
Bacillus subtilis	Increased α-amylase production	Tang et al. (1973)
Claviceps purpurea	Increased alkaloid content	Ostrovskii et al. (1961)
Saccharomyces carlsbergensis	Stimulation of respiration	Rados (1970)

al., 1985), and an extensive review (Mander, 1982). Though stereospecific chemical synthesis of GA_3 is possible, if involves a series of steps and the use of expensive reagents. Therefore, chemical synthesis is not employed for the production of GA_3 anywhere in the world. In fact, attention given to the chemical synthesis of GA_3 is from the academic point of view only, as all attempts to chemically synthesize GA_3, at comparable cost to the fermentation technique, were not successful.

B. Extraction from Plants

The presence of GAs in higher plants was first reported by West and Phinney (1956) and subsequently their presence in a variety of plants was recorded by many workers (Lona, 1957; Murakami, 1957; MacMillan and Suter, 1958). The level of GAs in the vegetative parts of plants is generally much lower (a few micrograms per kilogram of fresh weight) as compared to the level of 10–100 mg/kg fresh weight in the reproductive parts, such as stamens and immature seeds (Graebe and Ropers, 1978). The exception is Phaseolus seeds, which were reported to contain as much as 16 mg/kg fresh weight (MacMillan et al., 1960). The GAs are present in higher plants mostly in bound forms, along with a small quantity of free GAs (Russell, 1975). The bound GAs are either neutral or more highly polar than free GAs. The procedure followed for the extraction of GAs from macerated plant tissues involved methanolic extraction (Kende, 1967) and subsequent partition of free GAs in ethyl acetate at pH 2.5. The leftover aqueous phase contains GA_{32} and bound GAs. The latter were purified by liquid–liquid partition (Sembdner et al., 1964; Yokota et al., 1971a,b). It is universally accepted that the extraction of GAs from plant tissue on a commercial scale is not economically feasible due to extremely low concentrations of GAs in plants.

C. Microbial Fermentation

Fermentation is the industrial method practiced for the manufacture of GA_3. Around 1954, ICI initiated the commercial production of GA_3 on a lab scale using G. fujikuroi to supply small quantities of GA_3 to scientists around the world. At present, a number of industries produce GA_3 by fermentation in a number of countries, e.g., the United States, England, Hungary, Poland, and Japan. The SmF technique is followed, although liquid surface fermentation was employed in earlier years.

It is interesting to follow the changes in the objectives of research on GAs in a country such as India. Up to the 1970s, the work in India on GAs was mostly confined to the isolation and identification of GAs from plant sources, characterization of Bakanae disease, effects of GAs on different plants, and chemical as well as biosynthetic aspects in plants (Prasad and Desai, 1952; Subramanian, 1951, 1952, 1954; Subba Rao, 1957; Nandi and Mondal, 1970). These investigations were more or less confined to meet basic research and academic interest rather than commercial pursuits. It was only in the late 1970s that efforts were made to develop indigenous technology for the production of GA_3 by fermentation. The work on the production of GAs by the SmF process was initiated at Saugar University, India, in 1976 (Thakur, 1981). Botryodiploidea theobromae, the strain selected for its ability to give the highest yield of GA_3 among all of the cultures isolated, gave about 0.31 and 0.50 g of GA_3 per liter in shake flasks and fermentors, respectively. At the Regional Research Laboratory, Jammu, India, the initial yield of 0.12–0.15 g of GA_3 per liter by G. fujikuroi GP-4 in SmF process was improved to 0.64 g of GA_3 per liter by optimizing cultural and nutritional parameters (Somal et al., 1978). The initial trials at CFTRI resulted in yields of 0.40–0.45 g of GA_3 per liter under SmF process conditions, even after the optimization of cultural, physical, and nutritional parameters [Central Food Technological Research Institute (CFTRI), 1981–1982). In subsequent attempts also, the yield of GA_3 obtained under SmF process conditions was 0.10 g/liter (Kumar, 1987).

VII. Microorganisms Producing Gibberellins

A. Fungi, Actinomycetes, Yeasts, and Bacteria

A number of microorganisms have been reported to produce GA_3 and GA-like substances. Among these, the fungal cultures (listed in Table IX) are able to produce GAs and GA-like activities in higher yields. In addition, a large number of actinomycetes and yeast cultures as well as bacteria are also known to produce these compounds (Table X). Several strains, isolated from a variety of fungus-affected plants, were tested by Borrow et al. (1955) but only nine cultures were able to show the Bakanae activity. Most of the work on production of GA_3 is based on the strains isolated from rice plants in Japan. Curtis (1957) studied fungi and 500 actinomycetes for their ability to produce GAs, and concluded that G. fujikuroi was the only microorganism that can produce GAs. Sanchez-Marroquin (1963) tested about 43 strains of Fusarium sp. and reported that F. moniliforme was able to give higher yields of GA_3 on a variety of media.

B. Industrially Used Cultures

The microorganisms preferred for commercial fermentation to produce GAs are G. fujikuroi or its imperfect stage, F. moniliforme (Borrow et al., 1955). None of the other microorganisms were able to produce GAs and GA-like substances at commercially feasible levels (Jefferys, 1970). The ability of various strains to produce GA_3 varied widely (Kumar, 1987; Riciova et al., 1960) and this fact illustrates the importance of strain selection. Even the yields from G. fujikuroi or F. moniliforme are low, and the optimization of the process as well as strain improvement by selection, mutation, and breeding techniques (Erokhina and Sokolova, 1966; Erokhina, 1967; Imshenetsky and Ul'Yanova, 1961, 1962a,b) are invariably undertaken to produce GAs in higher yields and at economic costs. For example, Imshenetsky and Ul'Yanova (1961, 1962a) have tested 120 mutants of G. fujikuroi and selected two mutants that produced 1.5–3 times more GAs at a faster rate than the parent culture. A process involving Sphaceloma manihoticola for the production of GAs was patented by Graebe and Rademacher (1979). However, the yield is very low (7320 µg/liter).

TABLE IX

Fungal Species Capable of Producing GAs and GA-like Substances

Strain	Reference
Agaricus biporous	Pegg (1973a)
Aspergillus flavus	Nair and Subba Rao (1977)
Aspergillus fumigatus	Aseeva and Barmenkova (1967)
Boletus eleqaus	Pegg (1973b)
Botryodiploidea theobromae	Thakur (1981)
Chaetomium sp.	Aube and Sackston (1965)
Clitocybe dicolor	Netien and Oddoux (1961)
Clitopilus pinsitus	Netien and Oddoux (1961)
Colletotrichum sp.	Aube and Sackston (1965)
Collybia conigena	Netien and Oddoux (1961)
Fusarium sp.	Krasilnikov et al. (1956)
Fusarium avenaceum	Manaka (1980)
Fusarium cucurbitae	Hirata (1958)
Fusarium culmonum	Manaka (1980)
Fusarium herbanum	Kurosawa (1926)
Fusarium heterosporum	Manaka (1980)
Fusarium moniliforme	Stodola et al. (1955), Zweig and DeVay (1959), Kefeli et al. (1969), Mertz and Henson (1967a,b), Maddox and Richert (1977a), El-Bahrawi (1977), Kurosawa (1926)
F. moniliforme var. *anthophilum*	Gordon (1960)
F. moniliforme f. *majus*	Hiroe (1969), Maddox and Richert (1977a)
F. moniliforme var. *subglutinans*	Gordon (1960)
Fusarium oxysporum	Hirata (1958), Malcolm and Fahy (1971)
Fusarium solani	Hirata (1958)
Gibberella fujikuroi	Ito and Kimura (1929), Sternberg (1962), Borrow et al. (1961, 1964a,b), Muromtsev and Globus (1976)
Gibberella lateritium	Hirata (1958)
Gibberella zeae	Hirata (1958)
Geophita fasicularis	Netien and Oddoux (1961)
Geopetalum geogenium	Netien and Oddoux (1961)
Grifola freudosa	Pegg (1973b)
Hypholama fasiculase	Pegg (1973b)
Lisea fujikuroi	Sawada (1917)
Nectria galligena	Taris and Clemencet (1970)
Neurospora crassa	Kawanabe et al. (1983)
Penicillium sp.	Aube and Sackston (1965)
Phallus impudicus	Pegg (1973b)
Phellinus pomaceous	Pegg (1973b)
Rhizoctonia solani	Peterson et al. (1963), Aube and Sackston (1965)
Schizophyllum commune	Kukharskaya (1968)
Spaceloma manihoticola	Rademacher and Graebe (1979)
Verticillium albo-atrum	Aube and Sackston (1965)
Verticillium lateritium	Domasch (1963)
Verticillium nigrescens	Aube and Sackston (1965)
Verticillium dahliae	Aube and Sackston (1965)

TABLE X

BACTERIA, ACTINOMYCETES, AND YEASTS CAPABLE OF PRODUCING
GAs AND GA-LIKE SUBSTANCES

Strain	Reference
Bacteria	
Azotobacter chroococcum	Vancura (1961)
Azotobacter vinelandii	Zarnescu and Nita (1964)
Agrobacterium tumefaciens	Galsky and Lippincott (1967)
Agrobacterium radiobacter	Katznelson and Cole (1965)
Bacillus licheniformis	Montuelle (1966)
Bacillus cereus	Montuelle (1966)
Bacillus pumilis	Montuelle (1966)
Bacillus megaterium	Montuelle (1966)
Bacillus polymyxa	Katznelson and Cole (1965)
Arthobacter sp.	Brown (1972)
Arthobacter globiformis	Katznelson et al. (1962)
Pseudomonas sp.	Montuelle and Cheminais (1964)
Pseudomonas sp. pigmented	Montuelle (1966)
Pseudomonas sp. nonpigmented	Montuelle (1966)
Pseudomonas fluorescens	Panosyan and Babayan (1966)
Pseudomonas liquefaciens	Panosyan and Babayan (1966)
Pseudomonas desmolytica	Panosyan and Babayan (1966)
Pseudomonas aeruginosa	Zarnescu and Nita (1964)
Flavobacterium sp.	Montuelle (1966)
Achromobacter sp.	Brown (1972)
Brevibacterium sp.	Brown (1972)
Alcaligenes sp.	Brown (1972)
Actinomycetes	
Actinomycetes (unidentified)	Krasilnikov et al. (1956)
Actinomyces sp.	Katznelson and Cole (1965)
Nocardia sp.	Brown (1972)
Yeasts	
Candida pulcherrima	Aseeva and Barmenkova (1967)
Torula pulcherrima	Krasilnikov et al. (1958)
Torulopsis sp.	Panosyan and Babayan (1966)

VIII. Biosynthesis Pathways

Progress in the elucidation of biosynthetic pathways of GAs has proceeded at a fast pace, mainly because of its straightforward nature. The results available so far suggest that the biosynthetic pathways are more or less similar in both higher plants and G. fujikuroi (Goodwin and Mercer, 1983). The elucidation of the biosynthetic pathways was mainly based on results of the feeding of radioactively labeled acetate or mevalonate to G. fujikuroi and their incorporation into GA_3 synthesized by the fungus (Birch et al., 1959a; Bearder et al., 1974). The other methods employed include the use of mutants of G. fujikuroi B1-41a with a blocked pathway between ent-kaurenal and ent-kaurenoic acid (Bearder et al., 1974), as well as the use of metabolic inhibitors (Ninnemann et al., 1964; Fall and West, 1971). The main sites for biosynthesis of GAs in higher plants were reported to be seedlings, shoots, root tips, young leaves, flower parts, mature seeds, and germinating embryos (Crozier and Reid, 1971; Stoddart, 1983; Phinney, 1983). Though it is not yet conclusively proven, the plastids were identified as the subcellular sites of GA synthesis in plants (Rogers et al., 1965). Spector and Phinney (1968) have also identified two genes in G. fujikuroi that control the production of GAs. The gene g_2 blocks the synthesis of GA_1 and GA_3, but has no effect on the synthesis of GA_4 and GA_7. The diffusion of GAs from the fungal system is either through the membrane-bound vesicles or as a result of general diffusional loss into the external medium (Dockerill et al., 1977; Stoddart, 1983).

A. Formation of Isopentenyl Pyrophosphate

Acetyl coenzyme A (CoA) is the basic compound in the biosynthesis of terpenoids and the pathway for formation of Δ^3-isopentenyl pyrophosphate (IPP) from acetyl CoA is outlined in Fig. 2. Two molecules of acetyl CoA under the action of acetyl CoA transferase forms acetoacetyl CoA (Birch et al., 1959a), which is further converted into hydroxymethylglutaryl CoA (HMG CoA) by the action of HMG CoA synthase. Subsequently, HMB CoA reductase catalyzes the formation of mevalonic acid via mevaldic acid. The mevalonic acid thus formed is converted into mevalonic acid–5-phosphate and mevalonic acid–5-pyrophosphate by the action of mevalonate kinase and phosphomevalonate kinase (Birch et al., 1959a). The enzyme, pyrophosphate mevalonate decarboxylase, further converts mevalonic acid-5-pyrophosphate into IPP (Popják and Cornforth, 1960).

B. Formation of Terpenes and Terpenoids

The different types of terpenes and terpenoid compounds obtained from IPP are shown in Fig. 3. IPP is first converted into dimethyl-allyl pyrophosphate (DMAPP) by the action of sulfhydryl enzyme, IPP isomerase (Graebe et al., 1965; Kandutsch et al., 1964; Nandi and

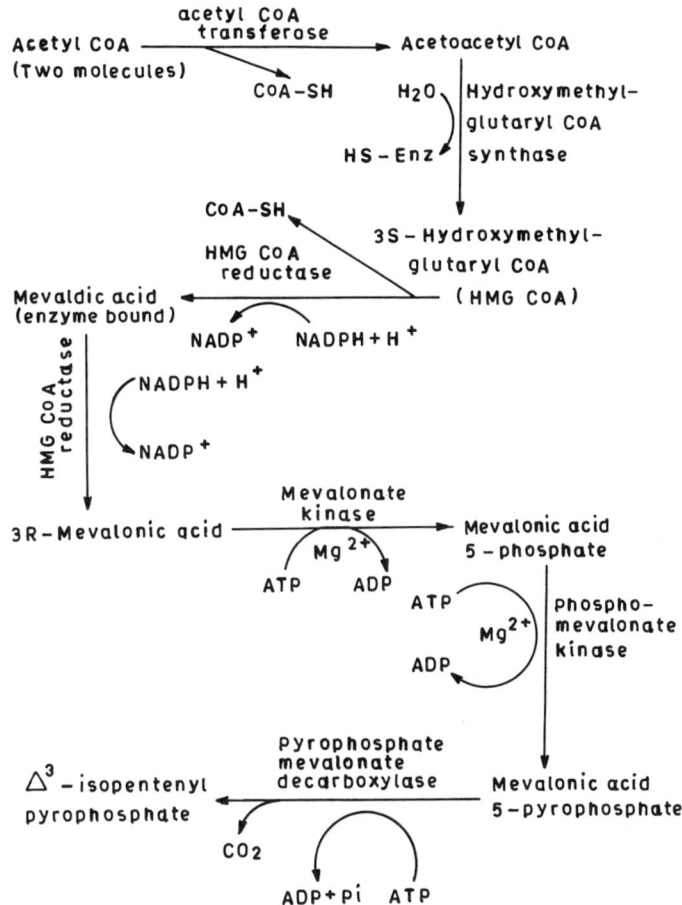

FIG. 2. Pathways for formation of isopentenyl pyrophosphate from acetyl CoA.

Porter, 1964). Either IPP or DMAPP is then converted into hemiterpenes. DMAPP can also act as a starter of chain elongation and, under the action of prenyl transferase, is condensed with an IPP molecule to form geranyl pyrophosphate (GPP), a C-10 compound (Cross, 1968). GPP is then channeled into either monoterpene biosynthesis or is condensed with one molecule of IPP to form farnesyl pyrophosphate (FPP), a C-15 compound. FPP can then undergo three different metabolic transformations, via., (1) channeling into sesquiterpenes (Kandutsch et al., 1964; Nandi and Porter, 1964) to undergo chain extension to form the C-20 compound, geranyl geranyl pyrophosphate (GGPP), (2)

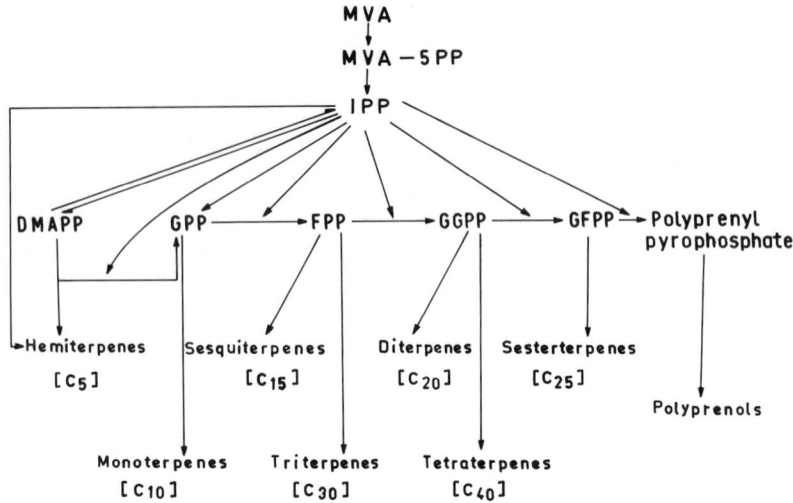

FIG. 3. Pathways for formation of terpenes and terpenoids from mevalonic acid. MVA, Mevalonic acid; MVA-5PP, mevalonic acid–5-pyrophosphate; IPP, Δ^3-isopentenyl pyrophosphate; DMAPP, dimethyl-allyl pyrophosphate; GPP, geranyl pyrophosphate; FPP, farnesyl pyrophosphate; GGPP, geranyl geranyl pyrophosphate, GFPP, geranyl farnesyl pyrophosphate.

tail-to-tail dimerization to form C-30 triterpenes, and (3) conversion to C-15 compounds, sesquiterpenes. Alternatively, GGPP can undergo similar types of metabolic transformations as FPP, viz., (1) diversion to diterpenes by chain elongation to geranyl farnesyl pyrophosphate, (2) the dimerization to C-40 tetraterpenes, or (3) formation of diterpenes. The enzymes catalyzing the formation of GGPP from mevalonic acid were detected in the soluble fraction of cell extract and were reported to originate from the cell organelles (Crozier, 1981).

C. Formation of ent-Kaurene

A cyclization process is involved in the formation of copalyl pyrophosphate from GGPP (Shechter and West, 1969). It is initiated by the electrophilic attachment of H$^+$ on the Δ^{14} double bond and is terminated by loss of H$^+$ from the C-7 methyl group, which thus becomes a methylene group. Further cyclization into ent-kaurene from copalyl pyrophosphate is initiated by a removal of the pyrophosphate ion, followed by cyclization of the resulting carbonium ion under the action of ent-kaurene synthase (Cross, 1968; Cross et al., 1963, 1964;

Fall and West, 1971). These steps are followed by the sequential oxidation of C-19 methyl group to *ent*-kaurenol, *ent*-kaurenal, and *ent*-kaurenoic acid (Shechter and West, 1969). Similarly, the hydroxylation at C-7 gave *ent*-7α-hydroxykaurenoic acid (West, 1973; Bearder *et al.*, 1975a). The enzymes that catalyze the steps from *ent*-kaurene to *ent*-7α-hydroxykaurenoic acid were found to be membrane bound and were identified as cytochrome P-450 mixed-function oxygenases (Dennis and West, 1967; Lew and West, 1971). The inhibition of the oxidation of *ent*-kaurene to *ent*-kaurenol, and that of *ent*-kaurenal to *ent*-kaurenoic acid by carbon monoxide, was also reported (Murphy and West, 1969). The generalized pathway for these steps is given in Fig. 4.

D. Formation of GA_{12}-Aldehyde

The contraction of ring B of *ent*-7α-hydroxykaurenoic acid forms a five-membered ring from one with six carbons by extrusion of C-7 as an aldehyde group and formation of new bond between C-6 and C-8, thereby leading to formation of GA_{12}-aldehyde (Birch *et al.*, 1959a). A generalized pathway (Goodwin and Mercer, 1983) is depicted in Fig. 4. Studies of *ent*-7α-hydroxykaurenoic acid sterospecifically labeled at C-6 with tritium showed that the 6α-hydrogen is lost in this process (Hanson *et al.*, 1972).

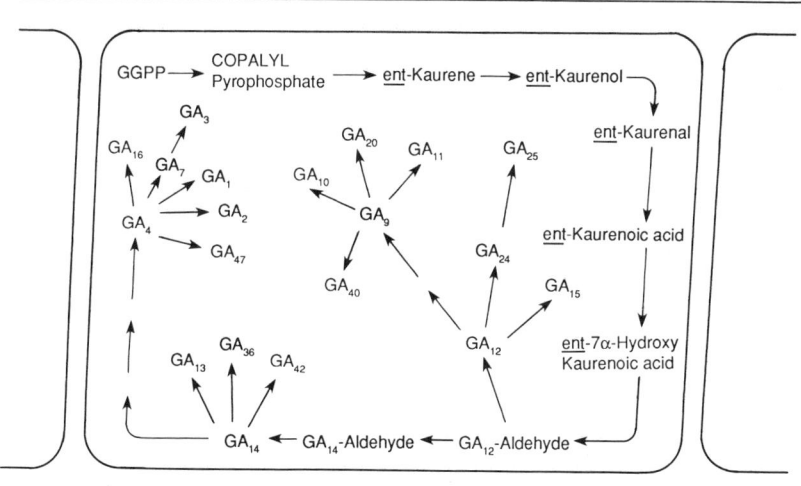

Fig. 4. Generalized pathway for formation of *ent*-kaurene, GA_{12}-aldehyde, and different GAs.

E. Pathways beyond GA_{12}-Aldehyde

The pathway for GA_3 biosynthesis from GA_{12}-aldehyde is presented in Fig. 4. This pathway has emanated from studies on the mutant of *G. fujikuroi* B1-41a, in which the pathway is blocked between *ent*-kaurenal and *ent*-kaurenoic acid but the mutant still possesses all the postblockage enzymes (Bearder et al., 1974).

It is well established that the biosynthetic pathway bifurcates at the GA_{12}-aldehyde stage to an early 3β-hydroxylation pathway and a non-3β-hydroxylation pathway (Bearder et al., 1975a,b; Bearder and MacMillan, 1973; Evans and Hanson, 1975; Cross, 1968). The former pathway leads to the formation of GA_3 and a range of other GAs, such as GA_1, GA_2, GA_3, GA_4, GA_7, GA_{13}, GA_{14}, GA_{16}, GA_{18}, and GA_{23} (Bearder et al., 1975a; Hedden et al., 1974), while the latter pathway yields GA_{12} and GA_9 as main products. Both of these pathways also operate in some higher plants and *G. fujikuroi*. The pathway operating beyond the formation of GA_4 is depicted in Table XI.

In the case of the non-3β-hydroxylation pathway, the oxidation of 7β-aldehyde (formyl) group of GA_{12}-aldehyde to a 7β-carboxyl group occurs to form GA_{12} (Evans and Hanson, 1975). The metabolites formed from GA_{12} are 3-deoxy GAs, such as GA_9, GA_{15}, GA_{24}, and GA_{25} (Bearder et al., 1975a). Further metabolism of GA_9 results in the formation of four minor metabolites, i.e., GA_{10}, GA_{11}, GA_{20}, and GA_{40} (Cross et al., 1964; Bearder et al., 1976). The studies with mutants of *F. moniliforme* and the use of precursors (Phinney and Spector, 1967) suggested that at least three separate pathways diverge. One pathway yielded GA_9 and then GA_{10}; the second yielded GA_1 and GA_9, which, on subsequent hydroxylation, yielded GA_1 and GA_3; while the third

TABLE XI

Pathways Operating beyond the Formation of GA_4

Starting metabolite	Reaction	Product
GA_4	α-Hydroxylation	GA_{16}
GA_4	2α-Hydroxylation	GA_{47}
GA_4	β-Hydroxylation	GA_1
GA_4	Introduction of Δ' double bond	GA_7
GA_7	β-Hydroxylation	GA_3
GA_{14}	Hydration at its Δ^{16} double bond	GA_{42}
GA_{14}	Oxidation at C-20	GA_{36} (10α-CHO) and GA_{13} (10α-COOH)

pathway provided C-19 GAs. The post-GA_{12}-aldehyde pathway in higher plants (Crozier, 1981) was derived from studies on several plant species, such as Cucurbita maxima (Graebe et al., 1974a–c; Graebe and Hedden, 1974), Phaseolus vulgaris (Hiraga et al., 1974a,b; Yamane et al., 1975; Sembdner 1968), and Pisum sativum (Frydman and MacMillan, 1975; Frydman et al., 1974).

Apart from the these two pathways, which are operating in some higher plants and also in G. fujikuroi, there is another pathway called the early 13-hydroxylation pathway (Durley et al., 1971, 1979; Frydman and MacMillan, 1975; Sponsel and MacMillan, 1977, 1978; Ropers et al., 1978). This pathway is absent in G. fujikuroi. Apart from the 13-hydroxylation pathway, the 2β-hydroxylation pathway also operates in some higher plant but it is absent in G. fujikuroi (Rappaport et al., 1974).

F. INHIBITORS OF THE BIOSYNTHESIS

A number of compounds, such as 2'-isopropyl-4'-trimethyl ammonium chloride-5'-methyl piperidine carboxylate (AMO-1618), β-chloroethyl trimethyl ammonium chloride (CCC), and tributyl 2,4-dichloro-benzyl phosphonium chloride (phosphon D), inhibit the biosynthesis of GAs by affecting the activity of ent-kaurene synthases (Ninnemann et al., 1964; Dennis et al., 1965; Lang, 1970). On the other hand, α-4-methoxyphenyl (ancymidol), α-cyclopropyl-5-pyrimidine methanol (EL 531), and closely related α-2,4-dichlorophenyl (triarinol) and α-phenyl-5-pyrimidine methanol (EL 273) inhibit all steps of the sequence, ent-kaurene→ent-kaurenol→ent-kaurenal→ent-kaurenoic acid (Coolbaugh and Hamilton, 1976). These are all oxidation steps catalyzed by cytochrome P-450 mixed-function oxygenases, and the inhibition is probably due to interaction with one of the nitrogen atoms of the pyrimidine in the heme of cytochrome P-450 and the consequent prevention of oxygen binding (Murphy and West, 1969). These compounds are also known to inhibit the growth of higher plants, probably by blocking GA formation (Leopold, 1971; Coolbaugh et al., 1978).

IX. Liquid Surface Fermentation

Liquid surface fermentation (LSF), often referred to as surface fermentation by many workers, was employed in earlier years for the production of GAs (Yabuta et al., 1934, 1939; Yabuta and Hayashi, 1936, 1937, 1939; Curtis and Cross, 1954; Kurosawa, 1934). The use of this fermentation technique was continued until 1955. Using a culture medium containing 3 g of glycerol, 3 g of NH_4Cl, 3 g of KH_2PO_4, and 1 liter of water, Yabuta et al. (1939) produced GAs at pH 3.2 by growing G. fujikuroi at 25°C for 30 days. The yield was 3.3–9.2 g of crude powdered GAs from 130 liters of medium (Yabuta et al., 1940). In Japan, glycerol was considered a most suitable carbon source at the optimum concentration at 1–3 ml of 88% glycerol in 100 ml of medium. The optimum initial pH for GAs production was 3.4 (Yabuta et al., 1939). Either HCl (Yabuta and Hayashi, 1939) or H_2SO_4 (Yabuta et al., 1939) was used for pH adjustment. It was also reported by Yabuta et al. (1934) that plant growth promoting GAs or plant growth retarding fusaric acid could be produced by *Fusarium heterosporium* Nees by altering the culture conditions. The use of glucose as a carbon source at pH 9.0 produced 0.5–1.0 g of fusaric acid/10 liters after 40–50 days of incubation. Glucose was also reported to be the best carbon source for production of fusarin at pH 9.0 in 15–30 days (Yabuta et al., 1939).

The production of GAs under LSF using 1 liter of medium in flat earthenware vessels at 25°C was reported by other workers. The incubation period employed was 10 days (Brian et al., 1957) or 15 days (Kitamura et al., 1953). The level of GAs in the medium beyond 15 days remained more or less constant. The LSF process using Raulin–Thom medium (Brian et al., 1946) containing 2.5% sucrose or glucose at an initial pH of 5.0 was also reported (Borrow et al., 1955). The inoculum was grown on potato dextrose agar in flat medicine bottles and the fermentation was carried out using 30 ml of medium in 100-ml flasks at 25°C. The yield of GAs was 40 mg/liter in sucrose medium but it was of lower magnitude in glucose medium (Borrow et al., 1955). The product from the broth, after removal of the fungal mat, was adsorbed on activated carbon and eluted with $MeOH–NH_4OH$ at room temperature (Yabuta and Hayashi, 1939). Further purification involved concentration of the extract, treatment with lead acetate, H_2S, and Na_2CO_3, and finally extraction with ether to obtain an almost colorless, amorphous powder of GAs consisting of GA A and B (Yabuta and Hayashi, 1939).

The LSF technique offers advantages, such as no foam formation and no mechanical damage to mycelial cells, as compared to the SmF process. It was pointed out by Kluyver and Perquin (1933) that the LSF technique is of value in exceptional cases in which the SmF technique cannot be used. The information on LSF for the production of GAs is of historical importance only. The use of LSF not only resulted in the production of a wide range of products along with GAs, the yields were also of very low magnitude and prolonged incubation periods were necessary. These disadvantages are inherently present in LSF, which is also known to be cumbersome, uneconomical, prone to contamination, labor intensive, and less reliable (Calam, 1969; Prescott and Dunn,

1949; Kluyver and Perquin, 1933; Johnson, 1954). The problems in precise control of temperature, the presence of nonuniform type of growth, the critical importance of the ratio of surface area to volume, and unreasonable space requirements have also been reported (Calam, 1969; Kluyver and Perquin, 1933). The level and nature of the inoculum are also most critical in LSF. The use of less inoculum will not cover the liquid surface completely and may also lead to a semisubmerged condition (Calam, 1969). The formation of islands of mycelial mass resulting from the tendency of spores to go to the edge of the culture vessel, loss to productivity in the case of inoculum that sinks in the liquid medium and inability to use inoculum from different stages of growth are also well known (Calam, 1969).

The LSF technique was also reported to be unfeasible for large-scale use, as the problems associated with the technique increased with the increase in the scale of operation (Kluyver and Perquin, 1933). It was even stated that the initiation of laboratory scale work based on the LSF technique was a waste of time, as it was useless for further development (Kluyver and Perquin, 1933). The comparatively lower contact of cells with nutrients leading to inefficient utilization, creation of O_2-limited and CO_2-enriched atmospheres due to the barrier formed by the mycelial mat, and inability to agitate the liquid are some of the other problems associated with LSF.

X. Submerged Fermentation

A. The Technique and Utility

In SmF, the liquid medium is employed in greater depth in vessels with a diameter:height ratio of 1:2 to 1:3. In most of the cases, all nutrients are dissolved in water except for those fermentations involving water-insoluble solid substrates or water-immiscible liquids. As the concentrations of the insoluble or immiscible substrates are quite low (1–10%), the medium used in the SmF process is free flowing in nature. This characteristic distinguishes it from the SSF technique, while the use of medium in greater depth makes it distinct from the LSF process. The employment of the medium in deep layers in the SmF process confers upon it many advantages, such as operational convenience, economy, reduced requirement of space, greater reliability in inoculation and growth pattern, improved contamination control, and precise control of parameters as compared to the LSF technique (Calam, 1969). However, it has been argued in recent years that the presence of the product in dilute form and the disposal of the large quantity of water in SmF processes is a major obstacle in developing biotechnological processes (Hahn-Hägerdal, 1986; Datar, 1986).

The SmF processes are carried out in shake flasks or aerated, agitated fermentors equipped for control of parameters (Banks, 1979; Aiba et al., 1973). It was only in the early 1940s that the stirred and aerated fermentor vessels were designed to achieve more uniform growth of the fungal strain under the SmF technique. The first report on the production of GA_3 by the SmF process appeared in 1953 (Kitamura et al., 1953). Subsequently, other workers in other countries have also reported the production of GA_3 by the SmF technique (Stodola et al., 1955, 1957; Borrow et al., 1955; Fuska et al., 1960; Darken et al., 1959).

Yields in the SmF and LSF Techniques

Borrow et al. (1955) have compared the yields of GA_3 by *G. fujikuroi* strain ACC 917 when grown under the SmF and LSF techniques. The LSF process involved the use of Raulin–Thom medium, initial pH adjustment to 5.0, temperature maintenance at 25°C, use of 1 liter of medium in a flat earthenware vessel, and a fermentation period of 15 days. The SmF technique involved the use of Raulin–Thom medium in 30 liters of fermentor, aeration by sterile, humid air at a rate of 0.3–0.5 vvm (volume of air/volume of medium/minute), temperature maintenance at 25°C, initial pH adjustment of 5.5, and a fermentation period of 18 days. The culture grew in the form of a thick uniform suspension of highly branched mycelia in the SmF process (Borrow et al., 1955) and in the form of the usual mycelial mat in the LSF process (Brian et al., 1946). The yield obtained by the SmF technique was 0.2 g/liter in the medium containing 4% glucose or sucrose as carbon sources, as compared to 0.04 g/liter in the medium containing 2.5% sucrose by the LSF process. It is interesting to note that any increase beyond 2.5% in sucrose concentration in the LSF technique did not increase the yield of GA_3. Also, sucrose was found to be a better carbon source than glucose in the LSF process, while both these substrates were equally effective in the SmF technique.

The production of pigment by *G. fujikuroi* strain ACC 917 was also found to be far greater in the SmF process than in the LSF technique (Borrow et al., 1955). It was also reported by Brian et al. (1957) that the fermentation time to achieve the maximum production of GA_3 and its derivatives, in the case of *G. fujikuroi*, was 3–5 days in the SmF process, in contrast to 10 days in the LSF technique. In addition, the LSF technique was also found to result in a wide range of products, usually in low yields, after prolonged periods of incubation (Jefferys, 1970).

B. Physical Factors

The SmF technique for production of GAs, like all other fermentation techniques (Munro, 1970; Patching and Rose, 1970), is influenced to a great extent by a variety of physical factors such as pH, temperature, light, and aeration as well as agitation rates. These factors were reviewed earlier by Jefferys (1970).

1. pH

The initial pH values generally employed by various workers were either around 5.5 or within the range 3.5–5.8 (Stodola et al., 1955; Hernandez and Mendoza, 1976; Shin et al., 1985; Borrow et al., 1964a). These were usually not controlled during fermentation (Holme and Zacharias, 1965) and thus resulted in a final pH of 3.9–5.2 or 1.8–1.9 or in slight alkalinization (Stodola et al., 1955; Sanchez-Marroquin, 1963). The effects of the pH of the fermentation medium on growth of the strain and production of GAs include (1) no significant differences in product yield and growth were reported when initial pH was in the range of 3.5–5.5 (Jefferys, 1970; Somal et al., 1978; Gohlwar et al., 1984); (2) the pattern of pH changes during fermentation were similar in media containing glycerol or glucose, although these patterns were different when the complete dose of glucose was added initially (Darken et al., 1959); (3) the specific growth rate was fairly constant in ammonium tartrate media in the pH range of 3.5–6.3 but it decreased beyond this range (Borrow et al., 1964a); (4) the inhibition of NH_3-N assimilation in presence of NO_3-N at pH 2.8–3.0 was observed until the pH increased to a value at which ammonia assimilation was resumed (Borrow et al., 1961, 1964a); (5) the yield constant for glucose was not affected by initial pH, although the yield constant for nitrogen was found to increase with increasing pH value (Borrow et al., 1964a); (6) the rate of production decreased when the pH was outside the range of 3.0–5.5 (Borrow et al., 1964a); (7) concomitant production of GA_1 and GA_3 was reported at low initial pH values (Stodola 1955; Kuhr et al., 1961; Fuska et al., 1962); and (8) concomitant production of GA_4, GA_7, GA_9, GA_{12}, GA_{14}, GA_{16}, and dibasic acid was reported at an initial pH around 7.0 and maintained during fermentation at that level (Cross, 1966; Galt, 1968; Sumiki et al., 1966; Cross and Norton, 1965; Cross et al., 1960a,b, 1962a).

2. Temperature

The effect of temperature on the production of GA_3 is dependent on the strain employed. The optimum temperatures reported for the production of GA_3 using *G. fujikuroi* or *F. moniliforme* include room temperature (Hitzman and Mills, 1963), 25°C (Shin et al., 1985; Stodola et al., 1955), 28.5–29.5°C (Borrow et al., 1964a,b), 30°C (Sanchez-Marroquin, 1963; Maddox and Richert, 1977a), 34°C (Kyowa Hakko Kogyo Co. Ltd., 1983), 29 ± 0.5°C (Holme and Zacharias, 1965), 27°C (Gohlwar et al., 1984), and 28°C (Darken et al., 1959). Kinetic studies showed that variation in the temperature of the fermentation broth affects the process both qualitatively and quantitatively (Borrow et al., 1964a,b). For example, phthalic acid was produced by the strain at only 12°C but it was absent when the culture was grown at other temperatures (Cross et al., 1963).

A process for the production of GA_3 by *G. fujikuroi* in higher yields by maintaining the temperature at 27.5–30°C in the final stage of fermentation was patented by Borrow et al. (1959c). This process emanated from the findings that optimum temperature for growth of the strain was between 31 and 32°C, while the production of GA_3 was maximal at ~29°C (Jefferys, 1970). At temperatures above 29°C, the reduction in the production of GA_3 was rather rapid. Improvement in the production of GA_4 and GA_7 at about 32°C was also reported by Sumiki et al. (1966).

3. *Aeration–Agitation*

Though the effect of aeration–agitation is pronounced in the production of GA_3 by the SmF process, no kinetic studies are available nor was it directly related to the rate of production of GA_3 (Jefferys, 1970). It is well recognized that a continuous supply of oxygen is required for production of GA_3 (Hitzman and Mills, 1963; Jefferys, 1970), as the biosynthesis progresses through compounds of increasing levels of oxidation (Geissman et al., 1966). The aeration rates employed by different workers were 0.25 (Stodola et al., 1955) or 0.3–0.5 vvm (Borrow et al., 1955) in production and seed fermentors. The other levels reported were 0.5 vol of air per minute in 5000–6300 gallons of medium (Bergman et al., 1962) and 0.5 liters of air per minute in a 3-liter fermentor or 5 liters of air per minute in a 20-liter fermentor (Holme and Zacharias, 1965). The agitation rates employed were 100, 150, 350, or 1400 rpm (Stodola et al., 1955; Holme and Zacharias, 1965; Bergman et al., 1962). The rates of aeration and agitation were not specified by Darken et al. (1959) and Sanchez-Marroquin (1963), while humidified air was used by Borrow et al. (1955). Oxygen uptakes of 85–165 and 90–340 mM O_2/liter/hr with an air supply of 5–20 liters/min in a 20-liter vessel and 0.5–2 liters/min in a 3-liter vessel, respectively, with the use of impellers of different diameters, were reported by Holme and Zacharias (1965). In continuous fermentation experiments, the medium was not agitated in the first 2–3 days, but aeration at the specific level was employed (Holme and Zacharias, 1965). The air was sterilized by heating to 250°C for 5 seconds in a sterilizer (Borrow et al., 1961).

An important aspect of the effect of CO_2 on the yield of GA_3 was patented by Borrow et al. (1958). CO_2-free air, when supplied to the fermentor, was found to result in a protracted lag phase, especially with low levels of inoculum. The elimination of the prolonged lag phase and increased overall growth rate and yield of GA_3 were obtained when air was supplemented with CO_2. For example, aeration of the medium with 40 liters of air and 1.5 liters of CO_2 per minute resulted in production of 0.62 g of GA_3 in contrast to 0.42 g when 40 liters of air per minute was supplied. Similarly, the yield of GA_3 was 0.63 g when 19 liters of air and 1 liter of CO_2 were supplied per minute, as compared to the production of 0.50 g with 20 liters of air per minute. In a two-stage process based on the use of a high percentage of inoculum for the production of GA_3, the culture was shown to generate sufficient CO_2 to meet its own demand. The interactions of different aeration and agitation rates and the consequent changes in oxygen and gas transfer processes were reported by some workers (Cross et al., 1963; Borrow et al., 1958, 1964a; Jefferys, 1970). These were caused mainly by the lower rate of oxygen supply or the conditions causing oxygen restriction and are (1) lower yields of acidic compounds, (2) diversion of metabolic pathways, (3) production of a new range of compounds, (4) development of estery smell in the medium, (5) unchanged yield constants for nitrogen, (6) introduction of linear growth phase, (7) lower utilization of glucose, (8) lower productivity, and (9) changes in biomass formation.

4. *Light*

The effect of GAs in counteracting light-induced dwarfism in pea seedlings (Brian, 1957; Gorter, 1961) and in a transient increase in extractable GA-like activity in seeds and leaf sections when exposed to red light (Koehler, 1966; Reid et al., 1968) are well known. In addition, the lower activity of GAs and its faster metabolism when grown during a photo-period as compared to the corresponding dark-grown controls have also been reported (Crozier and Audus, 1968; Bown et al., 1975). In contrast to the latter phenomena, the increased growth of the culture as well as the production of GA_3 due to the effect of light were reported (Zweig and DeVay, 1959; Mertz and Henson, 1967a,b; Mertz, 1970). The various effects of light on *G. fujikuroi* and *F. moniliforme* include

(1) stock cultures of F. moniliforme ATCC 1261, F-18, and ICI 917 when grown in darkness produced GAs after exposure to an 80-ft candle source for 48 hours; (2) inhibitors such as AMO 1618 and CCC acted differently on light- and dark-grown cultures; (3) over 60% more incorporation of leucine into GAs by G. fujikuroi grown under continuous illumination of 1800 lux as compared to the dark-grown controls; (4) GAs formation in cultures grown under light and in darkness was equal if the latter medium contained Ca^{2+} and a C:N ratio of 37.6; (5) the medium containing Ca^{2+} with a C/N ratio of 9.4 failed to increase GA synthesis in dark grown culture; and (6) the regulatory roles of light and Ca^{2+} were probably exerted at different sites. It was stated by Jefferys (1970) that absence of GA_3 in dark-grown cells is hard to reconcile with general experiences and that the phenomena are worthy of further investigations.

C. Nutritional Factors

The effect of nutritional factors on the production of GA_3 were earlier reviewed by Jefferys (1970).

1. Carbon Sources

Different workers have used a wide variety of carbon sources for the production of GAs (Table XII). A distinction is made in slowly and readily utilizable carbon sources, and both are often used at various ratios. The salient features of the effect of carbon sources on fermentative production of GAs include (1) an inhibitory effect of a higher concentration of glucose on the specific growth rate of G. fujikuroi strain ACC 917 was found (Borrow et al., 1964a); (2) a decreased rate of production of GA_3 and overall productivity with an increase in the initial concentration of glucose and its level at the time of exhaustion of nitrogen during the course of fermentation was found (Borrow et al., 1964a); (3) a combination of readily and slowly metabolizable carbon sources gave a higher yield of GA_3 (Darken et al., 1959); (4) use of molasses led to decreased but economically useful yields (Dietrich, 1960; De Conejos et al., 1975); (5) there was a 300–559% increase in the yield of GAs with the use of natural oils such as linseed oil, sunflower oil, olive oil, cottonseed oil, and ethyl palmitate as compared to the yields on sucrose (Muromtsev and Dubovaya, 1964) and a reported improvement in yield of 16.7% with linseed oil (Kumar, 1987); (6) there was a 49% decrease in the yield of GAs with stearic acid as compared to sucrose (Muromtsev and Dubovaya, 1964); (7) there was improved yield of GAs with the use of carbohydrate polymers such as

TABLE XII
Carbon Substrates Used in the Production of GAs

Substrate/combination	Concentration/ratio (g/liter)	Reference
Glucose or sucrose	25	Borrow et al. (1955)
Glycerol esters or glycerol	20–100	Borrow et al. (1959b)
Glucose or lactose	160	Holme and Zacharias (1965)
Sugar cane molasses	40	De Conejos et al. (1975)
Whey permeate	40	Gohlwar et al. (1984)
Molasses residue, whey, sulfite waste liquor, or skimmed milk	—	Dietrich (1960)
Sunflower oil	—	Agnistikova et al. (1966), Erokhina (1967), Muromtsev et al. (1968)
Linseed oil, olive oil, cotton seed oil, ethyl palmitate, or stearic acid	40	Muromtsev and Dubovaya (1964)
Hydrocarbons or diesel oil	—	Flippin et al. (1964), Hitzman and Mills (1963)
Plant meals, e.g., soya flour, soya meal, or cotton seed meal	—	ICI (1960a)
Olive oil, olive pulp, or by-products of olive oil extraction	200	Hernandez and Mendoza (1976)
Glucose	12.5–30	Stodola et al. (1955), Sanchez-Marroquin (1963)
Dextrin + sucrose	50:80	Bergman et al. (1962)
Glycerol + starch	20:30	Darken et al (1959)
Glycerol + glucose	20:50	Darken et al. (1959)
Glycerol + glucose + lactose	20:10:20	Darken et al. 1959)
Glycerol + lactose	20:50	Darken et al. (1959)
Corn steep liquor + sucrose	15:30	Darken et al. (1959)
Glucose + corn steep solids	200:1	Pitel et al. (1971)
Sucrose + corn steep solids + soybean meal	120:1:10	Pitel et al. (1971)

(continued)

TABLE XII (continued)

Substrate/combination	Concentration/ratio (g/liter)	Reference
Corn steep liquor + sucrose	25 : 10–40	Sanchez-Marroquin (1963)
Corn steep liquor + glucose	25 : 20–40	Sanchez-Marroquin (1963)
Glucose + methanol	30 : 30	Sanchez-Marroquin (1963)
Glucose + ethanol	30 : 35	Sanchez-Marroquin (1963)
Glucose + malt extract	30 : 10	Sanchez-Marroquin (1963)

plant meals or wheat bran in the media [Imperial Chemical Industries, Ltd. (ICI), 1960b; Kumar, 1987; Thakur et al., 1983]; (8) the yield with glucose was better than that with sucrose (Darken et al., 1959) but was equal to that with lactose (Holme and Zacharias, 1965); (9) the improvement of yield was marginal when 3% glucose in the medium was supplemented with a 3% methanol, 3.5% ethanol, 1% malt extract (Sanchez-Marroquin, 1963); (10) the yield of GAs was ~0.5 g/liter in molasses residue, sulfite liquor, or skimmed milk media (Dietrich, 1960); (11) use of dairy waste as a carbon source resulted in 0.75 g GA_3/liter in 12 days (Maddox and Richert, 1977a); (12) production of GA_3 was inhibited by rice bran oil, although the rate of GA_3 formation was slightly enhanced for up to 5 days of fermentation (Kumar, 1987); and (13) the production was completely inhibited by 1 ppm geraniol due to total inhibition of cell growth (Kumar, 1987; Thakur et al., 1983).

2. *Nitrogen Sources*

The involvement of nitrogen in growth, metabolism, and product formation is well known (Payne, 1980; Dunn, 1985; Ribbons, 1970). A variety of organic and inorganic nitrogen sources, including those from plant and animal origin (Table XIII), were evaluated by different workers to study their effect on the production of GAs. Although ammonia was shown to be utilized in preference to nitrate by Harhash (1966), no information is available on its use in the fermentation processes for the production of GAs. Some of the important results on the efficiency of different nitrogen sources in the production of GAs include (1) among various nitrogen sources tested, ammonium nitrate at 1.0–4.0 g/liter concentration was the most adequate and higher yields were obtained by supplementation with corn steep liquor (Sanchez-Marroquin, 1963); (2) incorporation of nitrogenous compounds in whey medium resulted in higher biomass formation, but yields of GA_3 were lower (Gohlwar et al., 1984); (3) increased total production of GAs was observed when corn extract in Raulin–Thom medium was replaced with soybean flour or any one of the nine fractions of soybean flour (Fuska et al., 1964); (4) addition of soybean and arachis flour to nutrient medium enhanced the production of GAs and shortened the fermentation period (Fuska et al., 1960); (5) the use of ammonium acetate resulted in poor productivity (Borrow et al., 1964a,b), however, good yields with ammonium acetate have also been reported (Nestyouk et al., 1961); (6) production of metabolites other than GAs was at a higher level in glycine-based media (Cross et al., 1963); (7) addition of thiourea in the medium of Borrow et al. (1955) gave slightly higher yields (Sanchez-Marroquin, 1963); (8) peanut or soybean meal were better sources than corn steep liquor (Fuska et al., 1960, 1962); (9) GA_3 was the only member of GAs to be synthesized in corn steep liquor medium (Podojil and Ricicova, 1965); (10) substitution of corn steep liquor by soybean flour led to production of GA_3 and GA_1 at 1:1 ratio and the concentration of GA_3 thus produced was equal to that produced in corn steep liquor medium (Podojil and Ricicova, 1965); (11) plant seed meals may contain some precursors for GAs (Podojil and Ricicova, 1965); (12) higher yields of GAs were obtained by substituting ammonium succinate for ammonium nitrate and by adding corn steep liquor to the medium (Pitel et al., 1971); and (13) a combination of ammonia nitrogen and natural plant meals, as well as the selection of their proper concentrations to match the gas transfer characteristics of the fermentor vessel, led to improved yield of GA_3 (Jefferys, 1970).

3. C:N Ratio

The existence of two phases in the fermentative production of GAs, i.e., the distinction between growth and production phases, is well recognized (Borrow et al., 1964a; Holme and Zacharias, 1965; Jefferys, 1970). The importance of the C:N ratio in the production of GAs by fermentation is indicated by (1) initial active mycelial growth in a nitrogen-limited, balanced medium; (2) low growth rate of the strain in the production phase on an unbalanced medium; (3) initiation of production of GAs after nitrogen exhaustion; (4) continued production of GAs in the presence of sufficiently available carbon substrate; (5) correlation between onset of the stationary phase and exhaustion of nitrogen as well as initiation of GA production; and (6) a direct proportional relationship between the initial concentration of nitrogen supplied in the medium and the rate of product formation as well as the amount of metabolite produced (Borrow et al., 1959a, 1961, 1964a,b; Hanson, 1967; Jefferys, 1970; Bu'Lock et al., 1974).

TABLE XIII
NITROGENOUS COMPOUNDS USED IN THE PRODUCTION OF GAs

Nitrogen compound	Concentration (g/liter)	Reference
Ammonium chloride	1.7, 3.0, 15.3	Sanchez-Marroquin (1963), Stodola et al. (1955), Gohlwar et al. (1984)
Ammonium sulfate	0.5–1.0, 4.3, 14.7	Maddox and Richert (1977a), Sanchez-Marroquin (1963), Gohlwar et al. (1984)
Ammonium phosphate	4.3, 13.8	Sanchez-Marroquin (1963), Gohlwar et al (1984)
Ammonium carbonate	3.1	Sanchez-Marroquin (1963)
Ammonium nitrate	0.48, 1.2, 1.0–4.0, 2.4, 3.0, 5.5	Graebe and Rademacher (1979), Kyowa Hakko Kogyo Co. Ltd. (1983), Sanchez-Marroquin (1963), Borrow et al. (1958), Holme and Zacharias (1965), Hernandez and Mendoza (1976)
Sodium nitrate	3.0	Thakur (1981)
Ammonium tartarate	3.6–9.2, 6.0	Shin et al. (1985)
Potassium nitrate	13.4	Gohlwar et al. (1984)
Glycine	1.88–4.69, 4.0	Cross et al. (1963), Bu'Lock et al. (1974)

Casein	13.1	Gohlwar et al. (1984)
Yeast extract	0.25–0.5, 5.0	Maddox and Richert (1977a), Kyowa Hakko Kogyo, Ltd. (1983)
		Thakur (1981)
Peptone	4.0	Gohlwar et al. (1984)
Gaur meal	15.3	Gohlwar et al. (1984)
Castor oil cake	19.4	
Peanut, soybean meal, soybean flour, cottonseed meal, peanut meal, wheat germ meal, linseed meal, or wheat flour	10.0–15.0	Fuska et al. (1960, 1962), ICI (1960b)
Corn steep liquor (CSL)	0.2–35	Sanchez-Marroquin (1963), ICI (1960b)
CSL + ammonium nitrate	25 + 2.6	Sanchez-Marroquin (1963)
CSL + ammonium nitrate	1.25 + 2.6	Sanchez-Marroquin (1963)
CSL + ammonium tartarate	1.25 + 9.5	Sanchez-Marroquin (1963)
CSL + ammonium sulfate	25.0 + 1.0	Sanchez-Marroquin (1963), Darken et al. (1959)
Corn steep solids + ammonium succinate	1.0 + 6.0	Pitel et al. (1971)
Sodium nitrate + tryptophan	2.0 + 0.02	Thakur and Vyas (1983)
Ammonium nitrate + corn extract	—	Ganchev et al. (1984)
CSL + defatted soybean flour	3.0 + 8.0	Fuska et al. (1960)
Inorganic N_2 + N_2 sources of plant and animal origin	—	Calam and Nixon (1958)

In the process patented by Borrow et al. (1959a) for production of GA_3 by *G. fujikuroi* in a two-stage process, the C:N ratios were used effectively to obtain higher yields. The first stage involved the use of a balanced medium with C:N ratios in the range of 10:1 to 25:1 and the concentration of NH_4NO_3 at a 0.2–0.5% level. In the final stage of fermentation involving an unbalanced medium, the C:N ratio used was in the range of 25:1 to 200:1, with the concentration of NH_4NO_3 at a 0.11–0.5% level (Borrow et al., 1959a). The C:N ratio used by Shin et al. (1985) was 30:1 for *F. moniliforme* strains I, II, and IV, while the ratio used by Mertz (1970) as 30:6 for the *G. fujikuroi* strain. The C:N ratios used in the production of GAs in the single-stage technique are calculated based on the total carbon and nitrogen present in the carbohydrates and nitrogen sources used; these are presented in Table XIV. It may be noted that the C:N ratios given in a chapter on gibberellic acid production by Hanson (1969) were based on the ratio of the quantities of carbohydrates and nitrogenous compounds.

4. *Mineral Salts, Trace Elements, and Growth Factors*

In spite of the pronounced effect of minerals and trace elements in the biosynthesis of secondary metabolites (Weinberg, 1970; Berry, 1975), negligible information is available on these aspects in the microbial production of GAs. The requirement for salts of Mg, K, and P in the production of GAs is well recognized, and in most cases the requirement was efficiently met by using the salt combinations from Czapek–Dox, Raulin–Thom (Borrow et al., 1955), or modified Raulin (Kawarada et al., 1955) media. Different combinations of KH_2PO_4, $K_2HPO_4 \cdot H_2O$, KCl, K_2SO_4, and $MgSO_4 \cdot 7H_2O$ along with various ratios of glucose and NH_4NO_3 were evaluated extensively by Borrow et al. (1961) to study the effect on growth and metabolism of *G. fujikuroi* in stirred culture. The data were analyzed in terms of morphology of mycelia, changes in pH, uptake of nutrients, and accumulation of mycelial fat, phospholipids, carbohydrates, and phosphorus-containing compounds. The results indicated the importance of mineral salts in the fermentative production of GAs. Eleven combinations of mineral salts, commonly used in production of GAs, are presented in Table XV.

Trace elements such as Fe, Cu, Mn, Mo, Zn, B, Al, and Ca are required in the fermentative production of GAs (Hanson, 1967; Krasilnikov et al., 1963; ICI, 1955; Mertz, 1970; Kyowa Hakko Kogyo Co. Ltd., 1983). These are usually added in excess or in combination, as in Raulin–Thom medium (Hanson, 1967). Alternatively, they were not added by relying on impurities present in other ingredients of the

TABLE XIV
C:N Ratios Used by Various Workers in Single-Stage Processes

Microbial strain	Substrates used	C:N ratio	Reference
Gibberella fujikuroi (Saw) Wr. strain BRL (ACC) [17	Glucose + ammonium nitrate	76:1	Holme and Zacharias (1965), Borrow et al. (1958, 1961)
G. fujikuroi NRRL 2633	Glucose + ammonium nitrate	95:1	Kyowa Hakko Kogyo Co. Ltd. (1983)
G. fujikuroi (Saw) Wr.	Dextrose + ammonium tartarate	20:1	Borrow et al. (1955)
Fusarium moniliforme NRRL 2284	Glucose + ammonium chloride	6:1–8:1	Stodola et al. (1955)
F. moniliforme IOC 3326	Glucose + ammonium sulfate	36:1	Sanchez-Marroquin (1963)
F. moniliforme	Dextrose + ammonium sulfate	65:1	Darken et al. (1959)
Sphaceloma manihoticola	Maltose + ammonium nitrate	188:1	Graebe and Rademachar (1979)

TABLE XV

DIFFERENT COMBINATIONS OF MINERAL SALTS USED BY VARIOUS WORKERS

Serial No.	Concentration used (g/liter)						Other additives supplying additional minerals[a]	Reference
	KH$_2$PO$_4$	MgSO$_4$	K$_2$SO$_4$	KCl	MgCl$_2$	K$_2$HPO$_4$		
1	5	1	—	—	—	—	Yeast extract (5)	Kyowa Hakko Kogyo Co. Ltd. (1983)
2	2	0.2	0.6	—	—	—	—	Borrow et al. (1955)
3	5	1	—	—	—	—	—	Borrow et al. (1961), Holme and Zacharias (1965)
4	3	3	—	—	—	—	—	Bergman et al. (1962), Stodola et al. (1955)
5	0.5	—	—	—	—	—	Corn steep liquor (25)	Darken et al. (1959)
6	—	5	—	—	—	—	Corn steep liquor (1.5), agar (2)	Sanchez-Marroquin (1963)
7	0.5	—	0.2	—	—	—	Corn steep liquor (0.2)	Sanchez-Marroquin (1963)
8	0.5	0.2	—	0.2	—	—	Corn steep solids (1)	Pitel et al. (1971)
9	—	—	—	0.43	0.3	—	Corn steep solids (1), soybean meal (10)	Pitel et al. (1971)
10	2.2	0.71	0.2	—	—	1.59	—	Borrow et al. (1961)
11	3.54	0.38	0.43	—	—	0.51	—	Borrow et al. (1961)

[a] Figures in parentheses give concentration in g/litre.

fermentation medium (Jefferys, 1970). The combination of trace elements used by some workers consisted of (g/liter of medium) 0.5 Al_2O_3, 0.5 $ZnCl_2$, and 0.01 $CuSO_4 \cdot 7H_2O$ (Kyowa Hakko Kogyo Co. Ltd., 1983), 0.05 $ZnSO_4 \cdot 7H_2O$, and 0.05 $FeSO_4 \cdot 7H_2O$ (Pitel et al., 1971) or 1–2 ml of the stock solution of trace elements (Borrow et al., 1955, 1961; Holme and Zacharias, 1965). The stock solution of the trace elements usually contained (g/liter) 1.0 $FeSO_4 \cdot 7H_2O$, 0.15 $Cu\text{-}SO_4 \cdot 5H_2O$, 1.0 $ZnSO_4 \cdot 7H_2O$, 0.1 $MnSO_4 \cdot 7H_2O$, 0.1 K_2MnO_4, 3.0 Na_2EDTA (ethylenediaminetetraacetic acid), 1 liter of distilled water, and hydrochloric acid sufficient to clarify the solution (Borrow et al., 1961; Bu'Lock et al., 1974). Some important results in the studies on the effects of trace elements on GAs production include (1) the production of GA_4 by G. fujikuroi NRRL 2633 was markedly increased by Al, Zn, and Cu (Kyowa Hakko Kogyo Co. Ltd., 1983); (2) Ca^{2+} increased the incorporation of leucine into GA_3 in dark-grown cells to essentially the same level as in light-stimulated incorporation (Mertz, 1970); (3) sodium molybdate at the 0.10–1.0 mg/liter level or $CaCO_3$ at 7 g/liter concentration depressed production of GA_3 (Sanchez-Marroquin, 1963); (4) $Al_2(SO_4)_3$ at the 0.1–0.25 mg/liter level or EDTA at the 5–50 μg/ml level had a negligible effect on the yield of GA_3 (Sanchez-Marroquin, 1963); (5) the overall yield of GA_3 was not affected by addition of 1.5% $CaCO_3$ in the idiophase (Darken et al., 1959); (6) the response to trace elements varied between the cultures (Krasilnikov et al., 1963); (7) Zn was far more effective than other trace elements in G. fujikuroi, while Cu was most effective in Fusarium (Krasilnikov et al., 1963); and (8) G. fujikuroi responded best to a combination of Zn, Mo, and Cu, while Fusarium culture gave best results with Cu, Co, and B (Krasilnikov et al., 1963).

Little information is available on the effect of growth factors on the production of GAs. It was reported by Elliott (1948) that cultures do not need any vitamin for the growth or production of GA_3. In contrast, improvement in the growth of the strain by incorporation of vitamin into the medium was reported by Ito and Kimura (1931). Yeast extract at the 5 g/liter level was also incorporated into the medium for production of GA_4 by Kyowa Hakko Kogyo Co. Ltd. (1983). The inhibition of GA_3 production by some heavy metals is also indicated by reduced levels of GA_3 obtained by the use of lined steel or stainless steel fermentation tanks (Kitamura et al., 1953).

5. Precursors

Secondary metabolic pathways branch off from primary metabolism at a relatively small number of points (Bentley and Campbell, 1968).

Genotypic diversity during a series of branching reactions (Bu'Lock, 1965a,b) allowed the increased production of secondary metabolites by inclusion of appropriate precursors in the medium (Rose, 1979). The precursor directed the formation of one specific product, and such directed biosynthesis or directed fermentations (Perlman, 1973; Wang et al., 1979) were used with considerable success in improving the yield of secondary metabolites (Queener and Swartz, 1979; Perlman, 1973). Claims for an increased production of GAs by appropriately incorporating precursors in the fermentation medium have been made by some workers in the patent literature. The addition of mevalonic acid or one of its salts or esters, such as mevaldic acid or isopentenol, formed the subject of a patent by Birch et al. (1959b) and involved (1) cultivation of G. fujikuroi at an 80-liter scale in the first stage at 26.2°C for 53 hours; (2) use of the mycelia thus grown to inoculate a second-stage fermentation medium; (3) allowing fermentation to proceed at 26.2°C for about 59 hours to reach nitrogen exhaustion and initiation of GA_3 production; and (4) addition of precursor and continuation of fermentation to achieve a 2.76 mg/liter/hour production rate of GA_3 as compared to a rate of 1.83 mg/liter/hour possible in the absence of the precursor. Similarly, the addition of 15 g of mevalonic acid to 80 liters of medium every 12 hours resulted in an increase of GA_3 production from 0.34 g/liter to 0.53 g/liter, due to a total addition of ~5.8 g of the precursor per liter of medium.

A process involving addition of (−)-kaurene, (−)-isokaurene, (−)-kaurenol, and (−)-β-epimanoyl oxide (olearyl alcohol) was also patented by ICI (1964a). The addition of alkali metal or alkaline earth salts of senecioic acid, e.g., sodium salt of senecioic (β,β-dimethylacrylic) acid in fermentations with F. moniliforme giving a two- to three-fold increase in the production of GAs was also patented by Redemann (1959). The precursor added was 122 mg/liter, while the level of GA_3 was increased from 22.5 mg/liter to 50 mg/liter. It was stated by Jefferys (1970) that the addition of precursors in fermentative production of GAs was not of any economic interest due to insufficient increases in GA levels.

6. Optimization Using Statistical Methods

Normally, the procedures varying one factor at a time are employed in biotechnology. This is time-consuming, dictates the availability of numerous analytical data, and is incapable of detecting interactions among two or more variables (Cheynier et al., 1983; Maddox and Richert, 1977b). The immense value of statistical methods such as response surface optimization (Cheynier et al., 1983; McDaniel et al.,

1976), factorial design (Auden et al., 1969; McDaniel et al., 1970), the supersimplex optimization program (Paquette and McKellar, 1986), and the multiple regression approach (Schutz, 1983) in optimization of microbiological media and fermentation parameters are well known, as these are capable of overcoming the above problem.

The use of response surface methodology for optimization of whey filtrate medium in the production of GA_3 by F. moniliforme strain ACC 917 was reported by Maddox and Richert (1977b). The methodology used involves determination of interactions of the factors under study in statistically designed experiments, estimation of coefficients in a mathematical model to find out the response surface, and checking the adequacy of the statistical inferences (Davies, 1967; Myers, 1971; Maddox and Richert, 1977b). The exercise indicated important results (Maddox and Richert, 1977b), such as (1) A maximum yield of 1.16 g/liter was predicted for supplementation of the basal medium with 0.1% ammonium sulfate only. However, the prediction was not valid since the level of nutrient supplementation was at the extreme edge of the experimental design, and the nutrient concentration used was predictably too high. (2) The yield of GA_3 is affected largely by nutrient supplementation. The combined effect of ammonium sulfate and yeast extract caused a marked change in the yield. (3) A maximum production of 0.83 g of GA_3 per liter predicted when the medium was supplemented only with 0.001% magnesium sulfate. This prediction was confirmed by experimental results although the observed value was 0.75 g of GA_3 per liter. The optimization of nutrient medium for the production of GAs by F. moniliforme 1M-11 by using mathematical methods was also reported by Ganchev et al. (1984). The interactions of five nutrients, i.e., sunflower oil, ammonium nitrate, corn extract, calcium dihydrogen phosphate, and potassium sulphate, were examined. Among these, sunflower oil played the major role. The maximum production of GAs was shown to occur when these nutrients were present at 10.5, 0.25, 0.15, 0.25, and 0.015% levels, respectively.

D. Growth Phases in Fermentor

The formation of negligible quantities of GAs before the exhaustion of nitrogen in nitrogen-limited medium, production of the bulk of the metabolite after exhaustion of nitrogen, and termination of production soon after the exhaustion of glucose in glucose-limited medium (Borrow et al., 1961, 1964a,b) indicated that GAs are secndary metabolites. The process for production of GA_3 by the SmF technique is, in fact, considered as a classical fermentation method by Borrow et al. (1961), as the phases of growth can be clearly distinguished and related to nutritional and environmental states operating in the fermentor (Vass and Jefferys, 1979). The exhaustive studies by Borrow et al. (1961, 1964a,b) on different types of media have established the existence of five phases in fermentations involving G. fujikuroi strain ACC 917. The salient features of these fermentation phases are described below.

1. Lag Phase

The conventional lag phase in nitrogen-limited medium is undetectable as the strain requires little or no adaptation, and growth in the fermentor commences quickly due to the use of vigorously growing mycelial cells as inoculum. In fact, the nonsporulating nature of the culture also necessitates use of vegetative cells as inoculum. However, the lag phase is detectable when either ammonium acetate media or carbohydrate-rich media containing >30% glucose are used.

2. Balance Phase

This phase extends until exhaustion of one of the nutrients occurs and involves periods of rapid growth and nutrient uptake. The dry cell weight increases exponentially and uptake of glucose, nitrogen, phosphate, magnesium, and potassium is almost constant per unit increase in dry weight. It involves rapid proliferation of the strain in nutritionally unlimited medium until the exhaustion of a limiting nutrient. The morphology and composition of mycelial cells, as well as their nutritional requirement, also remain constant. The growth is exponential in initial stages, subsequently becoming linear, and the cells are finally subjected to a deceleration stage due to severe oxygen restriction. The media used in the fermentation are designed to achieve either nitrogen or phosphorus exhaustion first. The continuation of the balanced phase, due to exhaustion of nutrient, is not possible and hence the mycelial cells are committed to biochemical differentiation. No GA is produced in this phase.

3. Transition Phase

In phosphate- or magnesium-limited media, the limitation of nutrient occurs before the onset of oxygen restriction. Consequently, the intracellular reserve of acid-soluble phosphate or magnesium allows further proliferation of the cells. Therefore, the mycelial composition in this phase is different from that in the balanced phase. The uptake of other nutrients continues and dry mycelial weight increases proportionately, although the rates of growth and nutrient uptake are lower.

The increase in carbohydrate and fat content of the cells is continued in this phase until exhaustion of glucose in the medium.

4. *Storage Phase*

The presence of excess glucose and the exhaustion of nitrogen cause cessation of proliferation of the cells, thus leading to formation of maximum fat and carbohydrates in the cells—45 and 32%, respectively, of dry mycelial weight. The uptake of phosphate, magnesium, potassium, and glucose is continued if these are present in the media, and eventually these uptakes, except for that of glucose, cease completely. The fat formed in the cells consists largely of relatively saturated triglycerides and smaller concentrations of free fatty acids and phospholipids. The production of GAs is initiated in this phase and is continued in the presence of available glucose. Although the time of initiation of this phase is distinct, the time at which it ends is not clear. An adequate criterion to mark this end is the time at which the dry mycelial weight is maximal.

5. *Maintenance Phase*

This phase is operative between the period of maximum mycelial formation and the onset of terminal breakdown of mycelial components. It is initiated only when glucose is present in the medium. Except for the continued uptake of glucose, no other nutrients are taken up by the cells and the dry mycelial weight remains constant. The intracellular reserve fat is broken down due to the exhaustion of glucose in the medium and thus results in a decreased level of triglycerides and dry mycelial weight. Continuation of the phase, even for several hundred hours with a concurrent production of GAs, if glucose is present in excess, is of industrial importance. Even after exhaustion of glucose in the medium, the production of GAs is continued until reserve fat is reduced to the level it had in the balanced phase.

6. *Terminal Phase*

The mycelial cells undergo many changes, such as increased vacuolation, loss of cell contents, and breakdown of mycelial cells, due to nonavailability of external and internal sources of utilizable carbon. Consequently, this leads to a return of ammonia, phosphate, magnesium, and potassium to the medium. It also increases the pH of the medium and leads to a decrease in dry mycelial weight. This phase is not allowed to occur in fermetations, as the fermentor run for production of GAs is terminated just prior to the onset of this phase.

7. Trophophase and Idiophase

The above five phases of fermentation in production of GAs were regrouped into two phases by Bu'Lock et al. (1965), based on respirometric and enzymatic studies as well as other metabolic activities related to secondary metabolites. For the exponential part of the balanced phase of Borrow et al. (1961, 1964a,b), the name "trophophase" (equivalent to commonly used terms such as "growth phase" or "logarithmic growth") was suggested. The idiophase (equivalent to production phase or stationary phase) includes the transition, storage, and maintenance phases of Borrow et al. (1961, 1964a,b). Thus, trophophase exists until one of the essential nutrients is exhausted. Consequently, GAs are produced in the idiophase. Due to the sharp transition period between these two phases, it is possible to distinguish between primary and secondary metabolites. These two phases described by Bu'Lock et al. (1965) are widely accepted in the fermentative production of secondary metabolites (Drew and Demain, 1977; Haslam, 1986; Campbell, 1983).

8. Phases A–F

Gibberella fujikuroi is known to produce many secondary metabolites (Jefferys, 1970). Among these, GAs and bikaverins were selected by Bu'Lock et al. (1974) in a study to establish that their production was successive rather than simultaneous. These studies established the existence of a total of six phases, designated as phases A–F, in the batch SmF process for production of GAs by *G. fujikuroi* strains ATCC 12616, CMI 58290, and CBS–BRL 917. These phases are described in Table XVI. The results obtained on mycelial weight, residual glycine, pH, bikaverins, GAs, residual glucose, and the mycelial content of nitrogen, protein, DNA, RNA, carbohydrates, and ergosterol were shown to agree well in cases of chemostat and batch SmF processes (Bu'Lock et al., 1974) with measurements reported by Borrow et al. (1961, 1964a,b).

9. Implications of the Phases

The in-depth studies by Borrow et al. (1961, 1964a,b) and Bu'Lock et al. (1965, 1974) provided many useful conclusions, such as (1) the production of GAs is controlled by the degree to which growth was limited; (2) the onset of limitations of different nutrients is controlled by the initial composition of the growth medium; (3) production of GAs

TABLE XVI

PHASES IN FERMENTATION AND THEIR CHARACTERISTICS[a]

Phase designation	Characteristics
A	Lag phase, short or absent
B	Nitrogen nonlimiting; rapid growth of mycelia; some synthesis of GAs and ergosterol, but not bikaverin
C	Glycine becomes limiting; bikaverin production begins rapidly and virtually ceases at the end of this phase; ergosterol synthesis slows down; GA_3 synthesis progresses slowly
D	Free glycine in medium not measurable; steady increase in dry weight; bikaverin no longer formed; GA_3 synthesis accelerated
E	Constant dry weight; consumption of remaining glucose; GAs accumulation slower than in phase D
F	Dry weight decreases; pH rises; autolytic phase

[a] Based on the studies carried out by Bu'Lock et al. (1974).

in nitrogen-limited medium is initiated by the exhaustion of nitrogen; (4) the rate of production of GAs and the quantity produced are proportional to the initial concentration of nitrogen; (5) the rate of production of GAs is a linear function of cell concentration until the oxygen-transfer capacity of aeration and agitation system was saturated; (6) exponential growth of the strain ceased with the exhaustion of assimilable nitrogen in the medium; (7) the onset of the rapid production of GAs is associated with the onset of the storage of fat and carbohydrate by cells, a decrease in the RNA content of cells, and the production of other secondary metabolites; (8) upon exhaustion of the assimilable carbon in the medium and before the maximum production of GAs, some further production continued at the expense of stored fat; and (9) GAs and bikaverins, the two secondary metabolites, are formed independently from the same metabolic intermediate, i.e., acetyl CoA.

E. REGULATION OF GA_3 PRODUCTION

The regulation of GA_3 production by the coordination of different microbial metabolic pathways and catabolic regulation was well established even in earlier work. Glycerol was used as a sole carbon source since it was utilized slowly by the strain. Subsequently, a combination of readily and slowly utilizable carbon sources, such as glucose–glycerol and glucose–glycerol–lactose, were used (Darken et al., 1959).

The latter combination, containing two slowly utilizable carbon sources, resulted in production of 0.88 g of GA_3 per liter, which was the highest among all of the experimental strategies employed by Darken et al. (1959). It was observed that rapidly utilizable carbon sources, such as glucose and sucrose, were involved in the establishment of growth and the achievement of maximum growth rate. Subsequently, the maximum production of GA_3 occurred due to slowly utilizable carbon sources, such as glycerol and lactose. The nonutilization of slowly utilizable carbon sources, in the presence of a readily utilizable carbon source, was also reported (Darken et al., 1959). In well-designed experiments in 500-gallon fermentors, containing a basic corn steep liquor–ammonium sulfate–monopotassium phosphate medium, the effect of the addition of glucose at a 2% level at the beginning, hourly feeding of glucose at a 0.02% level after 24 hours of fermentation, and the addition of 3% glucose after 24 hours of fermentation were examined (Darken et al., 1959). In the latter case, the glucose added was equivalent to the total amount used in the experiment involving hourly additions of 0.02% glucose. The degree of production of GA_3 was highest (0.62 g/liter) in the fermentor fed with hourly additions of glucose, as compared to the values of 0.13–0.32 g/liter attained in the other two fermentors.

Similarly, the incorporation of glucose at concentrations >30 g/100 ml in the initial period was shown to result in delayed production of GA_3 and a lag phase that is otherwise absent (Borrow et al., 1964a,b). The higher initial concentration of glucose was also reported to result in lower yields of GA_3. Moreover, the specific growth rate of the strain was found to increase with the increase in the initial concentration of glucose in the medium. These effects were successfully overcome by adding 10 g of glucose per 100 ml of medium at time zero and feeding glucose after 24 hours of fermentation in such a way that the concentration of glucose, at any time, was <4% in the medium. This was necessary, as a glucose concentration at the 4% level was stated to be an inhibitory concentration (Borrow et al., 1964a,b). The increased production of GA_3 by feeding sucrose after 3 days at a 1%/day level on 8 consecutive days (Bergman et al., 1962) or by feeding 0.5 g of sucrose per 200 ml of medium every 12 hours after the 9th day of incubation (Serzedello and Whitaker, 1960) was also documented. These results indicate substrate inhibition of the production of GA_3.

It is well known that the production of GA_3 in the idiophase requires a carbon source, for production of metabolite as well as for maintenance of the biomass. However, since this carbon should not be readily utilizable, slowly utilizable carbon sources are used in batch fermentation. The slow feeding of readily utilizable carbon sources, however, eliminated the above need. The important role played by catabolic regulation, when the microorganism is supplied with two utilizable carbon sources, is well known. The formation of only those enzymes acting on the most favorable substrate until it is exhausted, with subsequent production of the enzymes for utilization of slowly utilizable carbon substrate, is a vital aspect of induction and catabolic repression under the catabolic regulation (Wang et al., 1979; Bu'Lock, 1975; Demain, 1968, 1971; Drew and Demain, 1977; Haslam, 1986). The overcoming of catabolic repression by continuous or intermittent feeding of lower concentrations of readily utilizable carbon substrate is also well known (Gallo and Katz, 1972; Pogell et al., 1976; Kominek, 1972).

Marked changes in the enzymatic composition of cells, the sudden appearance and accumulation of enzymes specifically related to the formation of secondary metabolites, depression of key enzymes involved in primary metabolism, and the release of the genes of secondary metabolism from catabolic repression were reported to occur at the end of trophophase in the production of many secondary metabolites (Bu'Lock, 1975; Demain, 1968, 1971; Drew and Demain, 1977; Haslam, 1986). Some of these changes were also observed in the production of GA_3 (Bu'Lock, 1975; Bu'Lock et al., 1974; Borrow et al., 1961, 1964a,b). The build-up of initiators (Bu'Lock, 1961) or prime precursors (Woodruff, 1966) at the end of the trophophase and their actions as initiators in inducing enzymes responsible for the production of secondary metabolites is well documented and forms a part of catabolic regulation (Demain, 1968, 1961; Wang et al., 1979). Mevalonic acid was considered to be one such initiator in the production of GA_3 (Demain, 1968). The role of acetyl CoA as an initiator for inducing secondary metabolism under catabolic regulation in the case of the production of GA_3 was also indicated by initiation of GA_3 production in the idiophase, accumulation of acetyl CoA at the end of the trophophase, and involvement of acetyl CoA in the synthesis of GA_3.

F. Kinetic Studies

Exhaustive kinetic studies on G. fujikuroi in 30-liter Hoover and conventional cylindrical fermentors in terms of growth and metabolism of the strain have been reported by Borrow et al. (1964a). In other studies, the effect of temperature, ranging from 8 to 40°C, on the kinetics of metabolism of G. fujikuroi was reported (Borrow et al., 1964b). The growth of the culture on agar in the temperature range

3–36°C was also studied by Stoll (1954). The salient features of the results on growth and production (Borrow et al., 1964a,b) include (1) presence of a lag phase only in ammonium acetate media or in the presence of high concentrations of glucose; (2) exponential early growth on all nitrogen sources, with a decreased specific growth rate in ammonium acetate media, and utilization of more NH_3–N than NO_3-N, with a concomitant decrease in pH in early exponential growth on ammonium nitrate media; (3) cessation of NH_3–N uptake and an increase in dry weight in the 2.8–3.0 pH range but with continued NO_3-N uptake, leading to an increase in pH causing NH_3–N uptake and a final resumption of exponential growth at a low specific growth rate; (4) continuation of exponential growth up to formation of ~7 mg dry cell weight per gram of whole, unfiltered sample on glycine, urea, and ammonium tartarate media and constant uptake of glucose, nitrogen, phosphate, and magnesium, which are unaffected by the rate of agitation or specific growth rate, with, however, the latter decreasing with increasing glucose concentration; (5) follow-up of the exponential phase by a period of linear growth that involves a greater contribution of glucose and a lesser contribution of phosphate and magnesium as compared to the exponential growth phase, with the rate of agitation influencing the dry weight, probably due to a response to oxygen restriction; (6) an increase in dry weight after exhaustion of nitrogen, due to accumulation of fat and carbohydrates in the cells, with constant specific rates of dry weight increase and glucose uptake rates if the initial nitrogen concentration was low, and a decrease in both of these rates if the initial nitrogen concentration was high; (7) initiation of GA_3 production at, or soon after, nitrogen exhaustion and a linear increase in GA_3 level with time; (8) decreased productivity with increased glucose concentration; (9) increased productivity followed by a decrease with high initial nitrogen; (10) a maximal amount of GA_3 produced proportional to initial nitrogen provided; and (11) no significant difference between nitrogen-limited media containing either 0.36 or 0.92 g of ammonium tartarate per 100 ml of medium in the early period of fermentation, but all specific rates during storage and maintenance phases were considerably lower in media containing a high initial concentration of ammonium tartarate.

The results of kinetic studies on the temperature effect include (1) the minimum temperature for growth is <8°C (Borrow et al., 1964a,b) or at 3°C (Stoll, 1954); (2) the growth ceased at 38°C after forming about 2 mg dry weight/g of unfiltered broth while it is absent at 40°C; (3) discontinuity in many parameters occurred between 17 and 20°C, while most of the optima were observed at 29–32°C; (4) the yield constant of nitrogen to dry weight is not affected by temperature, although it decreased slightly or markedly with increasing temperature in the case of glucose- and magnesium sulfate-containing media, respectively, and remained constant at 8–20°C in the case of phosphate media but decreased greatly with an increase in temperature in the range 20–36°C; (5) the relationship between rate constants and temperature showed a typical "skew" curve with "tail off" at the lower temperatures for specific growth rate, nitrogen and glucose quotients, specific rate of glucose uptake in the maintenance phase and GA_3 productivity; and (6) another relationship showed a marked discontinuity between 17 and 20°C for phosphate, magnesium, and carbohydrate quotients, linear growth rate, mycelial cell formation rate, fat and carbohydrate accumulation in cells, and rate of glucose, phosphate, and magnesium uptake in the storage phase.

G. Mathematical Models

A mathematical model to study the relationship between GA_3 production by *F. moniliforme* 1 in whey permeate medium and lactose utilization as well as other fermentation parameters was derived by Gohlwar et al. (1984). The model is based on a nonlinear multiple regression analysis of observed data with a computer program written in Fortran IV on an HP/1000 computer following the procedures designed by Sethi et al. (1981). Each iteration comprised (1) approximating the value of the function evaluated, using the value of the parameter, by a linear function of the parameter vector; (2) solving the resulting linear regression problem to get a new estimate of the parameter vector; and (3) passing the revised set of parameters to the next iteration. The model developed is capable of simulating various fermentation conditions and predicted precise values for various cultural conditions (Gohlwar et al., 1984).

The observed experimental data on the effect of temperature, pH, inoculum ratio, and fermentation time on the production of GA_3 and the values predicted by mathematical model gave ranges for these parameters that led to the maximum production of GA_3. These include a temperature range of 27–30°C, a pH range of 3.5–5.5, an inoculum level of 10–12.5% (vol/vol), and a fermentation time of 12 days. The mathematical model also confirmed various points, such as (1) product formation following active growth of the organism; (2) GA_3 synthesis depending on lactose utilization; and (3) the influence of incubation temperature, pH, and inoculum level on the synthesis of GA_3 and the consumption of substrate. No other report is available on mathematical modeling of the fermentation process involving other strains and substrates for production of GA_3 or other GAs.

H. PROCESS OPERATION STRATEGIES

1. *Fermentors Used*

The data on the production of GA_3 and other GAs reported in the literature are based on the use of fermentors of different types and sizes. These include 3-liter (Holme and Zacharias, 1965) and 5-gallon air-agitated bottles (Hitzman and Mills, 1963), a 30-liter-capacity Hoover washing machine modified into a fermentor, an 80-liter conventional unit (Borrow et al., 1955), and 60 liters (Sanchez-Marroquin, 1963), 300 gallons (Stodola et al., 1955), 50, 650, and 1000 gallons (Darken et al., 1959), and 5000–6300 gallons of medium (Bergman et al., 1962) in fermentors of appropriate capacity. The fermentor was operated at atmospheric pressure (Hitzman and Mills, 1963) while positive pressure was maintained in the seed fermentor (Stodola et al., 1955). Shake flasks were employed in laboratory-scale studies (Maddox and Richert, 1977a; Gohlwar et al., 1984; Pitel et al., 1971; Thakur and Vyas, 1983) using a 180 stroke per minute shaker (Shin et al., 1985).

2. *Single-Stage Process*

A typical single-stage fermentation technique as used by Stodola et al. (1955) for the production of GA_3 involves (1) the use of glucose–ammonium chloride–potassium dihydrogen phosphate–magnesium sulfate medium at a 150-gallon volume in a 300-gallon-capacity stainless steel fermentor; (2) continuous sterilization of the medium at 135°C for 3 minutes; (3) cooling of the medium to 25°C; (4) preparation of the inoculum progressively in a 9-liter-capacity aerated bottle and a 30-gallon seed tank; (5) inoculation of production medium using 4% inoculum; (6) agitation at 100 rpm; (7) aeration at 0.25 vvm; (8) automatic foam control by 0.75% octadecanol in 96% alcohol; (9) no control of pH (initial pH after sterilization was 5.1–6.0); (10) temperature control at 25°C; (11) air sterilization by passing through a column of activated charcoal; (12) 65-hour fermentation time; and (13) downstream processing to obtain the product in crystalline form. The yield obtained was 12 g of crystalline GA_3 per 160 gallons of culture liquor (Stodola et al., 1955).

A similar one-stage fermentation technique was also used by many other workers (Borrow et al., 1955; Gohlwar et al., 1984; Darken et al., 1959; Sanchez-Marroquin, 1963; Bergman et al., 1962). Some of these processes employed a combination of readily and slowly utilizable carbon sources (Darken et al., 1959; Sanchez-Marroquin, 1963) in such a way that the readily available carbon susbstrate was exhausted at a specific time in the medium. Subsequently, the carbon needed for the production of GA_3 was supplied by the slowly utilizable carbon substrate. A combination of dextrin and sucrose with addition either initially or during the course of fermentation was also reported (Bergman et al., 1962). Another variation in the technique used by some workers (Borrow et al., 1955, 1961) was the use of a readily utilizable carbon source such as glucose as the sole source of carbon. The concentration employed was such that growth in the nitrogen-limited medium and production of GA_3 after exhaustion of nitrogen were favored.

3. *Two-Stage Process*

Higher production of GA_3 in the presence of slowly utilizable carbon sources, initiation of production only after exhaustion of nitrogen in the medium (Darken et al., 1959), and the involvement of different environmental parameters in the production of GA_3 (Holme and Zacharias, 1965) are well known. The low-nitrogen medium is needed in the first phase to achieve controlled but rapid mycelial growth and to impart desired characteristics to the mycelial cells. On the other hand, low growth rate and maintenance of optimum conditions for rapid and maximum production of GA_3 are essential in the second phase. These requirements were met satisfactorily in the two-stage fermentation techniques used by many workers (Borrow et al., 1958; Birch et al., 1959b).

In a two-stage process patented by Borrow et al. (1958), the culture was grown in the first stage using nitrogen-limited medium with a C:N ratio in the range of 10:1 to 25:1. After achieving nitrogen exhaustion and maximum growth in the first stage, the mycelial cells were transferred to another medium with a C:N ratio in the range of 30:1 and 55:1. This second-stage medium was an unbalanced medium and it allowed only negligible growth. A comparison of the yields of GA_3 under single- and two-stage techniques indicated better yields in the latter case. The GA_3 production was 0.308 g/liter in 501.7 hours in a single-stage process against 0.413 g/liter in 407.1 hours in a two-stage process (Borrow et al., 1958). In another patented two-stage process, the precursors of GA_3 were added at the time of initiation of GA_3 production in the second-stage (Birch et al., 1959b). The average rate of production of GA_3 was found to increase to 2.76 mg/liter/hr as compared to 1.83 mg/liter/hr in the absence of precursors. Borrow et al.

(1958) have also patented a modified two-stage process for the production of GA_3 at a yield of 1.0 g/liter in 450 or 500 hours. It involved a periodic or continuous addition of carbon in the second stage and was thus a combination of a two-stage process and a fed-batch culture.

4. Multistage Process

A multistage process for production of GA_3 by *G. fujikuroi* was also patented by Borrow et al. (1959a) to achieve higher yields than those attainable in the single- and two-stage techniques. It consisted of (1) the first stage, i.e., the stage of active growth of the culture, involving use of a balanced medium containing glucose and ammonium nitrate with a C:N ratio in the range of 10:1 to 25:1; (2) the culture grown at least twice in balanced medium, involving active growth in both the first and intermediate stages; (3) the final stage of the process carried out in an unbalanced medium in which the C:N ratio was in a range of 25:1 to 200:1; (4) the active growth of the culture being halted in the final stage, mainly due to the exhaustion of nitrogen; and (5) in the final stage, GA_3 production was achieved at an increased rate as compared to single- and two-stage processes. A yield of 1 g of GA_3 per liter in a multistage system was reported by Grove (1963) and Anke (1986).

5. Fed-Batch Culture

Fed-batch culture techniques were employed by some workers even before the term was introduced into fermentation technology and its benefits became widely known (Edwards et al., 1970; Yoshida et al., 1973). The fed-batch technique, sometimes termed semibatch or extended culture, refers to the constant or intermittent supply of one or more nutrients to the fermentor and confinement of the product in the fermentor until the end of the run (Yamane and Shimizu, 1984; Whitaker, 1980). It is accepted as a superior technique as compared to the conventional batch SmF technique, especially in cases in which higher concentration of one or more nutrients affects the yield of the product. It overcomes the problems posed by substrate inhibition, high cell concentration, glucose effect, catabolic repression, feedback inhibition, end-product repression, and higher viscosity of the broth. It also offers advantages in allowing extension of operation time and replacement of water loss by evaporation (Yamane and Shimizu, 1984).

As discussed earlier, the production of GA_3 in the batch SmF process is controlled by catabolic regulation and involves catabolic repression, induction, and substrate inhibition (see Section X,E). These problems were overcome by employing an appropriate feed policy to achieve higher yields of GA_3 (Darken et al., 1959; Borrow et al., 1955, 1958,

1961, 1964a,b; Abbott Laboratories, 1963; Serzedello and Whitaker, 1960; Bergman et al., 1962). The policies followed were a periodic or continuous addition of nutrient without any feedback control. These different policies and schedules for periodic or continuous addition of carbon substrate have resulted in improved yields of GA_3.

6. *Continuous Culture*

The continuous culture technique is used successfully in fermentations involving filamentous molds (Pirt and Callow, 1959, 1960; Sikyta et al., 1959; Sikyta, 1964; Bartlett and Gerhardt, 1959). The prolongation of trophophase and idiophase was achieved in continuous cultivation either in single-step (Fiechter, 1981; Bartlett and Gerhardt, 1959; Knorre, 1980) or two-step operations (Harte and Webb, 1967; Herbert, 1964; Málek and Fencle, 1966). The production of GA_3 using single-step continuous fermentation was also successfully carried out by Holme and Zacharias (1965) using G. *fujikuroi* (Saw) Wr. strain BRL (ACC) 917 in 3- and 20-liter fermentors. The system was found to maintain a high rate of synthesis of GA_3 as compared to two-stage batch cultivation. The yield of GA_3 in the latter case involving 7 days in the first stage and 15–16 days in the second was 0.8 mg/liter/hr or 0.02 mg/hr/mM mycelial nitrogen. On the other hand, continuous culture gave a yield of GA_3 at the 0.049 mg/hr/mM mycelial nitrogen level with a 200-hour holding time.

The production of GA_3 and other GAs in media containing aqueous and liquid hydrocarbon phases in a 5-gallon air-agitated bottle at room temperature and atmospheric pressure in batch fermentation was also reported by Hitzman and Mills (1963). The culture used was G. *fujikuroi* and it was reported to grow in the aqueous phase during a total fermentation time of 35 days. The authors suggested that this process could also be performed under continuous fermentation (Hitzman and Mills, 1963). However, cultivation of F. *moniliforme* on n-alkanes by the SmF technique was considered unsatisfactory by Heinrich and Rehm (1981). The results of chemostat fermentation in a 3-liter-capacity fermentor using G. *fujikuroi* at different dilution rates in glycine-limited medium are also reported by Bu'Lock et al. (1974). The values for maximum specific growth rate (μ_{max}) and Michaelis constant for substrate (K_s) were 0.18 ± 0.02 hours^{-1} and 1.0 ± 0.2 g of glycine per liter. Secondary metabolite formation was better at low dilution rates, probably because of lower growth and RNA and DNA content of cells and higher accumulation of carbohydrates in cells. The production of bikaverin was 2.05 mg/g dry weight at the 0.05 hours^{-1} dilution rate, but it was reduced to 1.84 mg at the 0.02 hours^{-1} dilution rate. On the other hand, production of GAs was lower (0.45 mg/g dry weight) at the 0.05 hours^{-1} dilution rate, but it increased to 2.05 mg/g dry weight at a 0.02 hours^{-1} dilution rate. These values were shown to be comparable to those in the batch SmF process occurring in phases C and D, respectively, for bikaverin and GAs.

Continuous fermentation offers many advantages, such as superiority in productivity, uniformity of operation, ease in automation, and economics as compared to the batch SmF technique (Hospodke, 1966). It is more advantageous especially with fast growing cells and growth-associated products (Aiba et al., 1973). In spite of these advantages, the fermentation industry is reluctant to adopt single-step continuous processes (Stanbury and Whitaker, 1984). In fact, this is even more true in the case of multistage systems, particularly due to their complexity. In addition, continuous fermentation processes are often associated with certain problems such as high susceptibility to contamination (Hospodke, 1966), risk of instability of stationary and quasistationary cultivation, age-dependent loss of productivity, disturbances to steady state due to the presence of young producing and aged nonproducing cells, difficulty in maintaining dilution rates, and genetic instability of cells (Hasegawa et al., 1985). In spite of extensive work on continuous fermentation for the production of antibiotics during the past 20 years, most of the antibiotics are still being produced by batch SmF processes by the fermentation industry. In the case of GA_3 also, the continuous fermentation process is not used by the industry.

7. Inoculum Development

The development of inoculum is one of the most important operations in the successful completion of fermentation runs at both laboratory and industrial scales. It is usually developed in stages in seed fermentors, and dosage is determined with a view to shorten the initial growth phase and to effectively use the available fermentation capacity (Banks, 1979). For example, the procedure used by Stodola et al. (1955) consisted of (1) growth of culture on potato dextrose agar (PDA) slants at 24°C for a week; (2) scraping growth from three slants into 10 ml of 1:10,000 solution of sodium lauryl sulfate in distilled water; (3) inoculation of a 9-liter aeration bottle containing 4 liters of medium with the entire suspension; (4) fermentation at 28°C for 75 hours with aeration; (5) using 1 liter of these contents to inoculate a 30-gallon seed fermentor containing 20 gallons of medium; (6) cultivation for 30 hours under optimum parameters (aeration at 0.25 vvm by sterile air, agitation at 150 rpm, no pH control, 25°C); and (7) use of the inoculum thus developed in the seed fermentor for inoculation of 154 gallons of medium in a 300-gallon-capacity production fermentor.

The procedure used by Borrow et al. (1961) involved growth of G. fujikuroi ACC 917 for 14–21 days on solid medium in 8-ounce medicine bottles, scraping the growth into sterile water, shaking it with glass beads, and using it as inoculum at a rate of one bottle per 40 liters of medium. Alternatively, the mycelia from one slope were used to inoculate a 10-liter-capacity bottle containing 4 liters of liquid medium, which was then employed at 30 ml to seed a 30-liter fermentor. The growth from a PDA slant was suspended in 10 ml of water and used at 1 ml/liter of production medium by Sanchez-Marroquin (1963). The growth of inoculum in shake flasks for 7 days (Holme and Zacharias, 1965), 24 hours (Hitzman and Mills, 1963), or 75 hours (Gohlwar et al., 1984) was the other variation followed. Similarly, the inoculum ratios used included 10% (vol/vol) (Gohlwar et al., 1984), 5% (Shin et al., 1985), 5 ml/18 liters of production medium (Hitzman and Mills, 1963), 2% (Holme and Zacharias, 1965), 4% by volume in a flask experiment, and 10% inoculum in two-stage fermentation studies (Darken et al., 1959).

Recently, a wide range of inoculum ratios was studied for their effect on the production of GA_3 by F. moniliforme 1 in whey permeate medium (Gohlwar et al., 1984). It was found that the inoculum ratio not only affected production of GA_3 but also the degree of growth and substrate utilization. The mathematical model predicted that 12.5% inoculum would be optimum for production of GA_3, whereas other equations indicated that 10% inoculum would be optimum for growth and lactose utilization (Gohlwar et al., 1984).

8. Foam Control

Foam control is essential in SmF processes, as it adversely affects many different aspects, including productivity (Ghildyal et al., 1987). However, little information is available on its formation and control in the production of GAs. In spite of the use of 50- to 650-gallon and 1000-gallon fermentors in their studies, Darken et al. (1959) did not mention foam control. Stodola et al. (1955) added 1 g of octadecanol per 4 liters of medium before sterilization for foam control in the inoculum development process, while in the production fermentor, foam was controlled automatically by the addition of a solution of 0.75% octadecanol in 95% alcohol. Comparatively little foam formation was reported in steady-state continuous culture with an air supply of 5 liters of air per minute in a 20-liter fermentor or 0.5 liters of air per minute in a 3-liter fermentor by Holme and Zacharias (1965). However, the formation of a 10–30 mm deep foam layer on the surface of the culture was stressed by these authors. A decrease in the rate of nutrient uptake and of mycelial accumulation due to the use of antifoam agents was reported by Borrow et al. (1961). Therefore, no antifoam agent was used by them, but a hole was provided in the lid of the fermentor for foam discharge.

I. Concomitant Products

A large number of products of varied nature are reported to be concomitantly produced along with GAs by G. fujikuroi and F. moniliforme. Some of these, such as gibberellenic acid and Sumuki's acid, are even present in commercially available solid GAs preparations at 4–10% and varying levels, respectively (Holbrook et al., 1961). The list of these concomitant products is quite extensive and was reviewed earlier in detail by Jefferys (1970).

1. *Products Related to GAs*

The biosynthesis of GAs from acetyl CoA is a complex process involving a large number of intermediates. Most of these and the different GAs are already reported in Sections III and VIII. In addition to these, some other related compounds have also been reported and are presented in Table XVII. Among these, fujenal was characterized as a diterpenoid that is possibly connected with the biosynthesis of GAs (Cross et al., 1963), although it was considered as an extraction artifact by Jefferys (1970). These metabolites were mostly present in neutral fractions of the fermentation broth when glucose–ammonium tartarate or glucose–ammonium nitrate media were used (Cross et al., 1963).

2. *Degradation Products of GAs*

Quite a large number of degradation products were found in fermentation broth as well as in final products; these are presented in Table XVIII. The presence of gibberic acid, dehydroallogibberic acid, allogibberic acid, gibberellenic acid, 2,3,7-trihydroxy-1-methyl-8-methylenegibb-4-ene-1,10-dicarboxylic acid, and 1-3-lactone in fermentation extracts or the final product was considered by Jefferys (1970) to be extraction artifacts.

3. *Plant Growth Repressors*

The production of plant growth suppressor by Bakanae fungi was reported by Kurosawa (1930). Subsequently, production of the suppressor at 35°C in contrast to that of the stimulator at 20°C, inhibition of

TABLE XVII

GAs-Related Metabolites Produced along with GA$_3$ in Fermentation

Metabolite	Melting point (°C)	Reference
5-Hydroxymethylfuran-2-carboxylic acid (Sumiki's acid)	—	Kawarada et al. (1955)
5-Hydroxymethyl-2-furoic acid	—	Cross et al. (1963)
2-O-Acetyl gibberellic acid	230–232	Schreiber et al. (1966)
Fujenal	169–170	Cavel et al. (1967) Cross et al. (1963), Brown et al. (1967)
Fujenoic acid	205–206	Cross et al. (1963)
7,16,18-Trihydroxy kaurenolide	250–255	Cross et al. (1963)
7,18-Dihydroxy kaurenolide	211–214	Cross et al. (1963), ICI (1964b)
13-Epi(−)-manoyl oxide (olearyl acid)	98–99.5	Cross et al. (1963)
$C_{19}H_{26}O_4$ oxolactone	—	Wierzchowski and Wierzchowska (1961)
$C_{20}H_{26}O_4$ oxolactone	—	Wierzchowski and Wierzchowska (1961)
$C_{20}H_{28}O_5$ (dibasic)	169–170	Cross et al. (1963)
Isogibberellins A$_3$ and A$_7$	210–212	Pitel et al. (1971)

TABLE XVIII

Degradation Products of GA$_3$ Present in Fermentation Broth or the Final Product

Degradation product	Biological activity	Reference
Gibberic acid	None	Kuhr (1962)
Isogibberic acid	Positive in some bioassays	Holbrook et al. (1961), Pryce (1973)
Dehydroallogibberic acid	Positive in lettuce hypocotyl tests	Cross et al. (1962b), Pryce (1973)
Allogibberic acid	Positive in maize for dwarf leaf sheath extension	Kuhr (1962), Brian et al. (1967)
Gibberellenic acid	Positive in some bioassays	Kuhr (1962), Pryce (1973)
Gibberene	Positive in corn fragments	Sternberg (1962)
2,3,7-Trihydroxyl-1-8-methylenegibb-4-ene-1,10-dicarboxylic acid, 1,3-lactone	—	Kuhr (1962), Muromtsev et al. (1966)

production of the suppressor at pH 3.0, and crystallization of the suppressor were reported (Kurosawa, 1932, 1934; Yabuta et al., 1934). This compound, with the trivial name "fusaric acid," was identified as picolinic acid and thereafter as 5-n-butyl picolinic acid (Yabuta et al., 1934; Yabuta and Hayashi, 1940). The production of fusaric acid and dihydrofusaric acid by G. fujikuroi ETH M82 on Richard's medium was also reported by Stoll and Renz (1957). At a concentration of 10^{-3} M, fusaric acid was reported to inhibit both oxygen consumption and coupled phosphorylation in mitochondria isolated from tomato hypocotyls and cauliflower buds (Sanwal and Waygood, 1961). However, only phosphate esterification was inhibited at lower concentrations, e.g., 5×10^{-4} and 2.5×10^{-4} M. The production of fusarinic acid, a growth-restraining substance by F. heterosporium was reported by Yabuta et al. (1934). The production of an unidentified plant growth inhibitor by Fusarium sp. II was also reported by Thakur and Vyas (1983).

In the commercial production of GAs, it is essential that these plant growth repressors, which are concomitantly produced along with GAs, be eliminated from the final product. The suppression of their production by employing selective parameters, such as lower pH and other appropriate cultural conditions, has been shown (Yabuta et al., 1934; Kurosawa, 1934). However, it is probably difficult to completely inhibit their production, and hence these are present in small amounts in some of the commercial products. The high dilution required in the commercial use of GAs reduces fusaric acid to a nonphytotoxic level (Jefferys, 1970).

4. *Plant Growth Promoters Other Than GAs*

In addition to the production of various GAs and O(2)-acetyl gibberellin (see Section III,E), strains of G. fujikuroi, F. moniliforme, and other related cultures were also reported to produce other plant growth promoters in varying amounts. A literature survey indicated the production of eight chemically defined plant growth promoters, other than GAs, and these are presented in Table XIX. The growth-stimulating action of fujic acid was reported to be 20 times lower than that of GA_3 (Sternberg, 1962). The presence of 3-indoleacetaldoxime and its conversion to 3-indole acetonitrile by G. fujikuroi, presumably due to action of 3-indoleacetaldoxime hydrolase, was reported by Mahadevan and co-workers (Mahadevan, 1963; Kumar and Mahadevan, 1963). Further conversion of 3-indole acetonitrile to indole-3-yl-acetic acid by an intracellular nitrilase of washed mycelia of G.

TABLE XIX

Plant Growth Promoters Other Than GAs Produced by the Strains

Plant growth promotor	Microorganism	Reference
Fujic acid	Gibberella fujikuroi	Sternberg (1962)
3-Indolacetaldoxime	G. fujikuroi	Mahadevan (1963)
3-Indoleacetonitrile	G. fujikuroi	Thimann and Mahadevan (1964)
Anthranilic acid	Fusarium moniliforme	Bakalivanov (1968)
Indolepropionic acid	F. moniliforme	Kefeli et al. (1969)
Ethylene	F. moniliforme	Swart and Kamerbeek (1977)
Indolyl ethanol (tryptophol)	F. moniliforme	Thakur and Vyas (1983)
3-Indolylpyruvic acid	F. moniliforme	Thakur and Vyas (1983)
Other unidentified indoles	F. moniliforme	Thakur and Vyas (1983)

fujikuroi was also reported (Thimann and Mahadevan, 1964). The ability of F. *moniliforme* to produce anthranilic acid and indolepropionic acid and that of F. *moniliforme* strains I and II to produce indole ethanol (tryptophol), indole-3-yl-pyruvic acid, and other unidentified indoles is also well known (Bakalivanov, 1968; Kefeli et al., 1969; Thakur and Vyas, 1983; Thakur, 1981). All of these indole related compounds are presumably precursors or intermediates in the biosynthesis of indole-3-yl-acetic acid (Mahadevan, 1984; Schneider and Wightman, 1978). Ethylene, a fruit-ripening hormone, was also detected by gas–liquid chromatography in culture filtrates of F. *moniliforme* by Swart and Kamerbeek (1977).

5. *General Metabolites*

A vast number of general metabolites have been reported by various workers and these are classified into groups such as organic acids, amino acids, enzyme systems, coenzymes, and other compounds in a classical review by Jefferys (1970). All of these metabolites, except for the enzyme systems and coenzymes, were considered to be shunt products or compounds that are normally expected in general metabolism of microorganisms (Jefferys, 1970). The general metabolites of G. *fujikuroi* and F. *moniliforme* that do not appear in the review by Jefferys (1970) are presented in Table XX.

TABLE XX

General Metabolites Other Than Those Reported in the Review by Jefferys (1970)

Metabolite	Melting point (°C)	Microorganism	Reference
Bikaverin	320–325	Gibberella fujikuroi	Balan et al. (1970), Kjaer et al. (1971)
Nor-bikaverin	>350	G. fujikuroi	Kjaer et al. (1971)
N-Jasmonyl isoleucine	147–149	G. fujikuroi	Cross and Webster (1970)
N-Dihydrojasmonyl isoleucine	140–141.5	G. fujikuroi	Cross and Webster (1970)
O-Dimethylanhydrofusarubin	198–200	G. fujikuroi	Cross et al. (1970a)
Phenethyl alcohol	107.5–108	G. fujikuroi	Cross et al. (1963)
Dimethyl phthalate	—	G. fujikuroi	Cross et al. (1963)
Tyrosol	82–92	G. fujikuroi	Cross et al. (1963)
Nitrilase	—	Fusarium moniliforme	Thimann and Mahadevan (1964)

J. Downstream Processing

At the end of the fermentation, when the product is at its maximum level, the fermentation broth is subjected to downstream processing operations. It is also essential to eliminate toxic and inhibitory substances and to prevent degradation or destruction of desirable properties of the product during downstream processing. Some loss of the product is inherent in all of these unit operations of downstream processing due to handling, spillage, and other related factors. The efficiency of the unit operation in terms of recovery also leads to some losses. In industrial fermentations, concentration of GA_3 in the fermentation medium is low compared to that of many other secondary metabolites. The cell density is also lower as compared to that in fermentations for primary metabolites. Thus, the recovery of GA_3 from fermentation broth involves handling of a large volume of liquid for separation of a comparatively small amount of GA_3. It is a well-established fact that downstream processing involves high expenditure, amounting to ~35% of the cost of the product in many cases (Kalk and Langlykke, 1986). Even mere centrifugation of cells from the fermentation broth is known to constitute ~48% of the cost of the product (Datar, 1986). Moreover, centrifugation is known as a highly problematic and poorly understood unit operation (Datar, 1986). Most of the information available in the literature on downstream processing strategies in the case of the GA_3 process is the outcome of laboratory-scale research work and thus probably is without due consideration to economics.

The separation of mycelial cells or pellets from the culture broth by centrifugation (Carl-Gustaf, 1984; Axelsson, 1985) or filtration (Belter, 1985; Henry, 1972; Henry and Allred, 1972) is the first unit operation in downstream processing of GA_3 by the SmF process. The use of such a cell-free culture filtrate, after lyophilization, was recommended by Bolgarev et al. (1962) as a marketable product for agricultural application. However, such a product may not be acceptable in an international market that deals with the pure product. The cell-free extract is then subjected to the unit operations involving adsorption on animal charcoal, ion-exchange resins, alkaline metal hydroxides or their salts, diatomaceous earth, etc. Alternatively, GAs are extracted in various solvents and then purified by repeated liquid–liquid partition and concentrated under a vacuum. Finally, an amorphous powder or crystalline product is obtained. Different strategies employed for downstream processing of cell-free broth by various workers are depicted in Table XXI. The technical aspects for isolation of GA_3 or GAs from spent end liquors left after crystallization of the product are

TABLE XXI

DOWNSTREAM PROCESSING OF CELL-FREE BROTH ADOPTED BY VARIOUS WORKERS

(A) Russell (1975)	(B) Borrow et al. (1955) Stodola et al. (1955)	(C) Probst (1961)	(D) Société d'Etudes et d'Applications Biochimiques (1963)	(E) Yabuta and Hayashi (1939)	(F) Calam and Curtis (1960)	(G) Stodola et al. (1957)	(H) Hitzman and Mills (1963)	(I) Pitel et al. (1971)	(J) Heropolitanski et al. (1981)	(K) Benedict-Ratz and Olah (1985)	(L) Ganchev et al. (1984)	(M) Mabelis (1984)
Acidification to pH 2.5 →	Charcoal absorption →	Adsorption on strongly basic anion exchange resin →	Adsorption of acidic nongibberellic-like substances on alkaline earth metal hydroxides or their salts →	Charcoal adsorption →	pH adjustment to 3.0 →	Adsorption on diatomaceous earth and impregnated with phosphate buffer at pH 6.2 →	Charcoal adsorption →	Stirring with charcoal for 5 hours →	Adsorption on Amberlite XAD-4 column (at a 4-vol broth per volume of resin) →	Ethyl acetate extract →	Treatment with 10% $ZnSO_4$ and 105 $KFe(CN)_6$ under continuous stirring for 3 hours →	Extraction in ethyl acetate →
Ethyl acetate extraction →	Charcoal washing →	Elution with aqueous solution containing electrolyte →	Adsorption of GA on anionic resin	Elution with MeOH–NH_4OH at room temperature →	Repeated extraction with ethyl acetate →	Elution with ether →	Elution with acetone →	Drying the charcoal cake to ~40% water →	Elution from resin by 2 vol of 90% acetone →	Reextraction with K_2HPO_4 →	Centrifugation of protein →	Concentration of extract →
1% Na carbonate extraction →	Elute with ammonical methanol or acetone →	Extraction with		Solvent concentration and basic lead acetate addition	Concentration under vacuum →	Concentration	Concentration →	Extraction with acetone →	Evaporation of the	Treatment of extract with NaCl, sodium sulfolaurate filtration →	Sorption on Amberlite	Treatment with propyl alcohol and N-benzyl-N-methyl amine →
Acidification to pH 2.5 →	Elute concentration under reduced pressure				Gibberellins		Extraction with ethyl acetate →	Concentration →		Precipita-		
Ethyl acetate extraction							Extraction with phos-	Reextraction with 0.5 M sodium carbonate →				
								Reextraction with ethyl acetate →				
								Extraction				

→ Evaporation under reduced pressure → Concentration → Crystallization	to aqueous solution → Extraction with ethyl acetate as per Scheme A	water-immiscible ketone → Adsorption on solid alkali metal bicarbonate salt → Recovery as salt of alkali metal	→ Elution with ammonia or buffered solution of alkaline earth metal salts → Repeated liquid–liquid partition → Crystallization	→ Filtration → Treating the filtrate with H_2S, Na_2CO_3 → Extraction with ether → Treating filtrate with H_2S, Na_2CO_3 → Aqueous solution reextracted with ether after acidification with H_2SO_4 → Gibberellins	under vacuum → Amorphous powder	phate buffer from ethyl acetate → Reextraction in ethyl acetate at acidic pH → Concentration under vacuum → GA_3
					with ethyl acetate → Concentration under vacuum	after acidification → Concentration under vacuum
					solvent → GA_3	tion of GA_4 and GA_7 at pH 2.5–3.2 → Filtration → pH adjustment to 3.5 → Extraction with ethyl acetate → Concentration under vacuum → Extraction with KH_2PO_4–KOH buffer at pH 6.2 → Ethyl acetate extraction at pH 3.1 → GA_3
						IRA-401 S or Wofatite SBW/MB in –OH form → Elution by N NH_4Cl at pH 5.5 → Reextraction in ethyl acetate → Crystallized product
						Precipitation of salt of GA_4, GA_7, and a small amount of GA_3 → Treatment of filtrate with citric acid → Free GA_3

also available (Heropolitanski et al., 1981). They involve extraction with ethyl acetate or adsorption on a nonionic resin. Further liquid–liquid partitioning and other purification steps are reported to recover 87.3% of GAs present in the spent liquor.

XI. Use of Immobilized Whole Cells

The immobilization of enzymes or whole cells offers many advantages, chiefly repetitive use, improved stability, enzyme contamination-free product, and easy stoppage of the reaction by removing the enzyme or the cells from the reaction mixture (Ghildyal et al., 1979). The successful use of immobilized enzymes and immobilized whole cells in laboratory- and industrial-scale processes has evoked worldwide attention, leading to application of these techniques to other products as well as to improve the techniques (Mohan and Li, 1974; Zaborsky, 1973; Chang, 1964; Durand and Novarro, 1978; Abbott, 1978; Messing, 1980; Linko and Linko, 1983; Vojtisek and Jirku, 1983). Among the immobilized systems, immobilization of whole cells provides a means for entrapment of multistep and cooperative enzyme systems present in the intact cell (Fukui and Tanaka, 1982). Three different types of microbial cells are employed in whole cell immobilization; these include dead or treated cells, resting cells, and growing cells. The latter is of commercial importance, as the cells are kept in the growing state within a gel matrix by constantly supplying suitable nutrients(Fukui and Tanaka, 1982).

Immobilized growing cells have been shown to offer advantages, such as (1) superior stability due to protection of cells by physicochemical interactions between gels and cells; (2) protection of growing cells against unfavorable environmental factors; (3) changed permeability of cells favoring high penetration of substrate; (4) faster removal of end products from fermentation vessels; and (5) the renewable, self-regenerating, or self-proliferating nature of the biocatalytic system (Fukui and Tanaka, 1982; Kahlon and Malhotra, 1986; Jack and Zajic, 1977; Holcberg and Margalith, 1981). It also obviates expensive and difficult procedures for the isolation and purification of complex multienzyme systems. The other advantages are (1) smaller reactor volume, (2) reduced inhibition by substrate as well as product, (3) retention of enzyme activity for a very long time, (4) possible reuse, and (5) economics (Fukui and Tanaka, 1982).

To date, there are only three reports on the use of immobilized whole cells for the production of GA_3 (Heinrich and Rehm, 1981; Kahlon and Malhotra, 1986; Kumar and Lonsane, 1988a). Although no specific information on the nature of cells was given except by Kumar and Lonsane (1988a), the operational details, results, and composition of the media used suggest involvement of the immobilized growing cell technique. The first report on the production of GA_3 by immobilized mycelia of F. moniliforme was based on the use of n-alkanes as a carbon source (Heinrich and Rehm, 1981). The cells and n-alkanes were adsorbed onto a glass carrier while a fixed bed system was used for the production of GA_3. These carriers were subsequently used for the production of GA_3 by circulating ammonia-free medium, containing gas oil or glucose as a carbon source, in fixed-bed reactors. The results indicated equal production of GA_3 from both substrates, although the requirement for oxygen was nearly double in the case of glucose. However, the yield of GA_3 was poor in both cases. Heinrich and Rehm (1981) have also reported advantages in the use of cell-covered carrier in fixed-bed system. These include (1) elimination of difficult immobilization procedures such as gel-entrapment of cells, (2) no damage of cells by shear forces, (3) absence of cells in the product, and (4) the necessity for only low-level purification of the product. The possibility of using a two-stage process by simple medium replacement in the same reactor was also pointed out.

Recently, results on the production of GA_3 in shake flasks using fungal mycelia entrapped in sodium alginate have been reported (Kahlon and Malhotra, 1986). The culture used was F. moniliforme 1 in paneer (cheese) whey, while the mycelial cells were immobilized using the method of Marwaha and Kennedy (1984). The maximum yield of GA_3 obtained was 680 mg/liter, and the utilization of lactose at a 78.16% level was achieved under optimized parameters, i.e., pH 5.5, 12 days' incubation at $25 \pm 1°C$ on a rotary shaker. The comparison of yields from free or immobilized mycelia indicated insignificant differences.

Both of the above studies were with F. moniliforme. Recently, studies on the production of GA_3 by immobilized growing cells of G. fujikuroi P-3 in diluted wheat bran (WB) extract medium in batch and semicontinuous cultures were reported (Kumar and Lonsane, 1988a). The performance of the immobilized growing cell beads in the production of GA_3 was found to be significantly affected by various factors such as selection of the gel entrapment agent, the nature and the age of the cells, the size of the beads, cell concentration, and inclusion of linseed oil or a mixture of linseed oil and $MgSO_4 \cdot 7H_2O$. The production of GA_3 by immobilized growing cells of G. fujikuroi P-3 was 70.4% and 80.7% at 72 hours and 120 hours of fermentation, respectively, in batch culture as compared to that by free mycelial cells under identical fermentation parameters. The rate of production of GA_3 in an inverted conical fluidized bioreactor (Wilson, 1976) of 1-liter capacity was in the range of 0.56–0.66 mg/liter/hr, while the total production of GA_3 at the end of a 5-day fermentation was between 69 and 79 mg/liter in the first eight cycles in semicontinuous culture. The productivity was found to decrease to 83% and 57% in the ninth and tenth cycles. The yields of GA_3 achieved so far by using immobilized growing cells were very low even in semicontinuous culture (Heinrich and Rehm, 1981; Kumar and Lonsane, 1988a). Better yields in whey medium were reported by Kahlon and Malhotra (1986), but the studies were limited to batch culture. It was speculated by Fukui and Tanaka (1982) that immobilized growing cells will be utilized extensively in the future, as they are more economical than conventional fermentation techniques. However, the immobilized growing cell systems have not proven their potential so far for the production of GA_3, probably due to the involvement of many complex steps and regulatory mechanisms in converting carbon substrates into GA_3.

XII. Solid-State Fermentation

A. The Technique and Its Potential

The solid-state fermentation (SSF) technique has been used from ancient times in the production of fermented foods, cheese, and composting, even without understanding the scientific principles involved (Lonsane et al., 1982). In fact, these processes were considered some sort of mystical phenomena in those early days (Aiba et al., 1973). The moist water-insoluble substrate is fermented by microorganisms in the absence of any free water in SSF processes (Lonsane et al., 1985). The moist solid substrate usually acts as a source of carbon, nitrogen, minerals, and other nutrients although, in some cases, it was enriched with additional nutrients for improved productivity (Lonsane et al., 1982). In addition, the solid substrate particles also provide an anchorage for the microorganisms. In the SSF technique, the growth of microorganisms on moist solids was made faster by using appropriate physicochemical and nutritional parameters and also by controlling parameters at optimum values during the course of fermentation (Lonsane et al., 1982). The industrial exploitation of these techniques has been confined to countries in the Orient (Yamada, 1977). By using SSF processes, enormous quantitites of fermented foods are being produced in Japan, Thailand, Indonesia, China, and other countries in Southeast Asia. Other successful products or processes involving use of the SSF technique are mold bran, enzymes, cheese, mushrooms, single-cell protein, citric acid, protein-enriched beans and barley, ensilaged fodder, toxins, compost, spores, antibiotics, sterols, curing of tobacco, and dehairing of skins and hides (Lonsane et al., 1982).

The SSF technique involves growth of molds on moist solid substrate in an appropriate fermentor wherein optimum growth conditions are provided to obtain the product at highest possible yields. In laboratory- and bench-scale fermentations, growth was allowed to occur in flasks (Nagaraja Rao, 1976), pot fermentors (Hao et al., 1943), rotating drums (Underkofler et al., 1939), four-section fermentors (Lindenfelser and Ciegler, 1975), cooker-cum-incubators (Takamine, 1914), wooden cells (Underkofler et al., 1947), covered pan fermentors (Underkofler et al., 1947; Hao et al., 1943), incubating chambers (Underkofler et al., 1947), butler-type corn storage bin fermentors (Silman et al., 1979), column fermentors (Raimbault and Alazard, 1980), and vertical incubation cells (Underkofler et al., 1947). For work at the pilot plant and commercial scale, vertical incubation cells (Schulze, 1962), inclined incubation cells (Underkofler et al., 1947), tray fermentors (Underkofler et al., 1946, 1947), conveyer belt tunnels (Jeffries, 1948), or koji rooms (Hesseltine, 1972; Ahmed et al., 1987) have been used.

The SSF process offers many advantages over the SmF technique; these include (1) simplicity of growth media since a single substrate provides almost all required nutrients; (2) greater compactness of the fermentor vessel due to the use of lower water volume with higher substrate concentration; (3) simpler hardware and control systems; (4) a simplified procedure for inoculum build-up; (5) easier scale-up of processes; (6) reduced solvent requirement for product recovery; (7) greater yields; and (8) easier control of bacterial contamination due to low moisture levels in the systems.

Other advantages, offered in individual cases, include (1) elimination of costly centrifugation or extensive dewatering in case of rye grass straw fermentation, (2) the possible use of impure sugars and molasses in citric acid production, (3) the imparting of typical flavors to mushrooms, and (4) an increase in the protein content of low-protein foodstuffs (Hesseltine, 1977; Ralph, 1976). Moreover, there are reports stating that products lose their characteristic flavors or that productivity is lower when produced by the SmF process (Eddy, 1958; Ghildyal et al., 1985).

Some disadvantages are inherent in the SSF process; these include (1) the need for pretreatment of the solid substrates; (2) difficulty in maintaining moisture at optimum levels; (3) limitations with regard to the microorganisms that can be used; (4) maintenance of the critical relationshp between the surface area of solids and the yield as well as between the optimal amount of substrate and the size and shape of the fermentor; (5) the demand for a high amount of spore inoculum of desired characteristics; (6) the need for simple, direct methods for cell biomass determination; (7) the high power requirement for rotating drum fermentors; (8) the lack of design and engineering development for various operations of bulk fermentation; (9) inadequate development of methods for inoculations; and (10) an exacting demand for the monitoring and control of parameters such as temperature, humidity, air flow, and free O_2 transfer as well as removal of CO_2 generated in the fermentor (Hesseltine, 1972, 1977; Lonsane et al., 1982; Ralph, 1976).

1. Suitability of the SSF Technique for Fungal Strains

The type of microbial growth that occurs on moist solid substrate particles in the SSF processes is generally similar to that which takes place in nature (Lonsane et al., 1985). The nature of growth reported in the literature includes feltlike mycelia on the inner and coarse sides of curled wheat bran particles (Takamine, 1914), impregnation of the entire surface of starch granules (Raimbault and Alazard, 1980) or cellulosic waste (Chahal, 1983), and growth on a small area of rice grain (Hesseltine, 1972). The broken and exposed ends of the cells of solids and mechanical breaks and spaces created by pretreatment of substrate serve as sites for the initial attachment of the mycelia and the entry of the hyphae into the cell lumen of the substrate (Hayes, 1977). The filamentous fungi have the ability to penetrate deeply into intercellular and intracellular spaces, even in the hardest of substrates by mechanical and/or enzymic means. The rigid cell wall behind the apex of the hyphae, firm anchoring of older, rearward parts of the hyphae in the substrate, exertion of considerable mechanical pressure on the hyphal tip as it extends under turgor, a complex system of heavy branching, and the creation of a base hole in the cell wall of the substrate by production of extracellular enzymes by the hyphae (Chahal, 1983; Bravery, 1975; Reese, 1959) play important roles in deep penetration. These growth characteristics are responsible for limiting the choice of microorganisms in SSF mainly to filamentous fungi because bacteria, yeasts, and single-cell fungi lack the ability to achieve deep penetration into the tissue of solid substrates (Chahal, 1983). Nevertheless, it is interesting to note that successful growth and metabolite production of bacteria (Beckord et al., 1945; Ramesh and Lonsane, 1987) and yeasts (Gibbons et al., 1984; Kirby and Mardon, 1980) have also been reported in SSF processes.

Thus, for successful use of the SSF process, it is essential that the microorganism penetrate or adhere to a solid particle for uptake of nutrients. Breakage of the attachment of cells to solid substrate particles was shown to reduce enzyme production in SSF processes involving agitation of the fermenting mass (Mudgett, 1986; Arima, 1964). All of these facts indicate that the ability of microorganisms to grow in a mycelial form and to penetrate into or adhere to the substrate particle is probably the most important characteristic for successful exploitation of microorganisms in the SSF technique. Another important requirement for successful use of the SSF technique is the ability of the culture to form spores. The mixing of a large body of moist solid medium with 10–20% inoculum is most efficient when spore inoculum is used. The mycelial growth or mycelial pellets will pose problems in uniform distribution of inoculum.

2. *Potential of the SSF Technique in GA_3 Production*

G. fujikuori and *F. moniliforme*, the species used in production of GA_3, are known to produce spores, indicating their potential for exploitation under the SSF technique. Surprisingly, no attempt has been made anywhere in the world, even in exploration of the possibility of using the SSF technique for production of GA_3, except for the pioneering work at CFTRI (Prapulla et al., 1983; CFTRI, 1981–1982; Prema et al., 1987; Kumar and Lonsane, 1986a,b, 1987a–c). In exploratory studies on the extension of the SSF technique to production of GA_3, *G. fujikuroi* FT-2 gave 1.14 g of GA_3 per kilogram of dry moldy bran (DMB) at 28°C after an incubation period of 10 days. However, the strain was found to give inconsistent results in subsequent studies (Prema et al., 1987; CFTRI, 1981–1982).

Work was subsequently revived in 1984 to obtain a consistent yield of GA_3 under the SSF process. Among ten strains screened, *G. fujikuroi* P-3, isolated from an infected rice plant, produced the highest and most consistent quantity of GA_3 at the end of a 7-day fermentation. The level of GA_3 production was ~0.35 g/kg of DMB but was improved 2.3-fold in preliminary optimization studies by supplementing basal wheat bran (WB) medium with 25% soluble starch or by using WB of 0.3- to 0.4-cm-particle size (Kumar and Lonsane, 1985b, 1987b). It is of interest to note that *F. moniliforme* NCIM 1100 gave lower yields of GA_3 under the SSF technique although it produced the highest quantity under SmF technique, as compared to other strains (Kumar, 1987). The results indicated that the capability of strains to grow and produce GA_3 under the SSF technique is strain dependent.

B. Physical Factors

Various physical factors were found to significantly affect the production of GA_3 by *G. fujikuroi* P-3 under the SSF technique (Kumar, 1987; CFTRI, 1981–1982). The production of GA_3 was highest at 30°C, while it was reduced to 55.5 and 34.7% at 35 and 25°C, respectively. The production was highest at 50% initial moisture and was reduced considerably at lower and higher initial moisture contents. The use of 0.3- to 0.4-cm WB particles was found to lead to about 2.45 times more production of GA_3 as compared to lower and higher particle sizes. However, commercial WB contains coarse particles of 0.3- to 0.4-cm size at only a 10–20% level and thus it is not practicable to use WB of 0.3- to 0.4-cm-particle size. At an initial pH of 3.5, the production of GA_3 was maximum as compared to lower and higher initial pH values. The production of GA_3 was found to increase slightly with an increase in the autoclaving time of the medium up to 45 minutes at 121°C. The yield of GA_3 was also reported to decrease from 1.116 to 0.916 g/kg of DMB, with an increase in the ratio of moist medium volume to flask volume from 0.024 to 0.1. The production of GA_3 was marginally higher when the flasks were incubated under 80 lux light but was slightly reduced in darkness. The production of GA_3 by *G. fujikuroi* P-3 was highest with the use of 15% inoculum (Kumar, 1987) grown for 7 days in liquid or solid media. However, the yield was reduced by 33% with the use of inoculum grown in liquid medium for 24 hours. In the case of *G. fujikuroi* FT-2, the use of fully sporulated DMB at a 10% level as inoculum gave better yields as compared to a 5% inoculum (CFTRI, 1981–1982). These results indicated that the inoculum ratio was specific to the strain.

C. Nutritional Factors

Commercial WB contains ~8.5 and 9.5% starch and protein, respectively (Fisher, 1973), and therefore its enrichment leads to enhanced production of GA_3. Detailed reports on the effect of various nutritional factors on the yield of GA_3 under SSF techniques are available (Kumar, 1987; CFTRI, 1981–1982). The enrichment of WB with 25% corn starch or soluble starch resulted in an increase in the production of GA_3 by 2.28- and 2.32-fold, respectively, in the case of *G. fujikuroi* P-3 (Kumar, 1987). The enrichment with rice bran resulted in lowering the yield, even below that obtained in the control. However, in the case of *G. fujikuroi* FT-2, the mixture of WB and rice bran at a 1:1 ratio gave higher yields, although the rate of formation of GA_3 in the initial 6 days of incubation was slower (CFTRI, 1981–1982). Enrichment of WB with acetate buffer, geraniol, and citronella oil individually also completely inhibited the production of GA_3 by strain FT-2. Geraniol, at the 1 ppm level, was also found to completely inhibit the biosynthesis of GA_3 by strain P-3 in the SSF and SmF processes due to the inhibitory effect on cell growth.

Substrate inhibition of the biosynthesis of GA_3 in *G. fujikuroi* P-3 under the SSF technique by corn starch and glucose was prominent and affected both the level and the rate of biosynthesis. The concentrations that did not affect the biosynthesis of GA_3 were 20 and 15%, respectively. The results on the directed biosynthesis of GA_3 in the SSF technique were also similar to those in the SmF process. About a 21% increase in the biosynthesis of GA_3 by strain P-3 was observed with the addition of mevalonate, a precursor. However, the increase was only 7% with the incorporation of HMG CoA (Fig. 2), an intermediate. The yield of GA_3 by strain P-3 was also improved by 2.01-fold with the enrichment of WB with urea to provide 70 mg/dl of nitrogen (Kumar and Lonsane, 1985b, 1987c). In general, the yield of GA_3 was higher with the enrichment of WB with ammonia compounds as compared to that containing nitrogen in the nitrate form. Enrichment with soybean meal or soya flour improved the yield by ~22% in the case of strain P-3, but the yield was reduced in the case of strain FT-2. The enrichment of WB with linseed oil at 1%, in the case of strain P-3, resulted in an increase in the yield by about twofold as compared to a 1.5-fold increased with olive or rice bran oil. The enrichment of WB with $MgSO_4$ at 0.007 g/dl enhanced the production of GA_3 by strain P-3, while the yield was lowered in the case of strain FT-2.

D. Large-Scale Trial

The production of GA_3 under the SSF process by *G. fujikuroi* FT-2 at flask and large-scale levels was found to be variable and inconsistent. In most of the cases, the yield was ~1 g/kg DMB, even at 96 and 150 tray levels (CFTRI, 1981–1982; Ahmed et al., 1987) but varied between 0.28 and 1.23 g of GA_3 per kilogram of DMB. *Gibberella fujikuroi* P-3 gave consistent production of GA_3 under the SSF process, but the yield was ~0.35 g/kg of DMB in basal WB medium, selected arbitrarily (Kumar and Lonsane, 1985b, 1987b). After standardization of various physical and nutritional parameters, an optimum medium was developed and it consistently gave 0.986–1.115 g of GA_3 per kilogram of DMB in flask-level experiments (Kumar, 1987). In a large-scale production of GA_3 by *G. fujikuroi* P-3 under the SSF technique at the 93 kg of moist WB medium level, a peak of 1.164 g/kg of DMB in GA_3 production was reached on the 6th day as compared to that in flask-level experiments on the 7th day. It remained more or less constant for the next 24 hours (Kumar and Lonsane, 1987b). These results established the feasibility of production of GA_3 under SSF technique by G. fujikuroi P-3 even on the large scale.

E. Fed-Batch Process

The application of fed-batch culture to the SSF process for production of GA_3 was found to lead to an increase of 18.2% in the yield of GA_3 at the end of a 7-day fermentation (Kumar and Lonsane, 1987a,c). It involved feeding of soluble starch at 72, 96, and 120 hours. However, the production of GA_3 was reduced, even to a level below that obtained in the batch SSF process, with the feeding of glucose. In the case of feeding moist WB medium, the productivity of GA_3 was between that of the batch SSF process and that of the fed-batch SSF process based on the feeding of soluble starch (Kumar and Lonsane, 1987a,c, 1988b). The replacement of soluble starch with corn starch and the extension of fermentation time to 8 days increased the productivity of GA_3 by 46.7% as compared to the batch SSF process. The substitution of soluble starch with corn starch also offers economic advantages. The data indicated the importance of selecting the correct feed substrate and feed policy in achieving higher yields of GA_3 under the fed-batch SSF process.

Comparative kinetic studies of the level and the rate of production of GA_3, dry cell biomass, and hydrolytic enzymes in batch and fed-batch SSF processes, the latter involving feeding of corn starch at 72, 96, and 120 hours of fermentation, indicated the complex nature of biochemical changes in both of the processes (Kumar and Lonsane,1988b). In the fed-batch SSF process, the rate and level of GA_3 production, dry biomass formation, and production of proteases were higher as compared to those in the batch SSF process. On the other hand, the level and the rate of production of amylases, cellulases, and xylanases were higher in the batch SSF process as compared to those in the fed-batch SSF process. Pectinases were absent in both of the systems throughout the fermentation period (Kumar and Lonsane, 1988b). The lower levels of cellulases and xylanases in the fed-batch culture indicated the possibility of improved yield by strain improvement. The results of the kinetic studies on protoplast formation in G. fujikuroi P-3 (Kumar and Lonsane, 1986b, 1988c) are thus of significant value.

F. Growth Pattern and GA_3 Biosynthesis

Gibberella fujikuroi P-3, after about 73 hours of the log phase in basal WB medium, remained in the stationary phase until 196 hours. The initiation of GA_3 biosynthesis coincided with the onset of the stationary phase. The peak in biosynthesis of GA_3 was reached at 168 hours and remained more or less constant for the next 24 hours. GA_3 was undetectable in the first 4 days, while dry biomass formation was ~9 mg/g of DMB at the end of the log phase. Total GA_3 produced at the end of 168 hours in basal WB medium was 0.35 g/kg of DMB (Kumar and Lonsane, 1987b). The trend of the relationship between cell growth and GA_3 biosynthesis in optimized WB medium was similar to that in basal WB medium; however, cell growth and GA_3 biosynthesis were at higher levels. A similar type of relationship was also observed in the SmF process, thereby indicating that the growth pattern and GA_3 biosynthesis were more or less similar in SmF and SSF processes. In optimized WB medium, G. fujikuroi P-3 was found to grow throughout the medium in the form of white mycelial growth after 24 hours, and the cells were found to adhere to WB particles. The medium acquired a cottony appearance by 96 hours. Spore formation was initiated at 96 hours and subsequently the medium became light purple at the peripheral parts due to pigment production (Kumar, 1987).

G. Concomitant Products

Gibberella fujikuroi P-3, during its growth on moist WB medium and production of GA_3 under the SSF technique, produced many concomitant products. Among these, four major and interesting concomitant products were purified and subjected to partial characterization. It is interesting to note that a variety of hydrolytic enzymes, such as amylases, cellulases, xylanases, and proteases, were also concomitantly produced. The kinetics of these enzymes are described in Section XII,E.

1. Compound with Sterol Configuration

The extraction of DMB by a supercritical fluid extraction technique gave a light red extract that yielded a pure white crystalline compound after purification (Kumar, 1987). Identification studies showed that the substance is a sterol that differs in melting point, specific rotation, and mass spectrum from known ergosterol type compounds. The white crystalline pure compound showed no bioactivity at the 20 ppm level in the dwarf rice seedling bioassay. However, the root system of all test plants and the rest of the plant structure, excluding the upper actively growing part, were found to be infected by a black aspergillus and a dark gray fungi. The rice seedlings became yellowish in color with a 50–150 ppm dose of the sterol (Kumar, 1987). Even a crude extract of DMB obtained by countercurrent leaching showed the above symptoms when applied at the 30 ppm level. However, these symptoms were absent when the crude extract was applied at the 14.58 ppm level.

2. *Polyketide Compound*

A dark purple pigment was concomitantly produced by G. fujikuroi P-3 when cultured under the SSF technique. However, the ultraviolet (UV) and infrared (IR) spectra (Kumar, 1987) do not match totally with those of the known polyketides such as bikaverin, nor-bikaverin, and O-dimethylanhydrofusarubin (Kjaer et al., 1971; Bu'Lock et al., 1974; Cross et al., 1970a).

3. *Fujic Acid*

A crude extract of DMB obtained by countercurrent leaching showed the presence of a fluorescent compound with an R_f value less than that of GA_3 (Kumar, 1987). The IR spectrum of the compound matched that of fujic acid exactly, a compound reported to be concomitantly produced with GA_3 in the SmF process (Sternberg, 1962).

4. *Brassinosteroid-like Compound*

A higher level of biological activity in the crude extract of DMB than accountable for by the GA_3 present in it (Kumar and Lonsane, 1985a, 1986a) indicated that another plant growth promoter is concomitantly produced by the strain. The brassinosteroid-like nature (Mitchell et al., 1970; Grove et al., 1979; Ishiguro et al., 1980) of this concomitant product is indicated by the splitting of second internodes of rice seedlings especially when a higher quantity of crude extract was bioassayed. It is further confirmed by results of the rice lamina inclination assay, which is known as a specific and highly sensitive test for brassinosteroids (Maeda, 1965; Wade et al., 1981, 1983; Adam and Marquardt, 1986). The concomitant production of brassinosteroid-like compound provides a valuable by-product. Its separation will provide additional economic value to the process for the production of GA_3.

H. Downstream Processing

In preliminary studies involving arbitrarily selected parameters, GA_3 was extracted from DMB with 25 times its weight of ethyl acetate in two steps and thus 23 liters of extract was obtained from 1 kg of DMB (Kumar and Lonsane, 1985a, 1987b). The concentration of DMB in the crude extract was 43.5 µg/ml, much lower than that in the SmF process. This defeated the advantage of lower expenditure in downstream processing in the SSF technique.

1. *Factors Affecting Extraction Efficiency*

The drying of moist fermented WB medium produced under the SSF technique by *G. fujikuroi* P-3 at 30°C in a cross-current dryer, use of an aqueous solution of 10% ethanol at 1 : 10 ratio of DMB to solvent and an extraction pH of 2.5 were found to give maximum recovery of GA_3 from DMB (Kumar and Lonsane, 1987d). The degree of extraction was found to be independent of contact time and temperature across the ranges evaluated. Maximum recovery was achieved by Prapulla *et al.* (1983) by using aqueous solution containing 5% ethanol, a 1 : 8 ratio of DMB to solvent, a 30-minute contact time, and a 28°C contact temperature. The extraction of DMB under the above parameters (Kumar and Lonsane, 1987d) also gave a too dilute extract containing ~0.13 mg of GA_3 per milliliter of extract. On the other hand, the extraction of GA_3 at a 1 : 3 ratio was ~56% as compared to that with 1 : 10 ratio, while the concentration of GA_3 in the extract was 0.41 mg/ml.

2. *Multiple-Contact Countercurrent Leaching*

Multiple-contact countercurrent leaching led to a higher concentration of product in the extract as compared to the percolation technique. Even lower DMB: solvent ratios can be used, and the vacuum concentration of the extract can be eliminated. Only one report is available in the literature on the extraction of GA_3 from DMB by multiple-contact countercurrent leaching (Kumar and Lonsane, 1987d).

A gradual increase in the concentration of GA_3 in the extract and in the percentage of recovery of GA_3 were related to an increase in the number of contact stages. The use of four contact stages leads to ~87% extraction efficiency and 0.90 mg/ml of GA_3 in the extract. The material balance and analytical studies showed that 3.6 contact stages were theoretically needed to achieve the extraction efficiency obtained by using four contact stages in the actual experimental work. This gave a contact stage efficiency of 90%. The application of a four contact stages countercurrent leaching technique to the extraction of GA_3 was reported to lead to handling only one sixth the volume of extract in further downstream processing as compared to a single-step extraction. Thus, the data indicated that four contact stage countercurrent leaching provided the key to reduced production cost (Kumar and Lonsane, 1987d).

3. Supercritical Fluid Extraction of DMB

Supercritical fluid extraction (SCFE), a technique based on the exploitation of the enhanced power of supercritical (SC) fluids at temperatures and pressures near the critical point, (Williams, 1981) is currently enjoying a surge of industrial interest (Randolph et al., 1985) in chemical industries (Kurzhals, 1982; Gardner, 1982; Basta and McQueen, 1985). It offers an efficient and powerful alternative to conventional distillation and solvent extraction processes for separation of heat-labile substances of low volatility at moderate temperature (Larson and King, 1986; Williams, 1981). In spite of the high initial capital investment, the tempo of worldwide research has sped up so much recently that the manufacture and sale of bench- or pilot-scale units has become a sizable business (Basta and McQueen, 1985). In the area of biotechnology, SCFE has simplified downstream separation steps, since the desired or undesired products can be removed from the reaction media by volatilization into or precipitation from an SC fluid phase (Wong and Johnston, 1986). Emphasis was also given by Mudgett (1986) to the potential use of SCFE in biotechnology for the recovery of products from DMB produced under the SSF technique. However, only two exploratory reports are available on its application to SSF processes.

The results on the extraction of DMB, produced by G. fujikuroi P-3 in the SSF process, by SCFE were reported by Kumar et al. (1989). A red extract was obtained, which showed the presence of 12 fluorescent compounds of different R_f values. Among these, only one spot was prominent and its purification led to the isolation of pure sterol. It does not possess any plant growth-promoting ability, but it is able to disturb the defense mechanism in plants as described in Section XII,G. Its presence in the final product, therefore, is objectionable and SCFE offers potential in its effective elimination, along with that of the 11 other fluorescent compounds coextracted by the SCFE technique (Kumar et al., 1989). Another report dealt with sterilization of rice koji with the simultaneous removal of lipids by the SCFE technique (Tokuda et al., 1986).

4. Purification of GA_3 Produced by the SSF Technique

The purification of GA_3 present in the crude extract of DMB, produced by using G. fujikuroi FT-2 in the SSF process, was reported by Prapulla et al. (1983) and involved sequential extraction of the dark brown extract by different solvents, elimination of proteins, and passage through a column of $NaHCO_3$ as in the procedure developed by

Kavanagh and Kuzel (1958). It gave a colorless product that gave one spot of GA_3 on thin-layer chromatography (TLC) plates. The methodology developed by Kumar (1987) for purification of GA_3 present in the crude extract of DMB, produced by *G. fujikuroi* P-3 in the SSF process, involved adsorption of GA_3 on a resin, silicic acid column chromatography, preparative TLC, and crystallization of the product. It was found to result in a product that gave a single spot corresponding to GA_3 on two-dimensional TLC (Kumar, 1987). Crystallized material was obtained in ethyl acetate.

5. *Identification of GA_3 Produced by the SSF Technique*

Purified crystalline GA_3, as obtained above, was subjected to IR spectroscopy, proton nuclear magnetic resonance, and mass spectroscopy (Kumar, 1987). The spectral characteristics are found to match totally those of authentic GA_3 samples and thus conclusively proved that the isolated compound is pure GA_3. It also showed biological activity and promoted the growth of the dwarf rice seedlings, similar to effects of authentic GA_3. The absence of brown lesions on the sheath of rice seedlings also indicated the absence of phytotoxins in the final purified material.

XIII. Analytical Methods

The analytical procedures for GA are of two types, i.e., biological assays and physicochemical methods.

A. Bioassays

Bioassays have played an important role in the discovery and identification of GAs. A total of 33 test systems involving various parts, organs, or functions of plants such as coleoptile, leaf, sheath, epicotyl, mesocotyl, bud dormancy, seed germination, induction of α-amylase synthesis, leaf expansion, senescence, and induction of flowers and cones were reported by Bailiss and Hill (1971). Bioassays are moderately selective though not a single bioassay is known to be entirely free from interaction with impurities. The accuracy of bioassays is always open to questions in spite of their repeatability (Graebe and Ropers, 1978). Moreover, bioassays are time-consuming, difficult, and cumbersome as compared to physicochemical detectors (Holbrook *et al.*, 1961; Graebe and Ropers, 1978). Those used most commonly are discussed below.

1. *Tan-ginbozu Dwarf Rice Microdrop Bioassay*

A dwarf variety of rice, deficient in endogenous synthesis of GAs, probably due to a blockage of biosynthetic pathways between mevalonate and kur-16-ene, is used (Suge and Murakami, 1968). The response to GAs is related to the length of the second leaf sheath. The method was reported to be extremely sensitive and capable of detecting 10^{-4} μg of GA_3 (Murakami, 1968, 1970b). However, the response was found to be inconsistent with a crude extract of DMB (Prapulla et al., 1983).

2. *Dwarf Pea Bioassay*

A seedling of *Pisum sativum* L. var. progress No. 9 is used and the methodology is similar to the dwarf rice bioassay (Koehler and Lang, 1963). GA samples are applied to seedlings and induced length is measured and compared to the controls. Although the assay was reported to be specific to GAs (Brian and Hemming, 1955), it also responded to α-tocopherol (Bruinsma, 1963), steroids (Kopcewicz, 1969), and indole acetic acid (Brian and Hemming, 1955). The response was reported to be inconsistent with a crude extract of DMB (Prapulla et al., 1983).

3. *Dwarf Maize Bioassay*

Single gene mutants of *Zea mays* such as d-1, d-2, d-3, d-5, and an_1 can be employed, but the most commonly used mutants are d-1 and d-5 (Phinney, 1956, 1961; Phinney and West, 1961). The samples are applied to the first leaf of the young seedling and the length between the first and second leaf sheaths are measured. The method was also reported to give weak response to α-tocopherol and precursors of GA (Bruinsma, 1963).

4. *Cucumber Hypocotyl Bioassay*

Cucumis sativum L. is generally used and the test sample is applied to the cotyledons after removing the seed coats (Brian et al., 1964). The length of hypocotyls showed the response. The assay also showed a response to other plant growth regulators such as helminthosporol and its derivatives and dihydroconiferyl alcohol (Kato et al., 1968; Sakurai et al., 1974). This method is also known to be nonsensitive to GA_3 (Brian et al., 1964).

5. *Lettuce Hypocotyl Bioassay*

This bioassay methodology is relatively simple and involves the use of lettuce seeds of *Lactuca sativa* L. var. Arctic (Frankland and

Wareing, 1960). GAs samples are applied to the leaves and the length of hypocotyls indicated the response. The assay is known to be sensitive to GAs but is prone to high inhibition by toxic compounds present in impure samples of plant extracts (Brian et al., 1964).

6. *Barley Aleurone Bioassay*

This assay is based on the synthesis and secretion of α-amylase in barley, *Hordeum vulgare*, by the action of GA (Nicholls and Paleg, 1963). The amount of enzymes produced is directly proportional to the amount of GA applied (Jones and Varner, 1967). The test samples were applied to embryoless half-seeds of barley. It is also found to respond to helminthosporol and helminthosporic acid (Briggs, 1973) and is shown to be prone to inhibition by abscisic acid (Chrispeels and Varner, 1966). The response is also poor in the case of a crude extract from DMB (Prapulla et al., 1983).

7. *Miscellaneous Bioassays*

Many other bioassays are also known (Radley, 1958; Wheeler, 1960; Hayashi and Murakami, 1953; Whyte and Luckwill, 1966), but these are not used for routine analyses, due probably to many shortcomings. Some of these are shown in Table XXII.

B. PHYSICOCHEMICAL INSTRUMENTATION METHODS

Physiocochemical instrumentation methods are comparatively simple and can be performed in less time and hence are employed routinely in fermentation industries. A variety of methods are being used in the identification and analysis of GAs and these are described below.

TABLE XXII

SOME MISCELLANEOUS BIOASSAYS

Method	Reference
Rice endosperm α-amylase assay	Murakami (1966), Ogawa (1966)
Barley endosperm phosphatase assay	Jones (1969a,b)
Cereal leaf segments	Radley (1958), Hayashi and Murakami (1954), Skene and Carr (1961)
Leaf discs of bean or apple	Wheeler (1960)
Epicotyls of pea seedlings	Hayashi and Murakami (1953), Phillips and Jones (1964)
Senescing leaves of *Rumex* and *Taraxacum*	Whyte and Luckwill (1966), Fletcher and Osborne (1966b)

1. Spectrophotometric Methods

These assays are based on the conversion of gibberellic acid into gibberellenic acid by the action of strong acids and the measurement of absorption maxima at 254 nm (Holbrook et al., 1961). The method is rapid and fairly specific for gibberellic acid. It is routinely followed in the analysis of fermentation broths produced by the SmF technique. The calibration graph between 1 and 4 mg of GA_3 was found to be reproducible and linear. However, this method was reported to give erratic results when used for estimating GA_3 in crude extracts produced by the SSF technique (Prapulla et al., 1983). It is also necessary to estimate Sumuki's acid under both neutral and acid conditions at 268 nm and to subtract the contribution of this acid from the absorption value of the sample before calculation of GA_3 content.

Shen and Zhang (1980) and Graham and Henderson (1961) have also reported a spectrophotometric method based upon the reaction of gibberellins with phosphomolybdic acid in the presence of 1.25–1.40 N H_2SO_4 or 1.68–1.92 N $HClO_4$. The optical density of the resulting product, i.e., molybdenum blue, was measured at 650 nm. It gave a linear calibration curve between 1 and 5 mg of GA_3 per 25 ml and the recovery was found to be 98.35%. The method is also known to give a positive response to degradation products of GA_3, amino acids, and sugars (Graham and Henderson, 1961). In addition, it also gives a positive response to methanol, ethanol, and acetonitrile, which are used for the recovery and purification of GA_3 produced by the SSF technique (Prapulla et al., 1983).

2. Fluorometric Assay

The fluorometric assay is used to estimate gibberellic acid and gibberellenic acid in fermentation broths, commercial formulations and purified materials. It is based on the conversion of these compounds into a fluorogen by 85% sulfuric acid (Kavanagh and Kuzel, 1958). The samples are purified on a potassium bicarbonate column before the conversion to fluorogen. However, the method did not distinguish between the above two acids. As compared to GA_3, the biological activity of gibberellenic acid is very low (Gerzon et al., 1957). The response due to the presence of gibberellenic acid can be corrected by eliminating the purification step on the potassium bicarbonate column (Kavanagh and Kuzel, 1958). Better accuracy was achieved by the inclusion of protein precipitation and removal steps to the above purification process (Prapulla et al., 1983). The method can also be used for estimation of GA_3 even in the presence of GA_1 as the latter does not form fluorogen (Thériault et al., 1961). However, it is less accurate as gibberellenic acid and other impurities present in the sample also give a positive response and the sensitivity range lies between 0.000625 and 3.2 µg per milliliter of gibberellic acid (Kavanagh and Kuzel, 1958).

3. *Spectrofluorodensitometric Method*

The spectrofluorodensitometric method is a recent addition to the list and offers a simple and rapid technique for the estimation of GA_3 in both crude and pure forms when produced by SmF or SSF processes (Kumar and Lonsane, 1985a, 1986a). It involves (1) separating GA_3 on silica gel plates by TLC using a solvent system consisting of chloroform : ethyl acetate : glacial acetic acid at a ratio of 5 : 4 : 1; (2) heating the plates in an oven at 100°C for 30 minutes; (3) marking the upper and lower limits of the GA_3 spot based upon the R_f value and emission of blue fluorescence at a 254-nm wavelength under UV light; (4) subjecting the demarked spot to the automatic recording of the spectrofluorodensitometric readings; and (5) calculating the level of GA_3 in the sample spot based upon the area under the peak with reference to a standard graph. The methodology eliminated the interference of all other compounds present in the crude extract and thus was capable of giving accurate results. The regression analysis showed a standard error as small as 0.0265, while the percentage of error is less than ±2. The sensitive range of GA_3 is between 2 and 10 µg/10-µl spot at 0.1 absorbance unit of spectrofluorodensitometer and it can be enhanced to 5–40 µg/10-µl spot, without affecting the reliability, if 2.0 absorbance units are used (Kumar and Lonsane, 1986a).

The spectrofluorodensitometric method is specific for GA_3. The cospectrofluorodensitometry of pure and crude GA_3 compared well with individual spectrofluorodensitometry, and the purity of the spot on TLC plates was also confirmed by two-dimensional TLC (Kumar and Lonsane, 1986a). The method and its response were also subjected to detailed studies on concept validity, optimization, and standardization aspects and therefore has emerged as a sensitive and accurate tool for rapid and reproducible estimation of GA_3 in both crude and pure forms (Kumar and Lonsane, 1985a, 1986a). The estimation methods based on spectrofluorodensitometry are also widely used for reliable estimation of other microbial products such as mycotoxins (Clalam and Stahr, 1979; Stubblefield, 1979; Debeaupuis and Lafont, 1978; Trucksess et al., 1975).

4. *Infrared Method*

This method is based on the absorbance of GAs in a 15% solution in pyridine at 12.86 and 10.85 µm and was used by Washburn et al.

(1959). These bands are very characteristic for GA_3 and GA A. Primarily, it is used for the estimation of GA_3 in minute quantities, especially in pure samples.

5. Gas Chromatography

Ikegawa et al. (1963) have used a gas chromatography method for the detection of gibberellic acid. The samples were first converted into their methyl esters or trimethylsilyl esters (Cavell et al., 1967), and unknown gibberellins were identified and quantified based upon their retention time as compared to a reference compound.

6. Other Miscellaneous Methods

Various other methods available for identification and quantification of GA_3 are listed in Table XXIII.

XIV. Economic Considerations

One of the most important objectives of fermentation process development is to manufacture the product at the most economic level. The fermentation industry gives highest priority to economics, and efforts are always made to reduce the cost of production, even in those processes used currently by the industry.

TABLE XXIII

Some Miscellaneous Physicochemical Instrumentation Methods

Method	Reference
Paper chromatography	Bird and Pugh (1958)
Thin-layer chromatography	Ikegawa et al. (1963), Kagawa et al. (1963), Aseeva (1963), Podojil and Ševčik (1960), Cavell et al. (1967), MacMillan and Suter (1963)
High performance thin-layer chromatography	Sackett (1984)
Radioimmunoassay	Atzorn and Weiler (1983)
Tracer techniques	Arison et al. (1958), Baumgartner et al. (1959, 1963)
Gas chromatography–mass spectroscopy	Binks et al. (1969)
High-pressure liquid chromatography	Heftmann et al. (1978), Roeben et al. (1985)
Use of monoclonal antibodies	Knox et al. (1987)

A. Liquid Surface Fermentation

No report is available on the economic considerations in the production of GA_3 by the LSF technique. The requirement for a longer incubation period and lower yields in the case of the production of GA_3 by the LSF process, as compared to the SmF technique, has been well documented (Hepner, 1985). The lower yield also means lower concentration of product in the fermented broth and the consequent necessity for subjecting higher volumes to downstream processing. These facts indicate that the economics of LSF processes will be comparatively poor as compared to processes involving the SmF technique. This is also confirmed by the replacement of LSF processes by SmF techniques throughout the world not only for the production of GA_3, but for many other fermentation metabolites (Hepner, 1985).

B. Submerged Batch Fermentation

The economic considerations related to the process for the fermentative production of GA_3 were analyzed in detail by Vass and Jefferys (1979). The data, collected over a period of 18 years on the commercial production of GA_3 in submerged fermentation by ICI was analyzed extensively. It was used as a model to determine relative fermentation costs and economic consequences of process improvement by changing nutritional and fermentation conditions or by improving the strain. The major findings related to GA_3 production (Vass and Jefferys, 1979) include (1) the improvements in strain and process technology resulted in increased yields and production rates, leading to an upgrading of the volumetric production rate by >60%; (2) the above process improvement was mainly due to using a controlled supply of assimilable nitrogen sources during the growth phase to obtain cell concentrations that nearly saturate the aeration–agitation system of the fermentor vessel and due to maintenace of a near-oxygen-depletion condition for a longer time to achieve an efficient production phase; (3) diversion of some carbohydrates into cell storage components such as cell fat and carbohydrates toward the end of fermentation and its control by adjusting proliferation and metabolism of the cell; (4) by a mere increase of 50% in total fermentation cost, the improved process showed a 1700% increase in gross yield per batch; (5) spending part of the income from the existing process for further improvement of the technology is better than the blind addition of extra capacity to the existing plant; (6) the capital investment in the latter is probably more than 20 times the total revenue costs of an 18-year period of development; (7) the market vulnerability and lack of technical knowledge for trouble-shooting are relevant; (8) a point is always reached in the process development, operation, and improvement, when the likelihood of material improvement in economics will diminish; (9) at such a point, any increased demand for the product can only be met by increasing the capacity of the plant; and (10) such a point is approaching for GA_3 production.

In addition, an array of general observations were also reported by Vass and Jefferys (1979) and these include (1) the measure of process efficiency is best represented by the cost of GA_3 produced as extractable material in the final broth from the fermentor; (2) the net profit per unit of capital employment or the cost per unit of purified product are not useful as a yardstick; (3) business considerations and internal adjustments are of critical importance in the assessment of the economics of fermentation plants and processes; (4) the actual fermentation time and time taken for vessel cleaning, batch make-up, and filling should be considered in determining volumetric production rate, i.e., an average rate of change of concentration of the product with time; (5) an increase in plant capacity to meet increased demand in the product involves a larger cost due to additional overhead charges; (6) a decision to develop processes and generate profit by improving efficiency of processing rather than by investing in more capacity or generating profit by larger throughput with small margins is of vital importance; (7) the changes that guarantee increased production and decreased costs can be contemplated; (8) introducing new strains can result in a significant improvement in fermentation and economics; (9) continuing reductions in cost make the process more attractive for existing applications and even previously uneconomic applications may become attractive; (10) increasing demand for the product tends to reduce cost; and (11) process improvement can be achieved by increasing the overall rate of production, hastening the onset of production, extending duration of production, achieving fewer by-products, and combining these conditions.

C. Continuous Submerged Fermentation

No data are available on the economics for the production of GA_3 by continuous SmF. The economics of continuous SmF, in general, is always considered to be better than batch SmF processes. However, even work on the production of GA_3 by continuous SmF is too meager to allow for speculation on economics.

D. GA_3 Production by Immobilized Whole Cells

In this case also, no information is available on the economics of production. Certain aspects of the process for production of GA_3 using immobilized whole cells suggest positive economic value (Heinrich and Rehm, 1981; Kahlon and Malhotra, 1986) and include (1) the technique leads to a cell-free medium that does not require further purification and hence expenditure on purification is eliminated, and (2) paneer (cheese) whey, a dairy by-product, is a cheap carbon source and its use in GA_3 production eliminates effluent treatment, which otherwise would be necessary.

E. Solid-State Fermentation

As work on the production of GAs using the SSF technique was initiated in the early 1980s and the initial studies gave variable yields of the product, no information was available on the economics of the process until early 1987. With the recent revival of work at CFTRI and the generation of preliminary cost calculations during the progress of the work (Kumar, 1987; Kumar and Lonsane 1987b,d), the economic advantages of the process have become apparent. Three cases were examined wherein the yields of GA_3 under the SSF technique were 0.825, 1.05, and 1.54 g/kg of DMB. The comparison was made with the medium cost involved in the production of GA_3 at a yield of 0.6 g/liter by the SmF process as developed by Somal et al. (1978). The expenses on medium constituents at the above yields were 79.5, 71.4, and 48.70%, respectively. The yield of GA_3 in the fed-batch SSF process was 1.54 g/kg of DMB (Kumar and Lonsane, 1988b) as compared to 1.05 g/kg of DMB in the batch SSF process (Kumar and Lonsane, 1987c). Thus, application of the fed-batch culture to the SSF process offers a 31.82% reduction in expenses on medium constituents as compared to the optimized batch SSF process.

The presence of the product in higher concentration in the SSF process as compared to the SmF technique and the resultant lower expenses on downstream processing and effluent treatment are among the well-known advantages of the SSF process (Arima, 1964; Hahn-Hägerdal, 1986; Rao et al., 1983; Steinkraus, 1984). Analysis of the data in these respects was also performed and gave interesting results (Kumar, 1987; Kumar and Lonsane, 1987b,d). The calculations were made for an assumed plant size of 500 kg of GA_3 per year and a batch time of 7 days. Thus, it is necessary to produce 11.63 kg of GA_3 per batch on the basis of 300 working days per year. The results were compared with the SmF process developed by Somal et al. (1978), which yields 0.6 g of GA_3/liter. The data indicate that merely 73.0, 57.2, and 38.8% volume of extract will be subjected to downstream processing, as compared to the SmF process, when the yields under the SSF process are 0.825, 1.05, and 1.54 g of GA_3 per kilogram of DMB. The above calculations are based on the assumption that 1 liter of extract will be obtained from 1 kg of DMB, without any vacuum concentration of the extract, by applying the multiple-contact countercurrent leaching technique. The data indicated tremendous economic gains in the production of GA_3 under the SSF process as compared to the currently used SmF technique, even when only the expense on medium constituents and downstream processing were considered.

F. Comparative Economics

It is of industrial interest to study the comparative economics of various fermentation techniques used or investigated for the production of GA_3. Although no comparative data are available on the production of GA_3, such comparisons are available for other fermentation products. A comparison for the production of citric acid by *Aspergillus niger* by LSF and SmF techniques for a production capacity of 0.5 and 3 tons/hr was made (Schierholt, 1976). The conclusions were (1) total investment costs were 25 and 15% lower in the SmF process for 0.5 and 3 tons/hr production capacity; (2) expenses on buildings were about double in the LSF process while those for equipment were ~63 and 70% in the LSF technique as compared to the SmF process; (3) expenses for personnel were more in the LSF technique by ~14–16% for both capacities; (4) expenses for energy were much higher in SmF processes, being merely 13.5–15% in the LSF technique; (5) similarly, expenditure on raw materials, pretreatment, and chemicals was higher in the SmF technique, being ~84% in the LSF technique as compared to the SmF process for both capacities; (6) total production cost was higher by 17.8 and 26.3% in the case of the SmF technique for 0.5 and 3 tons/hr capacity respectively; and (7) the SmF process was superior to the LSF process from an industrial management point of view, based mainly on considerations such as comparative values of space requirements and the number of assistants needed.

The strain of *A. niger* used by Schierholt (1976) was able to produce 70 kg of citric acid from 100 kg of sugar in a total fermentation batch time of 9 days by both LSF and SmF techniques. In the production of GA_3 by the LSF technique, the yield is lower and batch time is longer as compared to those in the SmF process. Hence, the economics for the production of GA_3 by the LSF method will be poorer as compared to these results reported for the production of citric acid.

Submerged versus Solid-State Fermentations

The data on the comparative economics of the SmF and SSF processes for the production of amyloglucosidase were reported by Ghildyal et al. (1985). The enzyme, amyloglucosidase, produced by A. niger CFTRI 1105 by the SSF technique was tenfold greater than in the SmF process, when compared on the basis of enzyme units per milliliter of SmF medium or enzyme units per milliliter of crude extract obtained from 1 g of DMB in the case of the SSF process. The major conclusions were (1) comparative total capital investments were 28, 23, 18%, while the total costs of production were 61, 52, and 58% for an SSF plant producing 9, 30, and 150 m^3 of enzyme concentrate per year, respectively; (2) capital investment on land, buildings, deposits, miscellaneous assets, and preliminary and preoperation expenses were more or less equal; (3) working capital, contingencies, and interest on loans during the construction period were about three times greater for the SmF technique; (4) comparative investments for plant, equipment, erection, and commissioning for 9, 30, and 150 m^3-capacity SSF plants were 17, 14, and 10%, respectively; and (5) the return on investment was negative for the lower capacities but ~8% for the 150 m^3 plant in the SmF process but the returns were 3, 24, and 102% for these capacities using the SSF technique.

The expenditure on downstream processing in the case of amyloglucosidase is negligible, as the enzyme is used in crude form in the industry. However, extensive downstream processing operations are involved in the production of GA_3. The ability of SSF techniques to produce GA_3 at a higher concentration as compared to the SmF process may therefore play an important role in the comparative economics of the SSF and SmF processes. The resultant need for disposal of lower volumes of effluent, in the case of the SSF process, is also of economic importance, as the cost of effluent treatment is the most significant cost element in many cases (Hepner, 1985).

XV. Epilogue

The state of the art that has emerged related to the fermentative production of GA_3 reveals several industrially and economically important facts, such as (1) constraint in the use of GA_3 due to high cost; (2) negligible scope for improvement in the yield and economics of the

SmF process; (3) lower yield of GA_3 with the use of immobilized growing cells; (4) practical difficulties in continuous cultivation on an industrial scale; and (5) potential feasibility of the SSF technique for better economics in the production of GA_3. With the fed-batch SSF process, the expenses on medium constituents work out to be 49% as compared to the SmF process. The concentration of GA_3 in the extract at 1.54 g/liter as against 0.6 or 1.0 g/liter in the SmF process, with or without the use of precursors, respectively, also leads to treating merely a 39% volume by downstream processing, as compared to the SmF process, yielding 0.6 g of GA_3 per liter. This constitutes a tremendous economic gain with respect to downstream processing and effluent treatment.

The technology available, at present, for the production of GA_3 under the SSF process is at the laboratory-scale level with a successful large-scale trial. Scale-up studies are essential for effective industrial exploitation. The concomitant production of a brassinosteroid-like compound by the SSF process also provides a valuable and powerful plant growth promotor, contributing to improved process economics. The research and development investigations on the optimization of the yield of the brassinosteroid-like compound, without affecting the yield of GA_3 by the SSF process, are of significant value. Other concomitant products, such as sterol and polyketide pigment, produced in DMB by G. fujikuroi P-3 by the SSF technique, also deserve further investigations, as they are different from the known compounds produced in the SmF technique.

The accurate and reliable estimation of GA_3, in crude and impure forms, by physicochemical and bioassay methods is practically impossible due to interference in the response by the impurities. The spectrofluorodensitometric method provides a valuable tool for reliable estimation of GA_3, both in pure and impure forms, especially in fermentation broths or extracts. Among the 71 GAs known to date, GA_3 has received the greatest attention, based on its commercial value, and may continue to enjoy this attention until other, better GAs emerge from microbial sources.

Aknowledgments

The authors are thankful to N. P. Ghildyal, Dr. N. G. Karanth, S. K. Majumder, and Dr. B. L. Amla for valuable suggestions and interest in the work. P.K.R.K. is grateful to the Council of Scientific and Industrial Research, New Delhi, India, for award of a fellowship.

References

Abbott, B. J. (1978). *Annu. Rep. Ferment. Processes* **2**, 91–123.
Abbott Laboratories (1963). British Patent 919,186.
Adam, G., and Marquardt, V. (1986). *Phytochemistry* **25**, 1787–1799.
Agnistikova, V. N., Dubovaya, L. P., Lekareva, T. A., Lupova, L. M., Muromtsev, G. S., Kucherov, V. F., and Serebryaov, E. P. (1966). *Microbiologiya* **35**, 1037–1043.
Agnistikova, V. N., Kobrina, N. S., Kucherov, V. F., and Serebryakov, E. P. (1974). In "Biochemistry and Chemistry of Plant Growth Regulators" (K. Schreiber, H. R. Schütte, and G. Sembdner, eds.), pp. 63–77. Academy of Science of the German Democratic Republic, Institute of Plant Biochemistry, Halle (Saale).
Ahmed, S. Y., Lonsane, B. K., Ghildyal, N. P., and Ramakrishna, S. V. (1987). *Biotechnol. Tech.* **1**, 97–102.
Aiba, S., Humphrey, A. E., and Millis, N. F. (1973). "Biochemical Engineering," 2nd ed., pp. 128–162. Academic Press, New York.
Albertini, E., Della, V. R., and Rezzesi, F. (1960). *Med. Exp.* **2**, 344–348.
Aldridge D. C., Grove, J. F., Speake, R. N., Tidd, B. K., and Klyne, W. (1963). *J. Chem. Soc.* pp. 143–154.
Aleksandrov, F. A. (1964). *Bot. Zh. (Leningrad)* **49**, 1056–1957.
Aleksandrowicz, J., Sasiadek, U., and Lisiewicz, J. (1975). *Prezegl. Lek.* **32**, 343–345.
American Chemical Society (1982–1986). "11th Collective Index. Chemical Abstracts," Vols. 96–105, p. 757G. Chemical Abstract Service, Columbus, Ohio.
Anke, T. (1986). In "Biotechnology" (H. Pape and R. J. Rehm, eds.) Vol. 4, pp. 611–622. VCH Verlagsges., Weinheim.
Arima, K. (1964). In "Global Impacts of Applied Microbiology" (M. P. Starr, ed.), pp. 277–294. Wiley, New York.
Arison, B. H., Speth, O. C., and Trenner, N. R. (1958). *Anal. Chem.* **30**, 1083–1085.
Aseeva, I. V. (1963). *Gibberelliny Ikh Deistvie Rast.* pp. 73–75.
Aseeva, I. V., and Barmenkova, R. A. (1967). *Nauchn. Dokl. Vyssh. Shk., Biol. Nauki* **2**, 123–126.
Atzorn, R., and Weiler, E.W. (1983). *Planta* **159**, 1–6.
Aube, C., and Sackston, W. E. (1965). *Can. J. Bot.* **43**, 1335–1342.
Auden, J. A., Gruner, J., Liersch, M., and Nueesch, J. (1969). *Pathol. Microbiol.* **34**, 240–242.
Axelsson, H. A. C. (1985). *Compr. Biotechnol.* **2**, 325–346.
Bailiss, K. W., and Hill, T. A. (1971). *Bot. Rev.* **37**, 437–479.
Bakalivanov, D. (1968). *Pochvozn. Agrokhim.* **3**, 81–86.
Balan, J., Fuska, J., Kuhr, I., and Kuhorva, V. (1970). *Folia Microbiol. (Prague)* **15**, 479–484.
Banks, G. T. (1979). *Top. Enzyme Ferment. Biotechnol.* **3**, 170–266.
Bartlett, M. C., and Gerhardt, P. (1959). *J. Biochem. Microbiol. Technol. Eng.* **1**, 359–377.
Basta, N., and McQueen, S. (1985). *Chem. Eng. (N.Y.)* Feb. 4, pp. 14–17.
Baumgartner, W. E., Lazer, L. S., Dalziel, A. M., Cardinal, E. V., and Varner, E. L. (1959). *J. Agric. Food Chem.* **7**, 422–425.
Baumgartner, W. E., Lazer, L. S., and Dalziel, A. M. (1963). *Adv. Tracer Methodol.* **1**, 257–262.
Bearder, J. R., and MacMillan, J. (1972). *Agric. Biol. Chem.* **36**, 342–344.
Bearder, J. R., and MacMillan, J. (1973). *J. Chem. Soc., Perkin Trans. 1* pp. 2824–2830.
Bearder, J. R., MacMillan, J., Wels, C. M., Chaffey, M. B., and Phinney, B. O. (1974). *Phytochemistry* **13**, 911–917.

Bearder, J. R., MacMillan, J., and Phinney, B. O. (1975a). *J. Chem. Soc., Perkins Trans. 1* pp. 721–726.
Bearder, J. R., Dennis, F. G., MacMillan, J., Martin, G. C., and Phinney, B. O. (1975b). *Tetrahedron Lett.* pp. 669–670.
Bearder, J. R., Frydman, V. M., Gaskin, P., Hotton, I. K., Harvey, W. E., MacMillan, J., and Phinney, B. O. (1976). *J. Chem. Soc., Perkin Trans. 1* pp. 178–183.
Becker, H., and Kempf, T. (1976). *Z. Pflanzenphysiol.* **80,** 87–91.
Beckord, L. D., Kneen, E., and Lewis, K. H. (1945). *Ind. Eng. Chem.* **37,** 692–696.
Beeley, L. J., and MacMillan, J. (1976). *J. Chem. Soc., Perkin Trans. 1* pp. 1022–1028.
Belter, P. A. (1985). *Compr. Biotechnol.* **2,** 347–350.
Bendana, F. E., and Fried, M. (1976). *Life Sci.* **6,** 1023–1033.
Benedict-Ratz, M., and Olah, B. (1985). Hungarian Patent 35,290.
Bentley, R., and Campbell, I. M. (1968). *Compr. Biochem.* **20,** 415–489.
Bergman, D. E., Denison, F. E., Jr., and Friedland, W. O. (1962). U.S. Patent 3,021,261.
Bernal-Lugo, I., Beachy, R. N., and Varner, J. E. (1981). *Biochem. Biophys. Res. Commun.* **102,** 617–623.
Berry, D. R. (1975). *Filamentous Fungi* **1,** 16–32.
Berry, M., and Sachar, R. C. (1981). *FEBS Lett.* **132,** 109–113.
Binks, R., MacMillan, J., and Pryce, R. J. (1969). *Phytochemistry* **8,** 271–284.
Birch, A. J., Richards, R. W., Smith, H., Harris, A., and Whalley, W. B. (1959a). *Tetrahedron* **7,** 241–251.
Birch, A. J., Nixon, I.S., and Grove, J. F. (1959b). British Patent 844,341.
Bird, H. L., Jr., and Pugh, C. T. (1958). *Plant Physiol.* **33,** 45–46.
Bolgarev, P. T., Muromtsev, G. S., and Nestyouk, M. N. (1962). *Bull. Invent.* No. 23. Certificate No. 762055/23-24, as cited by Jefferys, 1970.
Borrow, A., Brian, P. W., Chester, V. E., Curtis, P. J., Hemming, H. G., Henehan, C., Jefferys, E. G., Lloyd, P. B., Nixon, I. S., Norris, G. L. F., and Radley, M. (1955). *J. Sci. Food Agric.* **6,** 340–348.
Borrow, A., Jefferys, E. G., and Nixon, I. S. (1958). British Patent 803,591.
Borrow, A., Jefferys, E. G., and Nixon, I. S. (1959a). U.S. Patent 2,906,670.
Borrow, A., Jefferys, E. G., and Nixon, I. S. (1959b). U.S. Patent 2,906,671.
Borrow, A., Jefferys, E. G., and Nixon, I. S. (1959c). U.S. Patent 2,906,673.
Borrow, A., Jefferys, E. G., Kessell, R. H. J., Lloyd, E. C., Lloyd, P. B., and Nixon, I. S. (1961). *Can. J. Microbiol.* **7,** 227–276.
Borrow, A., Brown, S., Jefferys, E. G., Kessell, R. H. J., Lloyd, E. C., Lloyd, P. B., Rothwell, A., Rothwell, B., and Swait, J. C. (1964a). *Can. J. Microbiol.* **10,** 407–444.
Borrow, A., Brown, S., Jefferys, E. G., Kessell, R. H. J., Lloyd, E. C., Lloyd, P. B., Rothwell, A., Rothwell, B., and Swait, J. C. (1964b). *Can. J. Microbiol.* **10,** 445–466.
Bourne, P. M., Grove, J. F., Mulholland, T. P. C., Tidd, B. K., and Klyne, W. (1963). *J. Chem. Soc.* pp. 154–162.
Bown, A. W., Reeve, D. R., and Crozier, A. (1975). *Planta* **126,** 83–91.
Bravery, A. F. (1975). In "Biological Transformation of Wood by Microorganisms" (W. Liese, ed.), pp. 129–142. Springer-Verlag, Berlin and New York.
Brian, P. W. (1957). *Symp. Soc. Exp. Biol.* **11,** 166–182.
Brian, P. W. (1959). *Biol. Rev. Cambridge Philos. Soc.* **34,** 37–84.
Brian, P. W., and Hemming, H. G. (1955). *Physiol. Plant.* **8,** 669–681.
Brian, P. W., Curtis, P. J., and Hemming, H. G. (1946). *Trans. Br. Mycol. Soc.* **29,** 173–187.
Brian, P. W., Elson, G. W., Hemming, H. G., and Radley, M. (1954). *J. Sci. Food Agric.* **5,** 602–612.
Brian, P. W., Radley, M. E., Curtis, P. J., and Elson, G. W. (1957). British Patent 783,611.

Brian, P. W., Petty, J. H. P., and Richmond, P. T. (1959a). *Nature (London)* **183,** 58–59.
Brian, P. W., Petty, J. H. P., and Richmond, P. T. (1959b). *Nature (London)* **184,** 69.
Brian, P. W., Hemming, H. G., and Lowe, D. (1960). *Ann. Bot. (London)* [N.S.] **24,** 407–419.
Brian, P. W., Hemming, H. G., and Lowe, D. (1964). *Ann. Bot (London)* [N.S.] **28,** 369–389.
Brian, P. W., Grove, J. F., and Mulholland, T. P. C. (1967). *Phytochemistry* **6,** 1475–1499.
Briggs, D. E. (1973). In "Biosynthesis and Its Control in Plants" (B. V. Milborrow, ed.), pp. 219–277. Academic Press, New York.
Briggs, D. E. (1978). "Barley." Wiley, New York.
Brown, J. C., Cross, B. E., and Hanson, J. R. (1967). *Tetrahedron* **23,** 4095–4103.
Brown, M. E. (1972). *J. Appl. Bacteriol.* **35,** 443–451.
Bruinsma, J. (1963). *Chem. Weekbl.* **59,** 599.
Buller, D. C., Parker, W., and Grant Reid, J. S. (1976). *Nature (London)* **260,** 169–170.
Bu'Lock, J. D. (1961). *Adv. Appl. Microbiol.* **3,** 293–342.
Bu'Lock, J. D. (1965a). "The Biosynthesis of Natural Products." McGraw-Hill, London.
Bu'Lock, J. D. (1965b). In "Biogenesis of Antibiotic Substances" (Z. Vaněk and Z. Hošťálek eds.), pp. 61–72. Academic Press, New York.
Bu'Lock, J. D. (1975). *Filamentous Fungi* **1,** 33–58.
Bu'Lock, J. D., Hamilton, D., Hulme, M. A., Powell, A. J., Smalley, H. M., Shepherd, D., and Smith, G. N. (1965). *Can. J. Microbiol.* **11,** 765–778.
Bu'Lock, J. D., Detroy, R. W., Hošťálek, Z., and Munim-Al-Shakarchi, A. (1974). *Trans. Br. Mycol. Soc.* **62,** 377–389.
Burström, H. (1960). *Physiol. Plant.* **13,** 597–615.
Butcher, D. N., and Street, H. E. (1960). *J. Exp. Bot.* **11,** 206.
Calam, C. T. (1969). In "Methods in Microbiology" (J. R. Norris and D. W. Robbins, eds.), Vol. 1, pp. 255–326. Academic Press, New York.
Calam, C. T., and Curtis, P. J. (1960). U. S. Patent 2,950,288.
Calam, C. T., and Nixon, I. S. (1958). British Patent 839,652.
Campbell, I. M. (1983). *J. Nat. Prod.* **46,** 60–70.
Carl-Gustaf, R. (1984). In' "The World Biotech. Report 1984," Vol. 1, pp. 317–330. Online Publications, Pinner, Middlesex.
Cavell, B. D., MacMillan, J., Pryce, R. J., and Sheppard, A. C. (1967). *Phytochemistry* **6,** 867–874.
Central Food Technological Research Institute (CFTRI) (1981–1982). "Annual Report," p. 60. CFTRI, Mysore, India.
Chahal, D. S. (1983). *ACS Symp. Ser.* **207,** 421–442.
Chang, T. M. S. (1964). *Science* **146,** 524–525.
Cheynier, V., Feinberg, M., Chararas, C., and Ducauze, C. (1983). *Appl. Environ. Microbiol.* **45,** 634–639.
Chrispeels, M. J., and Varner, J. E. (1966). *Nature (London)* **212,** 1066–1067.
Ciferri, O., and Bertossi, F. (1957). *Boll. Soc. Ital. Biol. Sper.* **33,** 114–116.
Clalam, R. V., and Stahr, H. M. (1979). *J. Assoc. Off. Anal. Chem.* **62,** 570–572.
Coggins, C. W., Jr. (1969). *Proc. Int. Citrus Symp. 1st, 1968* Vol. 3, pp. 1177–1185.
Coggins, C. W., Jr. (1973). *Acta Hortic.* **34,** 469–472.
Coggins, C. W., Jr., Hield, H. Z. and Garber, M. J. (1960). *Proc. Am. Soc. Hortic. Sci.* **76,** 193–198.
Coolbaugh, R. C., and Hamilton, R. (1976). *Plant Physiol.* **57,** 245–248.
Coolbaugh, R. C., Hirano, S. S., and West, C. A. (1978). *Plant Physiol.* **62,** 571–576.
Corey, E. J., and Munroe, J. E. (1982). *J. Am. Chem. Soc.* **104,** 6129–6130.

Corey, E. J., and Smith, J. G. (1969). *J. Am. Chem. Soc.* **101,** 1038–1039.
Corey, E. J., Danheiser, R. L., Chandrasekaran, S., Siret, P., Keck, G. E., and Gras, J. L. (1978a). *J. Am. Chem. Soc.* **100,** 8031–8034.
Corey, E. J., Danheiser, R. L., Chandrasekaran, S., Keck, G. E., Gopalan, B., Larsen, S. D., Siret, P., and Gras, J. L. (1978b). *J. Am. Chem. Soc.* **100,** 8034–8036.
Cossey, A. L., Lombardo, L., and Mander, L. N. (1980). *Tetrahedron Lett.* **21,** 4383–4386.
Cross, B. E. (1954). *J. Chem. Soc.* pp. 4670–4676.
Cross, B. E. (1966). *J. Chem. Soc.* pp. 501–504.
Cross, B. E. (1968). *Prog. Phytochem.* **1,** 195–222.
Cross, B. E., and Norton, K. (1965). *J. Chem. Soc.* pp. 1570–1572.
Cross, B. E., and Webster, G. R. B. (1970). *J. Chem. Soc.* pp. 1839–1842.
Cross, B. E., Grove, J. F., McCloskey, P., Mulholland, T. P. C., and Klyne, W. (1959a). *Chem. Ind. (London)* pp. 1345–1346.
Cross, B. E., Grove, J. F., MacMillan, J., Moffatt, J. S., Mulholland, T. P. C., and Seaton, J. C. (1959b). *Proc. Chem. Soc., London* pp. 302–303.
Cross, B. E., Galt, R. H. B., and Hanson, J. R. (1960a). *Tetrahedron Lett.* **15,** 18–22.
Cross, B. E., Galt, R. H. B., and Hanson, J. R. (1960b). *Tetrahedron Lett.* **23,** 22–24.
Cross, B. E., Grove, J. F., and Morrison, A. (1961). *J. Chem. Soc.* pp. 2498–2515.
Cross, B. E., Galt, R. H. B., Hanson, J. R., and Klyne, W. (1962a). *Tetrahedron Lett.* **4,** 145–150.
Cross, B. E., Galt, R. H. B., and Hanson, J. R. (1962b). *Tetrahedron* **18,** 451–459.
Cross, B. E., Galt, R. H. B., Hanson, J. R., Curtis, P. J., Grove, J. F., and Morrison, A. (1963). *J. Chem. Soc.* pp. 2937–2943.
Cross, B. E., Galt, R. H. B., and Hanson, J. R. (1964). *J. Chem. Soc.* pp. 295–300.
Cross, B. E., Myers, P. L., and Webster, G. R. B. (1970a). *J. Chem. Soc.* p. 930.
Cross, B. E., Stewart, J. C., and Stoddart, J. L. (1970b). *Phytochemistry* **9,** 1065–1071.
Crozier, A. (1981). *Adv. Bot. Res.* **9,** 33–149.
Crozier, A., ed. (1983). "The Biochemistry and Physiology of Gibberellins," Vols. 1 and 2. Praeger, New York.
Crozier, A., and Audus, L. J. (1968). *Planta* **83,** 207–217.
Crozier, A., and Reid, D. M. (1971). *Can. J. Bot.* **49,** 967–976.
Crozier, A., Kuo, C. C., Durley, R. C., and Pharis, R. P. (1970). *Can. J. Bot.* **48,** 867–877.
Curtis, P. J., and Cross, B. E. (1954). *Chem. Ind. (London)* p. 1066.
Curtis, R. W. (1957). *Science* **125,** 646.
Dahlström, R. V., Gauger, G. W., and Martin, E. G. (1961). *Proc. Am. Soc. Brew. Chem.* pp. 98–102.
Darken, M. A., Jensen, A. L., and Shu, P. (1959). *Appl. Microbiol.* **7,** 301–303.
Datar, R. (1986). *Process Biochem.* **21,** 19–26.
Davies, O. L. (1967). "Design and Analysis of Industrial Experiments." Hafner, New York.
De Angelis, J. G. (1970). *Isr. J. Agric. Res.* **20,** 149–157.
Debeaupuis, J. P., and Lafont, P. (1978). *J. Chromatogr.* **157,** 451–454.
Debaska, H., and Urbanek, H. (1970). *Zesz. Nauk. Uniw. Lodz., Ser. 2* **37,** 53–58.
De Conejos, Rosa, L. F., De Campo, Gladys, G. (1975). *Arch. Bioquim. Quim. Farm.* **19,** 39–46.
Demain, A. L. (1968). *Lloydia* **31,** 395–418.
Demain, A. L. (1971). *Adv. Biochem. Eng.* **1,** 113–142.
Dennis, D. T., and West, C. A. (1967). *J. Biol. Chem.* **242,** 3293–3300.
Dennis, D. T., Upper, C. D., and West, C. A. (1965). *Plant Physiol.* **40,** 948–952.
Dietrich, K. R. (1960). German Patent 1,081,402.
Dockerill, B., Evans, R., and Hanson, J. R. (1977). *J. Chem. Soc., Chem. Commun.* pp. 919–921.

Dolan, S. C., Holdup, D. W., Hutchinson, M., and MacMillan, J. (1985). *J. Chem. Soc., Perkin Trans.* 1 pp. 651–654.
Domasch, K. H. (1963). *Z. Pflanzenkr. (Pflanzenpathol.) Pflanzenschutz* **70**, 470–476.
Donoho, C. W., Jr., and Walker, D. R. (1957). *Science* **126**, 1178–1179.
Dostal, H. C., and Leopold, A. C. (1967). *Science* **158**, 1579–1580.
Dostal, R. (1959). *Nature (London)* **183**, 1338.
Drew, S. W., and Demain, A. L. (1977). *Annu. Rev. Microbiol.* **31**, 343–356.
Dunn, G. M. (1985). *Compr. Biotechnol.* **1**, 113–126.
Durand, G., and Novarro, J. M. (1978). *Process Biochem.* **13**(9), 14–23.
Durley, R. C., MacMillan, J., and Pryce, R. J. (1971). *Phytochemistry* **10**, 1891–1908.
Durley, R. C., Sassa, T., and Pharis, R. P. (1979). *Plant Physiol.* **64**, 214–219.
Eddy, B. P. (1958). *J. Sci. Food Agric.* **9**, 644–649.
Edwards, V. H., Gottschalk, M. J., Noojin, A. Y., III, Tuthill, L. B., and Tannahill, A. L. (1970). *Biotechnol. Bioeng.* **12**, 975–990.
El-Bahrawi, S. (1977). *Zentralbl. Bakteriol., Parasitenkd., Infectionskr. Hyg., Abt. 2*, **132**, 178–183.
Elliott, E. S. (1948). *Proc. W. Va. Acad. Sci.* **20**, 65–68.
Erokhina, L. I. (1967). *Genetika* **7**, 77–82.
Erokhina, L. I., and Sokolova, E. V. (1966). *Genetika* **1**, 109–115.
Evans, R., and Hanson, J. R. (1975). *J. Chem. Soc., Perkin Trans.* 1 pp. 663–666.
Evins, W. H., and Varner, J. E. (1972). *Plant Physiol.* **49**, 348–352.
Fall, R. R., and West, C. A. (1971). *J. Biol. Chem.* **246**, 6913–6928.
Fiechter, A. (1981). *In* "Biotechnology: A Comprehensive Treatise" (H. J. Rehm and G. Reed, eds.), Vol. 1, pp. 453–501. Verlag-Chemie, Weinhein.
Fisher, N. (1973). *In* "Molecular Structure and Function of Food Carbohydrates" (G. G. Birch and L. F. Green, eds.), pp. 275–295. Applied Science Publ., London.
Filner, P., and Varner, J. E. (1967). *Proc. Natl. Acad. Sci. U.S.A.* **58**, 1520–1526.
Fletcher, R. A., and Osborne, D. J. (1965). *Nature (London)* **207**, 1176–1177.
Fletcher, R. A., and Osborne, D. J. (1966a). *Can. J. Bot.* **44**, 739–745.
Fletcher, R. A., and Osborne, D. J. (1966b). *Nature (London)* **211**, 743–744.
Flippin, R. S., Smith, C., and Mickelson, M. N. (1964). *Appl. Microbiol* **12**, 93–95.
Frankland, B., and Wareing, P. F. (1960). *Nature (London)* **185**, 255–256.
Frydman, V. M., and MacMillan, J. (1975). *Planta*, **125**, 181–195.
Frydman, V. M., Gaskin, P., and MacMillan, J. (1974). *Planta* **118**, 123–132.
Fukui, S., and Tanaka, A. (1982). *Annu. Rev. Microbiol.* **36**, 145–172.
Fuska, J., Kuhr, I., Podojil, M., and Sĕvčik, V. (1960). *Folia Microbiol. (Prague)* **6**, 18–21.
Fuska, J., Kuhr, I., Sĕvčik, V., Musilek, V., and Podojil, M. (1962). Czech. Patent 104,329.
Fuska, J., Kuhr, I., and Zajicek, I. (1964). *Mikrobiologiya* **33**, 783–786.
Gallo, M., and Katz, E. (1972). *J. Bacteriol.* **109**, 659–667.
Galsky, A. G., and Lippincott, J. A. (1967). *Plant Physiol., Suppl.* **42**, S-29.
Galt, R. H. B. (1965). *J. Chem. Soc.* pp. 3143–3151.
Galt, R. H. B. (1968). *Tetrahedron* **24**, 1337–1339.
Galun, E. (1959). *Phyton* **13**, 1–8.
Ganchev, K., Krachanov, M., and Popova, R. (1984). *Acta Microbiol. Bulg.* **15**, 43–49.
Gardner, D. S. (1982). *Chem. Ind. (London)* **19**, 402–405.
Garg, D. K., and Mehrotra, R. S. (1977). *Indian Phytopathol.* **30**, 546–548.
Gawienowski, A. M., and Chatterjee, D. (1980). *Life Sci.* **27**, 1393–1396.
Gawienowski, A. M., Standnicki, S. S., and Stacewicz-Sapuntzakis, M. (1977). *Life Sci.* **20**, 785–788.
Geissman, T. A., Vesbiscar, A. J., Phinney, B. O., and Cragg, G. (1966). *Phytochemistry* **5**, 933–947.

Gerzon, K., Bird, H. L., Jr., and Woolf, D. O. (1957). *Experientia* **13**, 487–489.
Ghildyal, N. P., Ramakrishna, S. V., Lonsane, B. K., and Ahmed, S. Y. (1979). In "Immobilized Enzyme Engineering Seminar," pp. 139–150. Jadavpur University, Calcutta.
Ghildyal, N. P., Lonsane, B. K., Sreekantiah, K R., and Murthy, V. S. (1985). *J. Food Sci. Technol.* **22**, 171–176.
Ghildyal, N. P., Lonsane, B. K., and Karanth, N. G. (1987). *Adv. Appl. Microbiol.* **33**, 173–222.
Gibbons, G. C. (1981). *J. Am. Soc. Brew. Chem.* **46**, 55–59.
Gibbons, W. R., Westby, C. A., and Dobbs, T. L. (1984). *Biotechnol. Bioeng.* **26**, 1098–1107.
Gohlwar, C. S., Sethi, R. P., Marwaha, S. S., Seghal, V. K., and Kennedy, J. F. (1984). *Enzyme Microb. Technol.* **6**, 312–316.
Goodwin, T. W., and Mercer, E. I. (1983). "Introduction to Plant Biochemistry," pp. 580–595. Pergamon, Oxford.
Gordon, W. L. (1960). *Nature (London)* **186**, 698–700.
Gordon, C. J. (1961). *Physiol. Plant* **14**, 332–343.
Gould, R. F., ed. (1961). "Gibberellins," Adv. Chem. Ser. No. 28. Am. Chem. Soc. Appl. Publ., Washington, D.C.
Graebe, J. E., and Hedden, P. (1974). In "Biochemistry and Chemistry of Plant Growth Regulators" (K. Schreiber, H. R. Schütte, and G. Sembdner, eds.) pp. 1–16. Academy of Science of the German Democratic Republic, Institute of Plant Biochemistry, Halle (Saale).
Graebe, J. E., and Rademacher, W. (1979). British Patent 79/30,585.
Graebe, J. E., and Ropers, H. J. (1978). In "Phytohormones and Related Compounds: A Comprehensive Treatise" (D. S. Letham, P. B. Goodwin, and T. V. J. Higgins, eds.), Vol. 2, pp. 107–204. Elsevier/North Holland Biomedical Press, Amsterdam.
Graebe, J. E., Dennis, D. T., Upper, C. D., and West, C. A. (1965). *J. Biol. Chem.* **240**, 1847–1854.
Graebe, J. E., Hedden, P., Gaskin, P., and MacMillan, J. (1974a). *Phytochemistry* **13**, 1433–1440.
Graebe, J. E., Hedden, P., Gaskin, P., and MacMillan, J. (1974b). *Planta* **120**, 307–309.
Graebe, J. E., Hedden, P., and MacMillan, J. (1974c). In "Plant Growth Substances, 1973," pp. 260–266. Hirokawa, Tokyo, as cited by Crozier, 1981.
Graham, H. D., and Henderson, J. H. M. (1961). *Plant Physiol.* **36**, 405–408.
Griggs, W. H., and Iwakiri, B. T. (1961). *Proc. Am. Soc. Hortic. Sci.* **77**, 73–89.
Grigorov, I., and Angelova, M. (1976). *Priroda (Sofia)* **25**, 68–70.
Grove J. F. (1961). *Q. Rev., Chem. Soc.* **15**, 56–70.
Grove J. F. (1963). In "Biochemistry of Industrial Microorganisms" (C. Rainbow and A. H. Rose, eds.), pp. 320–340. Academic Press, New York.
Grove J. F., Jeffs, P. W., and Mulholland, T. P. C. (1958). *J. Chem. Soc.* pp. 1236–1240.
Grove M. D., Spencer, G. F., Rohwedder, W. K., Mandava, N., Worley, J. F., Warthen, J. D., Jr., Steffens, G. L., Flippen-Anderson, J. L., and Cook, J. C., Jr. (1979). *Nature (London)* **281**, 216–217.
Hahn-Hägerdal, B. (1986). *Enzyme Microb. Technol.* **8**, 322–327.
Haissig, B. E. (1972). *Plant Physiol.* **49**, 886–892.
Hanson, A. M. (1969). In "Microbial Technology" (H. J. Peppler, ed.) pp. 222–250. Reinhold, New York.
Hanson, J. R. (1966). *Tetrahedron* **22**, 701–703.
Hanson, J. R. (1967). *Tetrahedron* **23**, 733–735.

Hanson, J. R. (1968). "The Tetracyclic Diterpenes," pp. 41–59. Pergamon, Oxford.
Hanson, J. R., Hawker, J., and White, A. P. (1972). *J. Chem. Soc., Perkin Trans. 1* pp. 1892–1895.
Hao, L. C., Fulmer, E. I., and Underkofler, L. A. (1943). *Ind. Eng. Chem.* **35**, 814–818.
Hardie, D. G. (1975). *Phytochemistry* **14**, 1719–1722.
Harhash, A. W. (1966). *Acta. Biol. Med. Ger.* **17**, 8–16.
Harrington, J. F. (1960). *Proc. Am. Soc. Hortic. Sci.* **75**, 476–479.
Harrison, D. M., and MacMillan, J. (1971). *J. Chem. Soc.* pp. 631–636.
Harte, M. J., and Webb, F. C. (1967). *Biotechnol. Bioeng.* **9**, 205–221.
Hartsuck, J. A., and Lipscomb, W. N. (1963). *J. Am. Chem. Soc.* **85**, 3414–3419.
Hasegawa, S., Shimizu, K., Kobayashi, T., and Matsubara, M. (1985). *J. Chem. Technol. Biotechnol.* **35B**, 33–42.
Hashimoto, T., and Rappaport, L. (1966a). *Plant Physiol.* **41**, 623–628.
Hashimoto, T., and Rappaport, L. (1966b). *Plant Physiol.* **41**, 629–632.
Haslam, E. (1986). *Nat. Prod. Rep.* **3**, 217–249.
Hayashi, T., and Murakami, Y. (1953). *J. Agric. Chem. Soc. Jpn.* **27**, 675.
Hayashi, T., and Murakami, Y. (1954). *J. Agric. Chem. Soc. Jpn.* **28**, 543–545.
Hayashi, T., Murakami, Y., and Matsunaka, S. (1956). *Bull. Agric. Chem. Soc. Jpn.* **20**, 159–164.
Hayes, W. A. (1977). In "Composting" (W. A. Hayes, ed.), pp. 1–20. Mushroom Growers Assoc., London.
Hedden, P. (1979). *ACS Symp. Ser.* **111**, 19–56.
Hedden, P., MacMillan, J., and Phinney, B. O. (1974). *J. Chem. Soc., Perkin Trans. 1* 587–592.
Heftmann, E., Saunders, G. A., and Haddon, W. F. (1978). *J. Chromatogr.* **156**, 71–85.
Heinrich, M., and Rehm, H. J. (1981). *Eur. J. Appl. Microbiol. Biotechnol.* **11**, 139–145.
Henke, O., and Schaller, G. (1965). *Bot. Mar.* **8**, 156–167.
Henry, J. D., Jr. (1972). In "Recent Developments in Separation Science." (N. N. Li, ed.), Vol. 2, pp. 205–225. CRC, Cleveland, Ohio.
Henry, J. D., Jr., and Allred, R. C. (1972). *Dev. Ind. Microbiol.* **13**, 177–190.
Henry, R. J. (1980). *HortScience* **15**, 613.
Hepner, L. (1985). *Eur. Congr. Biotechnol. Proc., 3rd,* 1984, Vol. 4, p. 465.
Herbert, D. (1964). *Contin. Cult. Microorg., Proc. Symp., 2nd,* 1962 pp. 23–32.
Hernandez, E., and Mendoza, M. D. (1976). *Rev. Agroquim. Tecnol. Aliment.* **16**, 357–366.
Heropolitanski, R., Kazmierczak, A., Gajewski, I., and Wosko, H. (1981). German (East) Patent DD 152,578.
Hesseltine, C. W. (1972). *Biotechnol. Bioeng.* **14**, 517–532.
Hesseltine, C. W. (1977). *Process Biochem.* **12**(6), 24–27.
Hill, T. A., and Wimble, R. H. (1969). *Planta* **87**, 20–25.
Hiraga, K., Yokota, T., Murofushi, N., and Takahashi, N. (1972). *Agric. Biol. Chem.* **36**, 345–347.
Hiraga, K., Yokota, T., Murofushi, N., and Takahashi, N. (1974a). *Agric. Biol. Chem.* **38**, 2511–2520.
Hiraga, K., Kawabe, S., Yokota, T., Murofushi, N., and Takahashi, N. (1974b). *Agric. Biol. Chem.* **38**, 2521–2527.
Hirata, S. (1958). *Bull. Fac. Agric., Miyazaki Univ.* **3**, 46–52.
Hiroe, I. (1969). *J. Fac. Agric., Tottori. Univ.* **5**, 1–5.
Hitzman, D. O., and Mills, A. M. (1963).U.S. Patent 3,084,106.
Ho, D. T. H., and Varner, J. E. (1974). *Proc. Natl. Acad. Sci. U.S.A.* **71**, 4783–4786.

Hoad, G. V. (1983). In "The Biochemistry and Physiology of Gibberellins" (A. Crozier ed.), Vol. 2, pp. 57–94. Praeger, New York.
Hoad, G. V., Phinney, B. O., Sponsel, V. M., and MacMillan, J. (1981). *Phytochemistry* **20**, 703–713.
Hoad, G. V., MacMillan, J., Smith, V. A., and Taylor, D. A. (1983). In "Plant Growth Substances 1982." (P. F. Wareing, ed.), p. 1. Academic Press, London.
Hodson, H. K., and Hamner, K. C. (1971). *Plant Physiol.* **47**, 726–728.
Holbrook, A. H., Edge, W. J., and Bailey, F. (1961). *Adv. Chem. Ser.* **28**, 159–167.
Holcberg, I. B., and Margalith, P. (1981). *Eur. J. Appl. Microbiol. Biotechnol.* **13**, 133–140.
Holme, T., and Zacharias, B. (1965). *Biotechnol. Bioeng.* **7**, 405–415.
Hook, J. M., Mander, L. N., and Rudolf, U. (1980). *J. Am. Chem. Soc.* **102**, 6628–6629.
Hori, S. (1898). *Mem. Agric. Res. Stn. (Tokyo)* **12**, 110–119.
Hori, S. (1903). "Bakanae Disease of Rice: Lectures on Plant Disease," 1st ed., pp. 114–121. Seibido, Tokyo.
Hospodke, J. (1966). In "Theoretical and Methodological Basis of Continuous Culture of Microorganisms" (I. Málek and Z. Fencl, eds.), p. 493. Academic Press, New York.
Humphreys, E. C., and Wheeler, A. W. (1960). *J. Exp. Bot.* **11**, 81–85.
Ikegawa, N., Kagawa, T., and Sumiki, Y. (1963). *Proc. Jpn. Acad.* **39**, 507–512.
Imperial Chemical Industries Ltd. (ICI) (1955). Commonwealth of Australia Patent 10190.
Imperial Chemical Industries Ltd. (ICI) (1960a). British Patent 838,032.
Imperial Chemical Industries Ltd. (ICI) (1960b). British Patent 839,652.
Imperial Chemical Industries Ltd. (ICI) (1964a). British Patent 957,634.
Imperial Chemical Industries Ltd. (ICI) (1964b). British Patent 967,596.
Imshenetsky, A. A., and Ul'Yanova, O. M. (1961). *Dokl. Akad. Nauk SSSR* **138**, 1204–1207.
Imshenetsky, A. A., and Ul'Yanova, O. M. (1962a). *Mikrobiologiya* **31**, 832–837.
Imshenetsky, A. A., and Ul'Yanova, O. M. (1962b). *Nature (London)* **195**, 62–63.
Ishiguro, M., Takatsuto, S., Morisaki, M., and Ikekawa, N. (1980). *Chem. Commun.* **20**, 962–964.
Ito, S., and Kimura, J. (1929). *J. Hokkaido Agric. Exp. Stn., Spec. Bull.* **86**, 6–8.
Ito, S., and Kimura, J. (1931). *Hokkaido Agric. Exp. Stn. Rep.* **27**, 1–99.
Jack, T. R., and Zajic, J. E. (1977). *Adv. Biochem. Eng.* **5**, 125–145.
Jackson, D. I., and Coombe, B. G. (1966). *Science* **154**, 277–278.
Jefferys, E. G. (1970). *Adv. Appl. Microbiol.* **13**, 283–323.
Jeffries, G. A. (1948). *Food Ind.* **20**, 688–690, 825–826.
Jennings, R. C. (1968). *Planta* **80**, 34–42.
Jennings, R. C. (1971). *Aust. J. Biol. Sci.* **24**, 1115–1124.
Jennings, R. C., and McComb, A. J. (1967). *Nature (London)* **215**, 872–873.
Johnson, K. D., and Kende, H. (1971). *Proc. Natl. Acad. Sci. U.S.A.* **68**, 2674–2677.
Johnson, M. J. (1954). In "Industrial Fermentations" (L. A. Underkofler and R. J. Hickey, eds.), Vol. 1, pp. 420–445. Chem. Publ., New York.
Jones, R. L. (1969a). *Plant Physiol.* **44**, 101–104.
Jones, R. L. (1969b). *Plant Physiol.* **44**, 1428–1438.
Jones, R. L. (1983). *CRC Crit. Rev. Plant Sci.* **1**, 23–47.
Jones, R. L., and Varner, J. E. (1967). *Planta* **72**, 155–161.
Jones, T. W. A. (1976). *Phytochemistry* **15**, 1825–1827.
Kachru, R. B., Singh, R. N., and Chacko, E. K. (1972). *Acta Hortic.* **24**, 206–208.
Kagawa, T., Fukinbara, T., and Sumiki, Y. (1963). *Agric. Biol. Chem.* **27**, 598–599.
Kahlon, S. S., and Malhotra, S. (1986). *Enzyme Microb. Technol.* **8**, 613–616.
Kalk, J. P., and Langlykke, A. F. (1986). In "Manual of Industrial Microbiology and

Biotechnology" (A. L. Demain and N. A. Solomon, eds.), pp. 363–385. Am. Soc. Microbiol., Washington, D.C.

Kandutsch, A. A., Paulus, H., Levin, E., and Bloch, K. (1964). *J. Biol. Chem.* **239**, 2507–2515.

Kato, J., Katsumi, M., Tamura, S., and Sakurai, A. (1968). In "Biochemistry and Physiology of Plant Growth Substances" (F. Wightman and G. Setterfield, eds.), pp. 347–359. Runge Press, Ottawa.

Katsumi, M. (1970). *Physiol. Plant.* **23**, 1077–1084.

Katznelson, H., and Cole, S. E. (1965). *Can. J. Microbiol.* **11**, 733–741.

Katznelson, H., Sirois, J. C., and Cole, S. E. (1962). *Nature (London)* **196**, 1012–1013.

Kavanagh, F. and Kuzel, N. R. (1958). *J. Agric. Food Chem.* **6**, 459–463.

Kawanabe, Y., Yamane, H., Murayama, T., Takahashi, N. and Nakamura, T. (1983). *Agric. Biol. Chem.* **47**, 1693–1694.

Kawarada, A., Takahashi, N., Kitamura, H., Seta, Y., Takai, M. and Tamura, S. (1955). *Bull. Agric. Chem. Soc. Jpn.* **19**, 84–86.

Kefeli, V. I., Muromtsev, G. S., Agnistikova, V. N., Saidova, S. A., and Drakina, T. I. (1969). *Dokl. Akad. Nauk SSSR* **188**, 1182–1185.

Kende, H. (1967). *Plant Physiol.* **42**, 1612–1618.

Kende, H., and Lang, A. (1964). *Plant Physiol.* **39**, 435–440.

Khan, A. A. (1975). *Bot. Rev.* **41**, 391–420.

Kim, W. K., and Greulach, V. A. (1961). *Am. J. Bot.* **48**, 534.

Kimura, E. T., Young, P. R., and Stamszewski, K. (1959). *J. Am. Pharm. Assoc.* **48**, 127–129.

Kirby, K. D., and Mardon, C. J. (1980). *Biotechnol. Bioeng.* **22**, 2425–2427.

Kitamura, H., Kawarada, A., Seta, Y., Takahashi, N., Otsuki,, T., and Sumiki, Y. (1953). *J. Agric. Chem. Soc. Jpn.* **27**, 545–549.

Kjaer, D., Kjaer, A., Pedersen, C., Bu'Lock, J. D., and Smith, J. R. (1971). *J. Chem. Soc.* pp. 2792–2797.

Kluyver, A. J., and Perquin, L. H. C. (1933). *Biochem. Z.* **266**, 82–95.

Knorre, W. A. (1980). In "Biophysikalishe Grundlagen der Medizin" (W. Beier and R. Rosen, eds.), pp. 132–138. Fischer, Stuttgart.

Knox, J. P., Beale, M. H., Butcher, G. W., and MacMillan, J. (1987). *Planta* **170**, 86–91.

Koehler, D. E. (1966). *Planta* **70**, 42–45.

Koehler, D. E., and Lang, A. (1963). *Plant Physiol.* **38**, 555–560.

Koehler, D. E., and Varner, J. E. (1973). *Plant Physiol.* **52**, 208–214.

Kominek, L. A. (1972). *Antimicrob. Agents Chemother.* **1**, 123–134.

Kopcewicz, J. (1969). *Naturwissenschaften* **56**, 287.

Kos, Y., and Loewenthal, H. J. E. (1963). *J. Chem. Soc.* pp. 605–611.

Krasilnikov, N. A., Chailakhyan, M. K., Skryabin, G. K., Khokholova, Yu.M., Ulezio, I. V., and Konstantinova, T. N. (1956). *Dokl. Akad. Nauk SSSR* **121**, 755–758.

Krasilnikov, N. A., Chailakhyan, M. K., Asseva, I. V., and Khlopenkova, L. P. (1958). *Dokl. Adak. Nauk SSSR* **123**, 1124–1127.

Krasilnikov, N. A., Shirokov, O. G., and Kuchaeva, A. G. (1963). *Gibberelliny Ikh Deistivie Rast.* pp. 39–44.

Krishnamoorthy, H. N., ed. (1975). "Gibberellins and Plant Growth." Wiley Eastern Limited, New Delhi.

Kuhr, I. (1962). *Folia Microbiol. (Prague)* **7**, 358–363

Kuhr, K., Fuska, J., Podojil, M., and Sĕvčik, V. (1961). *Folio Microbiol. (Prague)* **6**, 179–185.

Kukharskaya, L. K. (1968). *Issled.Lesakh. Sib.* **2**, 202–206.

Kumar, P. K. R. (1987). Ph.D. Thesis, University of Mysore, Mysore, India.
Kumar, P. K. R., and Lonsane, B. K. (1985a). In "26th Annual Conference of the Association of Microbiologists of India," Abstr. IND.0-6, p. 110, AMI, Madras.
Kumar, P. K. R., and Lonsane, B. K. (1985b). In "26th Annual Conference of the Association of Microbiologists of India," Abstr. IND.0-5, p. 110, AMI, Madras.
Kumar, P. K. R., and Lonsane, B. K. (1986a). *J. Chromatogr.* **369**, 222–226.
Kumar, P. K. R., and Lonsane, B. K. (1986b). In "27th Annual Conference of Microbiologists of India," Abstr. BB-4, p. 3. AMI, Nagpur.
Kumar, P. K. R., and Lonsane, B. K. (1987a). In "Convention of Food Scientist and Technologists," Abstr. APP-4, p. 81. Assoc. Food Sci. Technol. India, Mysore.
Kumar, P. K. R., and Lonsane, B. K. (1987b). *Biotechnol. Bioeng.* **30**, 267–271.
Kumar, P. K. R., and Lonsane, B. K. (1987c). *Biotechnol. Lett.* **9**, 179–182.
Kumar, P. K. R., and Lonsane, B. K. (1987d). *Process Biochem.* **22**, 139–143.
Kumar, P. K. R., and Lonsane, B. K. (1988a). *Appl. Microbiol. Biotechnol.* **28**, 537–542.
Kumar, P. K. R., and Lonsane, B. K. (1988b). *Process Biochem.* **23**, 43–47.
Kumar, P. K. R., and Lonsane, B. K. (1988c). *Folia Microbiol. (Prague)* **33**, 183–187.
Kumar, P. K. R., Udaya Sankar, K., and Lonsane, B. K. (1989). Unpublished data.
Kumar, S. A., and Mahadevan, S. (1963). *Arch. Biochem. Biophys.* **103**, 516–518.
Kurosawa, E. (1926). *Trans. Nat. Hist. Soc. Formosa* **16**, 213–227.
Kurosawa, E. (1930). *Trans. Nat. Hist. Soc. Formosa* **20**, 218–239.
Kurosawa, E. (1932). *Trans. Nat. Hist. Soc. Formosa* **22**, 198–201.
Kurosawa, E. (1934). *Ann. Phytopathol. Soc. Jpn.* **4**(1&2), 33–34.
Kurzhals, H. (1982). In "CO_2 in Solvent Extraction." Soc. Chem. Ind., London.
Kyowa Hakko Kogyo Co. Ltd. (1983). Japanese Patent 58,152,499.
Laidman, D. L. (1983). *Biochem. Soc. Trans.* **11**, 534–537.
Lang, A. (1970). *Annu. Rev. Plant Physiol.* **21**, 537–570.
Larson, K. A., and King, M. L. (1986). *Biotechnol. Prog.* **2**, 73–82.
Lenton, J. R. (1980). In "Gibberellins: Chemistry, Physiology and Use," Monogr. No. 5, pp. 1–143. British Plant Growth Regulator Group, Wantage.
Leopold, A. C. (1971). *Plant Physiol.* **48**, 537–540.
Lew, F. T., and West, C. A. (1971). *Phytochemistry* **10**, 2065–2076.
Lien, T., Pettersen, R., and Knutsen, G. (1971). *Physiol. Plant.* **24**, 185–190.
Lindenfelser, L. A., and Ciegler, A. (1975). *Appl. Microbiol.* **29**, 323–327.
Linko, P., and Linko, Y. Y. (1983). *Appl. Biochem. Bioeng.* **4**, 53–60.
Lockhard, R. G. (1975). *Malays. Agric. Res.* **4**, 19–20.
Lona, F. (1957). *Ateneo Parmense* **28**, 111–115.
Lonsane, B. K., Ghildyal, N. P., and Murthy, V. S. (1982). In "Technical Brochure, Symposium on Fermented Foods, Food Contaminants, Biofertilizer and Bioenergy," pp. 12–18. Association of Microbiologists of India, Mysore.
Lonsane, B. K., Ghildyal, N. P., Budiatman, S., and Ramakrishna, S. V. (1985). *Enzyme Microb. Technol.* **7**, 258–265.
Luckwill, L. C., Weaver, P., and MacMillan, J. (1969). *J. Hortic. Sci.* **44**, 413–424.
Mabelis, R. P. (1984). European Patent Appl. E.P. 112,629.
McCapra, F., McPhail, A. T., Scott, A. I., Sim, G. A., and Young, D. W. (1966). *J. Chem. Soc.* pp. 1577–1585.
McDaniel, L. E., Bailey, E. G., Ethiraj, S., and Andrews, H. P. (1970). Bacteriol. Proc. p. 7.
McDaniel, L. E., Bailey, E. G., Ethiraj, S., and Andrews, H. P. (1976). *Dev. Ind. Microbiol.* **17**, 91–98.
MacMillan, J., ed. (1980). "Encyclopaedia of Plant Physiology, New Series," Vol. 9, pp. 1–8. Springer-Verlag, Berlin and New York.

MacMillan, J., and Suter, P. J. (1958). *Naturwissenschaften* **45**, 46–47.
MacMillan, J., and Suter, P. J. (1963). *Nature (London)* **197**, 790.
MacMillan, J., and Takahashi, N. (1968). *Nature (London)* **217**, 170–171.
MacMillan, J., Seaton, J. C., and Suter, P. J. (1960). *Tetrahedron* **11**, 60–66.
Maddox, I. S., and Richert, S. H. (1977a). *Appl. Environ. Microbiol.* **33**, 210–202.
Maddox, I. S., and Richert, S. H. (1977b). *J. Appl. Bacteriol.* **43**, 197–204.
Maeda, E. (1960). *Physiol. Plant.* **13**, 214–226.
Maeda, E. (1965). *Physiol. Plant.* **18**, 813–827.
Mahadevan, A. (1984). "Growth Regulators, Microorganisms and Diseased Plants," pp. 269–303. Oxford and IBH, New Delhi.
Mahadevan, S. (1963). *Arch. Biochem. Biophys.* **100**, 557–558.
Malcolm, H. D., and Fahy, P. C. (1971). *Search* **2**, 141–142.
Málek, I., and Fencle, Z., eds. (1966). "Theoretical and Methodological Basis of Continuous Culture of Microorganisms." Academic Press, New York.
Manaka, M. (1980). *Acta Microbiol. Pol.* **29**, 365–374.
Mander, L. N. (1982). *Search* **13**, 188–190.
Mann, J. D. (1975). *In* "Gibberellins and Plant Growth" (H. N. Krishnamoorthy, ed.), pp. 239–287. Wiley Eastern Ltd., New Delhi.
Martin, G. C. (1983). *In* "The Biochemistry and Physiology of Gibberellins" (A. Crozier ed.), Vol. 2, pp. 395–444. Praeger, New York.
Marwaha, S. S., and Kennedy, J. F. (1984). *Enzyme Microb. Technol.* **6**, 18–22.
Mees, G. C., and Elson, G. W. (1978). *In* "Jealott's Hill-Fifty Years of Agricultural Research, 1928–1978" (F. C. Peacock, ed.), pp. 55–60. Kynoch Press, Birmingham, England.
Mertz, D. (1970). *Plant Cell Physiol.* **11**, 273–279.
Mertz, D., and Henson, W. (1967a). *Nature (London)* **214**, 844–846.
Mertz, D., and Henson, W. (1967b). *Physiol. Plant.* **20**, 187–199.
Messing, R. A. (1980). *Annu. Rep. Ferment. Processes* **4**, 105–121.
Mitchell, J. W., Mandava, N., Worley, J. F., Plimmer, J. R., and Smith, M. V. (1970). *Nature (London)* **225**, 1065–1066.
Modlibowska, I., and Wicken,, M. F. (1982). *J. Hortic. Sci.* **57**, 413–422.
Moffatt, J. S. (1960). *J. Chem. Soc.* pp. 3045–3049.
Mohan, R. R., and Li, N. N. (1974). *Biotechnol. Bioeng.* **16**, 513–523.
Money, T., Raphael, R. A., Scott, A. I., and Young, D. W. (1961). *J. Chem. Soc.* pp. 3958–3962.
Monselise, S. P., Weiser, M., Shafir, N., Goren, R., and Goldschmidt, E. E. (1976). *J. Hortic. Sci.* **51**, 341–351.
Montuelle, B. (1966). *Ann. Inst. Pasteur, Paris Suppl.* **3**, 136–146.
Montuelle, B., and Cheminais, L. (1964). *C. R. Hebd. Seances Acad. Sci.* **258**, 6016–6017.
Mudgett, R. E. (1986). *In* "Manual of Industrial Microbiology and Biotechnology" (A. L. Demain and N. A. Solomon, eds.), pp. 66–83. Am. Soc. Microbiol., Washington, D.C.
Munro, A. L. S. (1970). *In* "Methods in Microbiology" (J. R. Norris and D. W. Robbons, eds.), Vol. 2, pp. 39–89. Academic Press, New York.
Murakami, Y. (1957). *Bot. Mag.* **70**, 376–382.
Murakami, Y. (1966). *Bot. Mag.* **79**, 315–327.
Murakami, Y. (1968). *Bot. Mag.* **81**, 33–43.
Murakami, Y. (1970a). *Bot. Mag.* **83**, 211–213.
Murakami, Y. (1970b). *Jpn. Agric. Res. Q.* **5**, 5–9.
Murakami, Y. (1972). *In* "Plant Growth Substances 1970" (D. J. Carr, ed.), pp. 166–174. Springer-Verlag, Heidelberg.

Murashige, T. (1964). *Physiol. Plant.* **17**, 636–643.
Murofushi, N., Sugimoto, M., Itoh, K., and Takahashi, N. (1979). *Agric. Biol. Chem.* **43**, 2179–2185.
Murofushi, N., Sugimoto, M., Itoh, K., and Takahashi, N. (1980). *Agric. Biol. Chem.* **44**, 1583–1587.
Muromtsev, G. S., and Dubovaya, L. P. (1964). *Mikrobiologiya* **33**, 1048–1055.
Muromtsev, G. S., and Globus, G. A. (1976). *Dokl. Akad. Nauk SSSR* **226**, 204–206.
Muromtsev, G. S., Agnistikova, V. N., Lupova, L. M., Dobovaya, L. P., Lekareva, T. A., Serebryakov, E. P., and Kucherov, V. F. (1966). *Khim Prir. Soedin.* **2**, 114–120.
Muromtsev, G. S., Rakovskii, Y. S., Dubovaya, L. P., Temnikova, T. V., and Fedchenko, A. N. (1968). *Prikl. Biokhim. Mikrobiol.* **4**, 398–407.
Murphy, P. J., and West, C. A. (1969). *Arch. Biochem. Biophys.* **133**, 395–407.
Myers, R. H. (1971). "Response Surface Methodology." Allyn & Bacon, Boston, Massachusetts.
Nagaraja Rao, K. S. (1976). Ph.D. Thesis, University of Mysore, Mysore, India.
Nair, S. K., and Subba Rao, N. S. (1977). *Plant Soil* **46**, 511–519.
Nandi, B., and Mondal, S. (1970). *Sci. Cult.* **36**, 461–462.
Nandi, D. L., and Porter, J. W. (1964). *Arch. Biochem. Biophys.* **105**, 7–19.
Nestyouk, M. N., Dendze-Pletman, B. B., Ionova, N. B., Iofo, R. N., Kleiver, G. I., Kravchenko, B. F., Krutova, R. L., Muromtsev, G. C., and Rusanova, N. V. (1961). Bull. Invent. No. 18, USSR Certificate No. 141,352, as cited by Jefferys, 1970.
Netien, G., and Oddoux, L. (1961). *C. R. Hebd. Seances Acad. Sci.* **253**, 520–522.
Newman, J. C., and Briggs, D. E. (1976). *Phytochemistry* **15**, 1453–1458.
Nicholls, P. B. N., and Paleg, L. G. (1963). *Nature (London)* **199**, 823–824.
Nickell, L. G. (1980). In "Plant Growth Substances 1979" (F. Skoog, ed.), pp. 419–425. Springer-Verlag, Berlin and New York.
Nickell, L. G. (1982). "Plant Growth Regulators: Agricultural Uses." Springer-Verlag, Berlin and New York.
Ninnemann, H., Zeevaart, J. A. D., Kende, H., and Lang, A. (1964). *Planta* **61**, 229–235.
Nisikado, Y. (1931). *Ber. Oharo Inst. Landwirt. Forsch. Kurashiki Provinz Okayama, Jpn.* **5**, 87–106.
Nisikado, Y. (1932). *Jpn. J. Plant Prot.* **19**, 491–497.
Ogawa, Y. (1966). *Plant Cell Physiol.* **7**, 509–517.
Ohara, K., Tamaki, M., Kawahara, T., and Takada, H. (1975). *Nippon Kingakkai Kaiho* **16**, 247–252.
Ostrovskii, N. E., Shalagina, A. I., Kryukova, M. A., and Bankovskaya, A. N. (1961). *Sov. Plant Physiol. (Engl. Transl.)* **8**, 278–280.
Paleg, L. G. (1960a). *Plant Physiol.* **35**, 293–299.
Paleg, L. G. (1960b). *Plant Physiol.* **35**, 902–906.
Paleg, L. G. (1965). *Annu. Rev. Plant Physiol.* **16**, 291–322.
Paleg, L. G., Aspinall, D., Coombe, B., and Nicholls, P. (1964). *Plant Physiol.* **39**, 286–290.
Panosyan, A. K., and Babayan, G. S. (1966). *Biol. Zh. Arm.* **19**, 78–84.
Paquette, G. J., and McKellar, R. C. (1986). *J. Food Sci.* **51**, 655–658.
Patching, J. W., and Rose, A. H. (1970). In "Methods in Microbiology" (J. R. Norris and D. W. Ribbons, eds.), Vol. 2, pp. 23–38. Academic Press, New York.
Payne, J. W., ed. (1980). "Microorganisms and Nitrogen Sources." Wiley, Chichester.
Pegg, G. F. (1973a). *J. Exp. Biol.* **24**, 675–688.
Pegg, G. F. (1973b). *Trans. Br. Mycol. Soc.* **61**, 277–286.
Penner, J. (1960). *Planta* **55**, 542–572.
Perlman, D. (1973). *Process Biochem.* **8**(7), 18–20.
Peterson, L. J., DeVay, J. E., and Houston, B. R. (1963). *Phytopathology* **53**, 630–633.

Phillips, I. D. J., and Jones, R. L. (1964). *Planta* **63,** 269–278.
Phinney, B. O. (1956). *Proc. Natl. Acad. Sci. U.S.A.* **42,** 185–189.
Phinney, B. O. (1961) In "Plant Growth Regulators" (R. M. Klein, ed.), pp. 489–501. Iowa State Univ. Press, Ames.
Phinney, B. O. (1979). *ACS Symp. Ser.* **111,** 57–78.
Phinney, B. O. (1983). In "The Biochemistry and Physiology of Gibberellins" (A. Crozier, ed.), Vol. 1, pp. 19–52. Praeger, New York.
Phinney, B. O., and Spector, C. (1967). *Ann. N. Y. Acad. Sci.* **144,** 204–210.
Phinney, B. O., and West, C. A. (1961). *Encycl. Plant Physiol. New Ser.* **14,** 1185–1227.
Pirt, S. J., and Callow, D. S. (1959). *Nature (London)* **184,** 307–310.
Pirt, S. J., and Callow, D. S. (1960). *J. Appl. Bacteriol.* **23,** 87–98.
Pitel, D. W., Vining, L. C., and Arsenault, G. P. (1971). *Can. J. Biochem.* **49,** 185–193.
Podojil, M., and Ricicova, A. (1965). *Folia Microbiol. (Prague)* **10,** 55–59.
Podojil, M., and Ševčik, V. (1960). *Folia Microbiol. (Prague)* **5,** 192–197.
Pogell, B. M., Sankaran, L., Redshaw, P. A., and McCann, P. A. (1976). In "Microbiology 1976" (D. Schlessinger, ed.), pp. 543–547. Am. Soc. Microbiol., Washington, D.C.
Popják, G., and Cornforth, J. W. (1960). *Adv. Enzymol.* **22,** 281–335.
Prapulla, S. G., Thakur, M. S., Prema, P., Lonsane, B. K., Ravindranath, B., Ramakrishna, S. V., and Murthy, V. S. (1983). In "Third Indian Convention of Food Scientists and Technologists." Assoc. Food Sci. Technol., Mysore, India.
Prasad, N., and Desai, M. V. (1952). *Curr. Sci.* **21,** 17–18.
Prema, P., Thakur, M. S., Prapulla, S. G., Ramakrishna, S. V., and Lonsane, B. K. (1987). *Indian J. Microbiol.* **28,** 78–81.
Prescott, S. C., and Dunn, C. G. (1949). "Industrial Microbiology," p. 572–605. McGraw-Hill, New York.
Probst, G. W. (1961). U.S. Patent 2,980,700.
Pryce, R. J. (1973). *Phytochemistry* **12,** 507–514.
Queener, S., and Swartz, R. (1979). In "Secondary Products of Metabolism" (A. H. Rose, ed.), Vol. 3, pp. 35–122. Academic Press, New York.
Rademacher, W., and Graebe, J. E. (1979). *Biochem. Biophys. Res. Commun.* **91,** 35–40.
Radley, M. (1958). *Ann. Bot. (London)* [N.S.] **22,** 297–307.
Rados, G. (1970). *Soripar* **17,** 126–130.
Raimbault, M., and Alazard, D. (1980). *Eur. J. Appl. Microbiol. Biotechnol.* **9,** 199–209.
Ralph, B. J. (1976). *Food Technol. Aust.* **28,** 247–251.
Ramesh, M. V., and Lonsane, B. K. (1987). *Biotechnol. Lett.* **9,** 323–328.
Randolph, T. W., Blanch, H. W., Prausnitz, J. M., and Wilke, C. R. (1985). *Biotechnol. Lett.* **7,** 325–328.
Rao, M. N. A., Mittal, B. M., Thakur, R. N., and Shastry, K. S. M. (1983). *Biotechnol. Bioeng.* **25,** 869–872.
Rappaport, L., Davies, L., Lavee, S., Nadeau, R., Petterson, R., and Stolp, C. E. (1974). In "Isolation of Plant Growth Substances 1973," pp. 314–324. Hirokawa, Tokyo.
Ratsimamanga, A. R., and Boiteau, P. (1964). *Pathol. Biol.* **12,** 65–71.
Redemann, C. T. (1959). U.S. Patent 2,918,413.
Reese, E. T. (1959). In "Marine, Boring and Fouling Organisms" (D. L. Ray, ed.), pp. 265–300. Univ. of Washington Press, Seattle.
Reeve, D. R., and Crozier, A. (1975). In "Gibberellins and Plant Growth" (H. N. Krishnamoorthy, ed.), pp. 34–64. Wiley Eastern Ltd., New Delhi.
Reid, D. M., Clements, J. B., and Carr, D. J. (1968). *Nature (London)* **217,** 580–582.
Ribbons, D. W. (1970). In "Methods in Microbiology" (J. R. Norris and D. W. Ribbons, eds.), Vol. 3A, pp. 297–304. Academic Press, New York.

Ricicova, A., Podojil, M., Musilek, V., and Ševčik, V. (1960). *Folia Microbiol.* (Prague) **5,** 181–191.
Robbins, W. J. (1957). *Am. J. Bot.* **44,** 743–746.
Roeben, R., Echard, K., and Klaus, R. (1985). *Monatsschr. Branwiss.* pp. 160–163.
Rogers, L. J., Shah, S. P. J., and Goodwin, T. W. (1965). *Biochem. J.* **96,** 7p–8p.
Ropers, H. J., Graebe, J. E., and Gaskin, P., and MacMillan, J. (1978). *Biochem. Biophys. Res. Commun.* **80,** 690–697.
Rose, A. H. (1979). In "Secondary Products of Metabolism" (A. H. Rose, ed.), pp. 1–33. Academic Press, New York.
Roy, A. K. (1964). *Sci. Cult.* **30,** 242.
Russell, S. (1975). In "Gibberellins and Plant Growth (H. N. Krishnamoorthy, ed.), pp. 1–34. Wiley Eastern Ltd., New Delhi.
Sackett, P. H. (1984). *Anal. Chem.* **56,** 1600–1603.
Sakurai, N., Shibata, K., and Kamisaka, S. (1974). *Plant Cell Physiol.* **15,** 709–716.
Sanchez-Marroquin, A. (1963). *Appl. Microbiol.* **11,** 523–528.
Santoro, T., and Casida, L. E., Jr. (1962). *Mycologia* **45,** 70–71.
Sanwal, B. D., and Waygood, E. R. (1961). *Experientia* **17,** 174–175.
Saono, S. (1964). *Nature (London)* **204,** 1328–1329.
Sawada, K. (1912). *Formosan Agric. Rev.* **63,** 10 and 16.
Sawada, K. (1917). *Trans. Natl. Hist. Soc. Formosa* **7,** 131–136.
Schierholt, J. (1976). *Int. Ferment. Symp. 5th, 1976* Abstr. 3.11, p. 49.
Schneider, E. A., and Wightman, F. (1978). In "Phytohormones and Related Compounds: A Comprehensive Treatise" (D. S. Letham, P. B. Goodwin, and T. J. V. Higgins, eds.), Vol. 1, pp. 29–105. Elsevier/North-Holland Biomedical Press, New York.
Schreiber, K., Schneider, G., Sembdner, G., and Focke, I. (1966). *Phytochemistry* **5,** 1221–1225.
Schreiber, K., Weiland, J., and Sembdner, G. (1970). *Phytochemistry* **9,** 189–198.
Schulze, K. L. (1962). *Appl. Microbiol.* **10,** 108–122.
Schutz, H. (1983). *Food Technol.* **37,** 46–48.
Schwartz, E. (1962). *Biologia (Bratislava)* **12,** 650–653.
Schwartz, E. (1963). *Rozhl. Tuberk. Nemocech Plien.* **23,** 333–342.
Schwartz, E., and Laginova, V. (1966). *Strahlentherapie* **129,** 616–621.
Schwartz, E., Miklussak, S., Vincurova, M., Laczko, A., Krcova, M., and Zemkova, O. (1983). *Pharmazie* **38,** 716–718.
Scott, A. I., McCapra, F., Comer, F., Sutherland, S. A., and Young, D. W. (1964). *Tetrahedron* **20,** 1339–1358.
Selman, I. W., and Kandiah, U. (1971). *Bull. Entomol. Res.* **60,** 359–365.
Sembdner, G., Schneider, G., Weiland, J., and Schreiber, K. (1964). *Experientia* **20,** 89–90.
Sembdner, G., Weiland, J., Aurich, D., and Schreiber, K. (1968). *Soc. Chem. Ind. Monogr.* **31,** 70–86.
Sembdner, G., Weiland, J., Schneider, G., Schreiber, K., and Focke, I. (1972). In "Plant Growth Substances 1970" (D. J. Carr, ed.), pp. 143–150. Springer-Verlag, Heidelberg.
Sembdner, G., Schulze, C., Adam, G., Viogt, B., Hung, P. D., Weiland, J., and Schreiber, K. (1973). In "Wirkungsmechanismen von Herbiziden und Synthetischen Wachstumsregulatoren" (A. Banth, F. Jacob, and G. Feyerabend, eds.), Part 10. Wissenschaftliche Beitráge der Martin-Luther Universität, Halle, as cited by Hoad, 1983.
Sembdner, G., Gross, D., Liebisch, H. W., and Schneidner, G. (1980). *Encycl. Plant Physiol., New Ser.* **9,** 281–444.
Serzedello, A., and Whitaker, N. (1960). *Rev. Agric. (Piracicabo, Braz.)* **35,** 15–24.
Sethi, R. P., Sehgal, V. K., and Kapoor, A. (1981). In "International Conference on System

Theory and Application," pp. C51–55. Punjab Agricultural University, Ludhiana, India, as cited by Gohlwar et al. 1984.
Shechter, I., and West, C. A. (1969). *J. Biol. Chem.* **244**, 3200–3209.
Shen, N., and Chang, F. (1980). *Yao Hsueh Tung Pao* **15**, 46.
Shin, H. S., Bang, W. G., and Hong, B. G. (1985). *Nonglin Nonjip.* **25**, 45–53.
Sidahmed, O. A., and Kliewer, W. M. (1980). *Am. J. Enol. Vitic.* **31**, 149–153.
Sikyta, B. (1964). *Rozpr. Cesk. Akad. Ved, Rada Mat. Prir. Ved* **74**, 16–20.
Sikyta, B., Doskocil, J., and Kasparova, J. (959). *J. Biochem. Microbiol. Technol. Eng* **1**, 379–392.
Silman, R. W., Conway, H. F., Anderson, R. A., and Bagley, E. B. (1979). *Biotechnol. Bioeng.* **21**, 1799–1808.
Sironval, C. (1961). In "Plant Growth Regulators" (R. M. Klein, ed.), pp. 521–530. Iowa State Univ. Press, Ames.
Skene, K. G.M., and Carr, D. J. (1961). *Phyton* **16**, 97–115.
Société d'Etudes et d'Applications Biochimiques (1963). British Patent 936,548.
Somal, B. S., Shastri, K. S. M., Thakur, R. N., Pandotra, U. R., and Atal, C. K. (1978). *Proc. Ind. Ferment. Symp. 1978*, pp. 100–106.
Spector, C., and Phinney, B. O. (1968). *Physiol. Plant.* **21**, 127–136.
Sponsel, V. M., and Macmillan, J. (1977). *Planta* **135**, 129–136.
Sponsel, V. M., and MacMillan, J. (1978). *Planta* **144**, 69–78.
Sponsel, V. M., Hoad, G. V., and Beeley, L. J. (1977). *Planta* **135**, 143–147.
Stanbury, P. F., and Whitakar, A. (1984). In "Principles of Fermentation Technology," pp. 11–25. Pergamon, Oxford.
Steinkraus, K. H. (1984). *Acta Biotechnol.* **4**, 83–88.
Sternberg, M. (1962). *Arch. Biochem. Biophys.* **98**, 299–304.
Stoddart, J. L. (1972). *Planta* **107**, 81–88.
Stoddart, J. L. (1983). In "The Biochemistry and Physiology of Gibberellins" (A. Crozier, ed.), Vol. 2, pp. 1–55. Praeger, New York.
Stoddart, J. L., and Venis, M. A. (1980). *Encycl. Plant Physiol., New Ser.* **9**, 445–498.
Stoddart, J. L., Jones, T. W. A., and Tapster, S. M. (1978). *Planta* **141**, 283–288.
Stodola, F. H. (1958). *U.S., Agric. Res. Serv.*, **ARS-71-11**, 1–421.
Stodola, F. H., Raper, K. B., Fennell, D. I., Conway, H. F., Johns, V. E., Langford, C. T., and Jackson, R. W. (1955). *Arch. Biochem. Biophys.* **54**, 240–245.
Stodola, F. H., Nelson, G. E. N., and Spence, D. J. (1957). *Arch. Biochem. Biophys.* **66**, 438–444.
Stoll, C. (1954). *Phytopathol. Z.* **22**, 233–274.
Stoll, C., and Renz, J. (1957). *Phytopathol. Z.* **29**, 380–387.
Stork, G., and Newman, H. (1959). *J. Am. Chem. Soc.* **81**, 5518–5519.
Stowe, B. B., and Yamaki, T. (1957). *Annu. Rev. Plant Physiol.* **8**, 181–216.
Stowe, B. B., Stodola, F. H., Hayashi, T., and Brian, P. W. (1961). In "Plant Growth Regulators (R. M. Klien, ed.), pp. 465–471. Iowa State Univ. Press, Ames.
Stubblefield, R. D. (1979). *J. Assoc. Off. Anal. Chem.* **62**, 201–202.
Subba Rao, N. S. (1957). *Proc.—Indian Acad. Sci. Sect. B* **45B**, 91–94.
Subramanian, C. V. (1951). *Proc. Natl. Inst. Sci. India* **17**, 403–411.
Subramanian, C. V. (1952). *J. Madras Univ., Sect. B* **22**, 206–222.
Subramanian, C. V. (1954). *J. Madras Univ., Sect. B* **24**, 21–46.
Suge, H., and Murakami, Y. (1968). *Plant Cell Physiol.* **9**, 411–414.
Sukh Dev and Misra, R., eds. (1986). "CRC Handbook of Terpenoids diterpenoids," Vol. 4, p. 178 only. CRC Press, Boca Raton, Florida.
Sumiki, Y., Kagawa, T., and Fukanebara, T. (1966). Japanese Patent 16558.
Swart, A., and Kamerbeek, S. A. (1977). *Physiol. Plant.* **39**, 38–44.

Takahashi, N., Kitamura, H., Kawarada, A., Seta, Y., Takai, M., Tamura, S., and Sumiki, Y. (1955). *Bull. Agric. Chem. Soc. Jpn.* **19**, 267–277.
Takahashi, N., Seta, Y., Kitamura, H., and Sumiki, Y. (1957). *Bull. Agric. Chem. Soc. Jpn.* **21**, 396–398.
Takamine, J. (1914). *Ind. Eng. Chem.* **6**, 824–828.
Tang, Y. W., Hwang, Y. D., Lee, S. J., Hua, X. Z., and King, H. J. (1973). *Sci. Sin. (Engl. Ed.)* **16**, 512–518,
Taris, B., and Clemencet, Y. (1970). *C. R. Hebd. Seances Acad. Sci., Ser. D.* **270**, 1468–1471.
Tauriya, H., Morimura, Y., and Yokota, M. (1962). *Arch. Microbiol.* **42**, 4–16.
Taylor, I. E. P., and Wilkinson, A. J. (1977). *Phycologia* **16**, 37–42.
Thakur, M. S. (1981). Ph.D. Thesis, Saugar University, India.
Thakur, M. S., and Vyas, K. M. (1983). *Folia Microbiol. (Prague)* **28**, 124–129.
Thakur, M. S., Prema, P., Prapulla, S. G., Lonsane, B. K., Ramakrishna, S. V., Ahmed, S. Y., and Murthy, V. S. (1983). *In* "Third Indian Convention of Food Scientists and Technologists." Assoc. Food Sci. Technol., Mysore, India.
Thériault, R. J., Friedland, W. C., Peterson, M. H., and Sylvester, J. C. (1961). *J. Agric. Food Chem.* **9**, 21–23.
Thimann, K. V., and Mahadevan, S. (1964). *Arch. Biochem. Biophys.* **105**, 133–141.
Tokuda, K., Ito, K., Imamura, T., Taniguchi, M., Kobayashi, T., Aki, T., and Hara, S. (1986). *J. Brew. Soc. Jpn.* **81**, 194–198.
Trewavas, A. (1979). *Trends Biochem. Sci.* **4**, 199–202.
Trucksess, M. W., Stoloff, L., Pons, W. A., Jr., Cucullo, A. F., Lees, L. S., and Franz, A. O., Jr. *J. Assoc. Off. Anal. Chem.* **60**, 795–798.
Underkofler, L. A., Fulmer, E. I., and Schoene, L. (1939). *Ind. Eng. Chem.* **31**, 734–738.
Underkofler, L. A., Severson, G. M., and Goering, K. J. (1946). *Ind. Eng. Chem.* **38**, 980–985.
Underkofler, L. A., Severson, G. M., Goering, K. J., and Christensen, L. M. (1947). *Cereal Chem.* **24**, 1–22.
Vancura, V. (1961). *Nature (London)* **192**, 88–89.
Vass, R. C., and Jefferys, E. G. (1979). *In* "Secondary Products of Metabolism" (A. H. Rose, ed.), Vol. 3, pp. 421–433. Academic Press, London.
Vojtisek, V., and Jirku, V. (1983). *Folia Microbiol. (Prague)* **28**, 309–340.
Wada, K., Marumo, S., Ikekawa, N., Morisaki, M., and Mori, K. (1981). *Plant Cell Physiol.* **22**, 323–325.
Wada, K., Marumo, S., Mori, K., Takatsuto, S., Morisaki, M., and Ikekawa, N. (1983). *Agric. Biol. Chem.* **47**, 1139–1141.
Wakagi, S. (1958). *J. Ferment. Assoc. Jpn.* **16**, 150–164.
Walhood, V. T. (1958). *Proc. Beltwide Cotton Prod. Res. Conf.* p. 27. As cited by Nickell, L. G. (1983). *In* "Plant Growth Regulating Chemicals," Vol. 1, CRG Press, FL.
Wang, D. I. C., Cooney, C. L., Demain, A. L., Dunnill, P., Humphrey, A. E., and Lilly, M. D. (1979). "Fermentation and Enzyme Technology." Wiley, New York.
Warden, W. K., and Schaible, P. J. (1958). *Poult. Sci.* **37**, 490–491.
Washburn, W. H., Scheske, F. A., and Schenck, J. R. (1959). *J. Agric. Food Chem.* **7**, 420–422.
Weaver, R. J. (1958). *Nature (London)* **181**, 851–852.
Weaver, R. J. (1959). *Nature (London)* **183**, 1198–1199.
Weinberg, E. D. (1970). *Adv. Microbiol. Physiol.* **4**, 1–44.
Weindhmayer, J. (1964). *Z. Pflanzenkr. (Pflanzenpathol.) Pflanzenschutz* **71**, 317–331.
West, C. A. (1973). *In* "Biosynthesis and Its Control in Plants" (B. V. Milborrow, ed.), pp. 143–169. Academic Press, New York.

West, C. A., and Phinney, B. O. (1956). *Plant Physiol.* **31**,Suppl., XX.
Wheeler, A. W. (1960). *J. Exp. Bot.* **11**, 217–226.
Whitaker, A. (1980). *Process Biochem.* **15**(4), 10–15.
Whyte, P., and Luckwill, L. C. (1966). *Nature (London)* **210**, 1360.
Wierzchowski, P., and Wierzchowska, Z. (1961). *Naturwissenschaften* **48**, 653.
Wilhelm, R., and Graebe, J. E. (1979). *Biochem. Biophys. Res. Commun.* **91**, 35–40.
Williams, D. F. (1981). *Chem. Eng. Sci.* **36**, 1769–1788.
Wilson, G. (1976). *Ann. Bot. (London)* [N.S.] **40**, 919–932.
Wittwer, S. H., and Bukovac, M. J. (1958). *Econ. Bot.* **12**, 213–255.
Wong, J. M., and Johnston, K. P. (1986). *Biotechnol. Prog.* **2**, 29–39.
Woodruff, H. B. (1966). *Symp. Soc. Gen. Microbiol.* **16**, 22–46.
Yabuta, T. (1935). *Agric. Hortic.* **10**, 17–22.
Yabuta, T., and Hayashi, T. (1936). *Agric. Hortic.* **11**, 27–33.
Yabuta, T., and Hayashi, T. (1937). *Agric. Hortic.* **12**, 1073–1083.
Yabuta, T., and Hayashi, T. (1938). *Agric. Hortic.* **13**, 21–25.
Yabuta, T., and Hayashi, T. (1939). *J. Agric. Chem. Soc. Jpn.* **15**, 257–266.
Yabuta, T., and Hayashi, T. (1940). *J. Imp. Agric. Exp. Stn. (Jpn.)* **3**, 365–400.
Yabuta, T., and Sumiki, Y. (1938). *J. Agric. Chem. Soc. Jpn.* **14**, 1526.
Yabuta, T., and Sumiki, Y. (1944). *J. Agric. Chem. Soc. Jpn.* **20**, 52.
Yabuta, T., Kambe, K., and Hayashi, T. (1934). *J. Agric. Chem. Soc. Jpn.* **10**, 1059–1068.
Yabuta, T., Sumiki, Y., and Uno, S. (1939). *J. Agric. Chem. Soc. Jpn.* **15**, 1209–1220.
Yabuta, T., Sumiki, Y., Katayama, E., and Motoyama, H. (1940). *J. Agric. Chem. Soc. Jpn.* **16**, 1157–1158.
Yabuta, T., Sumiki, Y., Aso, K., and Hayashi, T. (1941). *J. Agric. Chem. Soc. Jpn.* **17**, 1001–1004.
Yamada, K. (1977). *In* "Most Advanced Industrial Technology and Industry: Academic Review," pp. 1–7. Int. Tech. Inf. Inst., Japan.
Yamaguchi, I., Miyamoto, M., Yamane, H., Murofushi, N., Takahashi, N., and Fujita, K. (1975). *J. Chem. Soc., Perkin Trans.* 1 pp. 996–999.
Yamane, T., and Shimizu, S. (1984). *Adv. Biochem. Eng. Biotechnol.* **30**, 147–194.
Yamane, H., Yamaguchi, I., Murofushi, N., and Takahashi, N. (1971). *Agric. Biol. Chem.* **35**, 1144–1146.
Yamane, H., Murofushi, N., and Takahashi, N. (1975). *Phytochemistry* **14**, 1195–1200.
Yamane, H., Yamaguchi, I., Kobayashi, M., Takahashi, M., Sato, Y., Takahashi, N., Iwatsuki, K., Phinney, B. O., and Spray, C. R. (1985). *Plant Physiol.* **78**, 889–896.
Yokota, T., Takahashi, N., Murofushi, N., and Tamura, S. (1969). *Tetrahedron Lett.* pp. 2081–2084.
Yokota, T., Murofushi, N., Takahashi, N., and Tamura, S. (1971a). *Agric. Biol. Chem.* **35**, 573–583.
Yokota, T., Murofushi, N., Takahashi, N., and Tamura, S. (1971b). *Agric. Biol. Chem.* **35**, 583–595.
Yomo, H. (1960a). *Hakko Kyoakishi* **18**, 600–602.
Yomo, H. (1960b). *Hakko Kyoakishi* **18**, 603–606.
Yoshida, F., Yamane, T., and Nakamoto, K. I. (1973). *Biotechnol. Bioeng.* **15**, 257–270.
Zaborsky, O. R. (1973). "Immobilized Enzymes." CRC Press, Boca Raton, Florida.
Zarnescu, A. and Nita, L. (1964). *An. Inst. Cercet Cereale Plante Ieh-Fundulea, Inst. Cent. Cercet. Agric., Ser. B* **32**, 443–445.
Zeevaart, J. A. D. (1969). *Neth. J. Agric. Sci.* **17**, 215–220.
Zucconi, F., and Bukovac, M. J. (1978). *Acta Hortic.* **80**, 159–162.
Zweig, G., and DeVay, J. E. (1959). *Mycologia* **51**, 877–886.

Microbial Dehydrogenations of Monosaccharides

MILOŠ KULHÁNEK

Research Institute for Pharmacy and Biochemistry
130 60, Prague, Czechoslovakia

I. Introduction
II. Dehydrogenases and Mechanisms of Dehydrogenations
   A. Membrane Dehydrogenases
   B. Cytosol Dehydrogenases
   C. Electron Transport Systems
III. Special Nature of Microbial Dehydrogenations of Monosaccharides
   A. Location of the Enzyme and Respiratory Systems
   B. Simplicity of the ETS
   C. Nonexistence of Any Respiratory Regulation
   D. Special Nature of *Gluconobacter oxydans*
IV. Relationships between Structure and Dehydrogenation of Monosaccharides
   A. Alditols
   B. Aldoses
   C. Aldonic Acids
   D. Other Structure–Dehydrogenation Relationships
V. Microorganisms and Fermentation Technique
   A. Dehydrogenations of Individual Groups of Monosaccharides
   B. Fermentation Technique
   C. Enzyme Inhibitors in the Fermentations
VI. Preparative and Industrial Applications
   A. Dehydrogenations of Alditols
   B. Dehydrogenations of Ketoses
   C. Dehydrogenations of Aldoses
   D. Dehydrogenations of Aldonates at C-2
   E. Dehydrogenations of Aldonates at C-(ω-1)
   F. Dehydrogenations of Aldonates at Both C-2 and C-(ω-1)
References

I. Introduction

Microbial dehydrogenations of monosaccharides are fermentation reactions catalyzed by enzyme systems of some bacteria or fungi, in which nonphosphorylated aldoses are dehydrogenated to aldonic acids or their lactones and eventually to ketoaldonic acids, while alditols are dehydrogenated to ketoses. All the mentioned metabolites accumulate in the fermentation medium. "Ketogenic fermentations" or "ketofermentations" are other suitable designations for these reactions. They are used for preparations of dihydroxyacetone, gluconate, ketogluconates, and other ketoderivatives. The sorbose fermentation is being increasingly applied in industrial production of vitamin C. Previous reviews (Bertrand, 1904; Bernhauer, 1938, 1950; Razumovskaya, 1956; Rao, 1957; Janke, 1960, 1962; Müller, 1960; Wolf, 1960; Rainbow, 1961; Touster and Shaw, 1962; Asai, 1968; Spencer and Gorin, 1968) surveyed microbial dehydrogenations of sugars up to 1968 and this chapter gives a survey of further development in this field. Conclusions on the mechanism of action of individual dehydrogenases (DHs) are presented, as well as configuration–dehydrogenation relationships in monosaccharides, along with some applied techniques of study.

II. Dehydrogenases and Mechanisms of Dehydrogenations

Through studies on the mechanisms of microbial dehydrogenations of monosaccharides, two groups of enzyme systems were found to be capable of dehydrogenation in bacterial cells. The two groups of enzyme systems differ in their location and functions within the cell and, consequently, represent two types of metabolism. The first type of DH system, located in cytoplasmic membranes, is found in particulate fractions following cell disintegration. These are extracellular enzymes and are NAD(P)-independent. They catalyze rapid dehydrogenations of monosaccharide products from which metabolites accumulate in the medium. The second type of DH is soluble and is found in the cytosol. It is NAD(P)-dependent and does not accumulate but is dissimilated immediately in general metabolic pathways. Location and functions of both types of DH are summarized in the Fig. 1.

The proportion of the these two types of metabolism may vary according to conditions of cultivation (saccharides, concentration, pH, etc.). *Pseudomonas aeruginosa* shifted its metabolism from the extracellular oxidation to the intracellular phosphorylation pathway when the carbon supply was diminished (Whiting et al., 1976). Low pH values of the medium result in inhibition of the pentose cycle and the whole pathway starting with the action of the cytosol enzymes.

In cells of *Aspergillus niger*, glucose oxidase and catalase have been found in special organelles called peroxisomes (Van Dijken and Veenhuis, 1980).

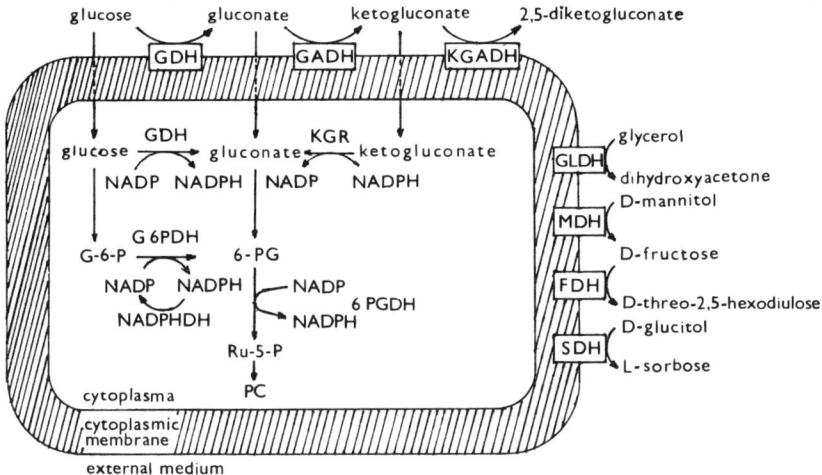

FIG. 1. A scheme of the location of the membrane and cytosol oxidoreductases and subsequent extra- and intracellular metabolic pathways in bacteria dehydrogenating monosaccharides. Membrane dehydrogenases: GDH, D-glucose DH; GADH, D-gluconate DH, e.g., EC 1.1.99.3; KGADH, keto-D-gluconate DH, e.g., EC 1.1.99.4; GLDH, glycerol DH; MDH, D-mannitol DH, e.g., EC 1.1.2.2; FDH, D-fructose DH; SDH, D-glucitol DH. Cytosol (intracellular) oxidoreductases: GDH, D-glucose DH, e.g., EC 1.1.1.118 or EC 1.1.1.119, or aldose DH, e.g., EC 1.1.1.121; KGR, keto-D-gluconate reductases, e.g., 5-keto-D-gluconate-5-reductase, EC 1.1.1.69 (similar is the action of, e.g., 5-keto-D-fructose reductase, EC 1.1.1.124); NADPHDH, NADPH-dehydrogenase, e.g., EC 1.6.99.1.

A. Membrane Dehydrogenases

After the whole system of D-glucose DH and D-gluconate DH, including hydrogen transmitters, had been found in submicroscopic particles obtained from the cells of *Pseudomonas fluorescens* (Wood and Schwerdt, 1953; Eagon, 1958), De Ley (1960) suggested that these active particles originated from cytoplasmic membranes by their mechanical disruption or by ultrasonic disintegration. Further research has resulted in the finding that the dehydrogenase systems catalyzing dehydrogenation of monosaccharides and associated with metabolite accumulation in the medium, caused by the absence of regulatory mechanisms, are located in the surface membranes and act extracellularly. Their substrates, in contrast to the substrates of the cytosol enzymes, do not need to penetrate into the cell nor do the generated metabolites have to be moved out of the cell. Their activity is not dependent on the presence of NAD(P). They may also function in resting cells. DHs are strongly bound to the electron transport system (ETS), and this has been the main difficulty confronted in studies on this type of enzyme system. It was clear that the particulate fractions obtained by disintegration of membranes and high-acceleration centrifugation were artifacts (De Ley and Dochy, 1960a,b; Birdsell and Cota-Robles, 1970), but studies on them brought valuable results. Solubilization of the particles for the purpose of isolation and identification of individual constituents of the respiratory chain resulted in the loss of activity in early studies (Ramakrishnan and Campbell, 1955; King and Cheldelin, 1957). Later studies using deoxycholate and nonionogenic tensides were successful. Three protein constituents were obtained from a series of dehydrogenase systems (see Section II,C). Known systems of membrane dehydrogenases have been reviewed (Kulhánek, 1984).

B. Cytosol Dehydrogenases

A series of DHs of monosaccharides has been isolated from the cytosol of ketogenic bacterial cells. From the viewpoint of preparative ketogenic fermentations, these enzymes result in the loss of monosaccharides by dissimilation. Besides their location in the cell, they differ from the membrane DHs in that they are NAD(P)-dependent and sensitive to sulfydryl reagents. Reactions catalyzed by these enzymes and subsequent dissimilation reactions proceed in the cells under the control of regulatory mechanisms (e.g., coupled to the supplies of NAD(P) and ADP), that is, similarly to mammalian tissues (Eagon, 1963). The accumulation of dehydrogenation products is thus avoided. Joining with the dissimilation pathway, starting with phosphorylation, enables the cells to utilize saccharides as a source of carbon and energy.

Stereospecificity of the cytosol DHs for saccharides is extremely variable. For individual enzymes, however, it may be expressed in the form of a simple relationship.

Many papers deal with cytosol DHs of alditols (Edson, 1953; Arcus and Edson, 1957; Shaw, 1956; Cummins et al., 1957; Bygrave and Shaw, 1961; Sasajima and Isono, 1968). Kersters et al. (1965) disintegrated cells of *Gluconobacter oxydans* by ultrasound. After separation of the particulate DH with stereospecificity according to the Bertrand–Hudson rule, they obtained a fraction of "soluble" enzymes by centrifugation at 150,000 g. The fraction contained at least five NAD(P)-dependent DHs of polyols of different specificities. Cytosol DHs of aldoses were also described (Cline and Hü, 1965; Anderson and Dahms, 1975).

The reduced coenzyme of NADP-dependent DHs of D-gluconate or DHs of aldehydes from the cytosol of G. oxydans is reoxidized by NADPHDH (Adachi et al., 1979, 1980a,b). NADP and NADPHDH also function in the subsequent oxidation step of phosphorylative metabolism, i.e., the oxidation of D-glucose-6-phosphate to 6-phospho-D-gluconate (Adachi, 1979). The production of ketogluconates by the particulate enzymes is about 100 times higher than their production by the cytosol enzymes. Equilibrium conditions of the reactions catalyzed by the particulate DHs of monosaccharides are shifted in favor of the formation of ketoderivatives to such an extent that no reversibility has been observed so far (Shinagawa et al., 1976). By contrast, equilibria of the analogous reactions catalyzed by the cytosol enzymes are shifted in the opposite direction, i.e., in favor of ketoderivative reduction, to such an extent that these enzymes are often designated reductases. A large series of such enzymes has been described. By hydrogenation of the ketoderivatives, they allow gradual intracellular utilization of these compounds.

C. Electron Transport Systems

Since the work of Weibull (1953), it has been known that bacterial membranes contain ETS. Efforts to separate the ETS into individual components were not as extensive as the studies with mitochondria. Only bacterial cytochromes were studied in detail. In fact, cytochromes of sugar oxidizing bacteria were studied much more extensively than complete respiratory systems. A large series of papers dealing with gradual elucidation of bacterial ETS has been reviewed recently (Kulhánek, 1984). This chapter will consider only the most recent findings.

By the action of nonionogenic tenside, the following membrane enzymes were solubilized: alcohol DHs from *Gluconobacter suboxydans* (Adachi et al., 1978a) and *Acetobacter aceti* (Adachi et al., 1978b), aldehyde DHs from *G. suboxydans* (Adachi et al., 1980b), D-fructose DHs from *Gluconobacter industrius* (Ameyama et al., 1981a), D-gluconate DHs from *Serratia marcescens* (Shinagawa et al., 1978b) and *P. aeruginosa* (Matsushita et al., 1979a,b), 2-keto-D-gluconate DHs from *Gluconobacter melanogenus* (Shinagawa et al., 1981), and D-glucitol DHs from *G. suboxydans* (Shinagawa et al., 1982). After purification, they were electrophoretically separated into three protein fractions. Fractions with a molecular weight of ~65,000 were flavoproteins, those with a molecular weight of ~50,000 were cytochrome c, which were autooxidable in some cases, and those with a molecular weight of ~20,000 were proteins of unknown function. Some authors indicated that sufficiently purified DH complexes manifested strict substrate specificities (Ameyama et al., 1978). D-Glucose DHs from G. *suboxydans* purified to homogeneity exhibited the presence of a new prosthetic group, pyrrole quinoline quinone (Ameyama et al., 1981b). Alcohol DHs from the same bacterium were isolated as a crystalline DH–cytochrome c complex (Adachi et al., 1982).

Membrane D-gluconate DHs from P. *aeruginosa* essentially remained in their native form after purification. The enzymes also contained the three protein components referred to above. They did not contain ubiquinone, which may have been lost during solubilization and purification (Matsushita et al., 1979a). In studying their properties, ubiquinone was found to be required for the oxidation of D-gluconate. It constitutes a component of the ETS, located at a specific site near an unknown protein or cytochrome c_1. As an acceptor of electrons, these DHs can also utilize pyocyanin, a redox dye released to the cultivation medium by P. *aeruginosa* (Matsushita et al., 1979b). Following aerobic cultivation, the ETS of these bacteria contained a series of respiratory components in their membrane particles (soluble flavin, coenzyme Q_9, heme c, heme b, cytochrome o). Cytochrome o and about half of the b-type cytochromes are supposed to function as the terminal oxidase (Matsushita et al., 1982a,b). The authors have proposed the following composition of the ETS: primary DH → cytochromes b, c_1, c, o → oxygen (Matsushita et al., 1980b). However, the ETS of P. *aeruginosa* has not yet been clarified in detail (Matsushita et al., 1982a,b). Membrane D-glucose DH from *Pseudomonas* and the *fluorescens* type contained a peptide. In this case, ubiquinone was a substantial component of the D-glucose oxidation system (Matsushita et al., 1982c). Other authors proposed a pteridine derivative to be the cofactor of the particulate D-glucose DHs from P. *fluorescens* (Imanoga et al., 1979).

The simplest single-stage system of dehydrogenation of monosaccharides is represented by aldose DHs (aldose oxidases), which oxidize aldoses to aldonic acids and with the simultaneous release of hydrogen peroxide. The best known of these enzymes, fungal glucose oxidase (EC 1.1.3.4), contains flavin adenine dinucleotide as the coenzyme (Nakamura and Ogura, 1962; Swoboda and Massey, 1965). The reaction itself is dehydrogenation. Hydrogen peroxide has been shown to originate from molecular oxygen and hydrogen originating in the aldoses. Hydrogen peroxide is decomposed immediately by catalase. D-Glucose is transformed to 1,5-D-gluconolactone, which is subsequently hydrolyzed.

III. Special Nature of Microbial Dehydrogenations of Monosaccharides

From the viewpoint of cellular metabolism, microbial dehydrogenations of monosaccharides seem to be useless reactions. They lead to the conversion of large amounts of monosaccharides to their oxidized metabolites without capture of released energy. Perhaps of some value is the fact that the oxidized derivatives are less utilizable for other microorganisms (e.g., membrane D-gluconate DH produces ketogluconates as an extracellular store of carbon sources, while the NAD(P)-dependent reductases of ketogluconates supply the cells with D-gluconate as a source of carbon).

The special nature of oxidizing bacteria, of which acetic bacteria have been known for ages, has always attracted attention. Their main representative, G. oxydans, originally called *Acetobacter suboxydans*, has been termed a "metabolic cripple" for its inability to catabolize the synthesized and accumulated oxidized metabolites (Cheldelin, 1961). According to Ameyama's group, who have isolated a series of enzymes from oxidizing bacteria, questions remain as to how and why the bacterial organisms accumulates acetic acid (*Acetobacter* genus), D-gluconic acid, or keto-D-gluconic acids (*Gluconobacter* genus) in high concentrations. Furthermore, why are acetic bacteria strictly aerobic? Why don't they grow quickly and produce plentiful populations of cells, even when cultivated in media rich in nutrients (Adachi et al., 1978a)?

Nevertheless, the present knowledge summarized in the preceding discussion helps to indicate some of the facts enabling these curious reactions to proceed. The existence of dehydrogenation of monosaccharides by oxidizing bacteria, although useless, rapid, and very prolific, is made possible by the location of the DH systems in the surface membranes of the cells and by the simplicity and operability of ETS under conditions of nonexistent respiratory regulation.

A. Location of the Enzyme and Respiratory Systems

All of the bacterial dehydrogenations of monosaccharides, accompanied by a rapid accumulation of large amounts of oxidized metabolites in the fermentation medium, are performed by the membrane systems. Within the cells, they are situated, or anchored, in the cytoplasmic membrane and act extracellularly in the medium. Thus, the substrate does not need to penetrate into the cell nor be moved out of the cell.

B. Simplicity of the ETS

In summarizing literature on the ETS of mitochondria and bacteria performing direct oxidations of sugars, a schematic illustration can be given as in Fig. 2. It is clear that direct oxidations of sugars are made possible by a simple, i.e., two-stage (flavoprotein–cytochromes) or single-stage (cytochromes), transfer of hydrogen or electrons, mediated by atypical cytochromes of the groups c and b, respectively, which can

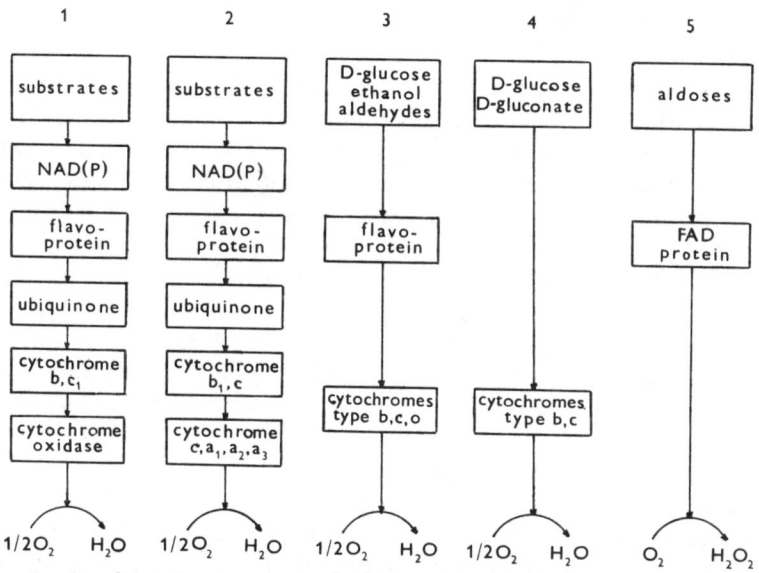

FIG. 2. A schematic presentation of the electron transport systems (respiratory chains) in mitochondria, common bacteria, and microorganisms oxidizing monosaccharides (Dolin, 1961). (1) Mammalian mitochondria (cytochrome oxidase represents a complex of cytochromes a, a_3). (2) Escherichia coli, Micrococcus denitrificans, Mycobacterium phlei, and Bacillus subtilis (Harold, 1972; Dickerson and Timchovich, 1975; Smith, 1968). (3) Acetic bacteria (Adachi et al., 1978a,b), Pseudomonas aeruginosa (Matsushita et al., 1980a). (4) Gluconobacter oxydans (Hauge, 1966; King, 1966; King and Cheldelin, 1958), Pseudomonas sp. (Wood and Schwerdt, 1953; Trutko et al., 1977). (5) Fungal glucose oxidase (Nakamura and Ogura, 1962; O'Malley and Weaver, 1972). Some of the cytochromes of groups 3 and 4 are autooxidable—they function as cytochrome oxidase (terminal oxidase).

be combined with molecular oxygen. The function of pyocyanin as an electron transmitter in cultures of P. *aeruginosa* is self-evident even in stationary cultures. In the course of saccharide dehydrogenation, the surface layer of the culture, suffused with oxygen, becomes green due to the oxidized form of pyocyanin. On mixing, the whole culture turns green, and after settling, the original equilibrium is reestablished. This depends on the rate and depth of oxygen diffusion.

C. Nonexistence of Any Respiratory Regulation

Aerobic phosphorylation has either not been proven (Widmer et al., 1956; Stouthamer, 1962) or it is documented at about one order of magnitude lower than in mitochondria (Klungsöyr et al., 1957). Our experience in dehydrogenations of monosaccharides using precultivated, washed cells in the presence of chloramphenicol as a preservative, i.e., in conditions of nongrowth, confirms the independence of dehydrogenation from growth (Uspenskaya and Loitsyanskaya, 1979). A low sensitivity to uncouplers of oxidative phosphorylation is also in agreement with this finding. These reactions are not dependent on ADP supplies either, and no respiratory regulation has been proven so far, the same being valid also for other bacteria (Forrest and Walker, 1971).

Very narrow specificity toward sugar substrates, defined by the Bertrand–Hudson rule for alditols and the gluconate rule for aldonic acids, is a disadvantage of these highly productive enzyme systems.

D. Special Nature of *Gluconobacter oxydans*

The special nature of G. *oxydans* is shown by requirements for a complex medium, a relatively slow growth rate, and a low concentration of cells in the mature culture after growth ceases. This holds true even in media containing adequate nutrients. Additionally, there is a higher level of transformation activity in the media with a lower content of organic sources of nitrogen. These are unusual features in common types of bacteria, but, according to our experience, quite typical for some laboratory-prepared auxotrophic mutants. Also unusual is the formation of cytosol membrane invaginations by continued synthesis of the membrane during the end phase of active differentiation of cells with enhanced dehydrogenase activity (Batzing and Claus, 1973; Claus et al., 1975; White and Claus, 1982). To my mind, all of the unusual features of acetic bacteria, and G. *oxydans* especially— including their disordered metabolism, resulting in noneffective production and accumulation of large quantities of metabolites in the expression of the pentose system as a single metabolic pathway (Eagon, 1963), and also resulting in demands for a complete culture medium, slow growth, low final concentration of the cells, etc.—can be best explained if these bacteria are considered naturally developing (spontaneous) auxotrophic mutants. Considerations on the possibiity of spontaneous transformations in acetic bacteria have been already published (De Ley and Dochy, 1960a,b).

IV. Relationships between Structure and Dehydrogenation of Monosaccharides

A. ALDITOLS

The relationship between configuration of alditols and their dehydrogenation to ketoses by the action of acetic bacteria *Acetobacter xylinum* was published by Bertrand (1898) at the end of the past century. This so-called Bertrand rule determined that dehydrogenations to ketoses proceeded with both alditols of the configuration D-*erythro*- at the next to last carbon and the adjacent asymmetric carbon, and the alditols of the configuration L-*erythro*-. In 1938, Hudson's group published (Hann et al., 1938) the relationship found between the configuration of alditols and their dehydrogenation to ketoses by the action of acetic bacterium *A. suboxydans*. The so-called Hudson rule determined that only the alditols of the D-*erythro*- configuration were dehydrogenated. The difference between the Bertrand and Hudson rules was explained by a narrower specificity of *G. oxydans*.

By analyzing papers by Bertrand and others, it was found that the dehydrogenation of alditols of the L-*erythro*- configuration, namely of L-arabinitol, by the action of *A. xylinum* to produce ketoses had never been proven at all. This was also confirmed by experiments with various strains of *A. xylinum* (Kulhánek and Tadra, 1973). It was proved that Bertrand had been wrong and that both bacteria were of the same specificity for alditols. We recommended continuing to designating this relationship the Bertrand–Hudson rule. Its history may be summarized as follows:

Acetobacter xylinum
The Bertrand rule (1898)

$$
\begin{array}{c}
\text{H}-\overset{|}{\text{C}}-\text{OH} \\
\text{H}-\overset{|}{\text{C}}-\text{OH} \\
\overset{|}{\text{CH}_2\text{OH}} \\
\text{D-}erythro\text{-}
\end{array}
\xrightarrow{-2\text{H}}
\begin{array}{c}
\text{H}-\overset{|}{\text{C}}-\text{OH} \\
\text{CO} \\
\overset{|}{\text{CH}_2\text{OH}} \\
\text{ketose}
\end{array}
$$

Acetobacter xylinum
The Bertrand rule (1898)

$$\begin{array}{c} | \\ HO-C-H \\ | \\ HO-C-H \\ | \\ CH_2OH \\ \text{L-}erythro\text{-} \end{array} \xrightarrow{-2H} \begin{array}{c} | \\ HO-C-H \\ | \\ CO \\ | \\ CH_2OH \\ \text{ketose} \end{array}$$

Gluconobacter oxydans
The Hudson rule (1938)

$$\begin{array}{c} | \\ H-C-OH \\ | \\ H-C-OH \\ | \\ CH_2OH \end{array} \xrightarrow{-2H} \begin{array}{c} | \\ H-C-OH \\ | \\ CO \\ | \\ CH_2OH \end{array}$$

Acetobacter xylinum
Gluconobacter oxydans
The Bertrand–Hudson rule (1973)

$$\begin{array}{c} | \\ H-C-OH \\ | \\ H-C-OH \\ | \\ CH_2OH \end{array} \xrightarrow{-2H} \begin{array}{c} | \\ H-C-OH \\ | \\ CO \\ | \\ CH_2OH \end{array}$$

Findings by Votoček et al. (1930) that dehydrogenations of rhodeite (now D-fucitol, 6-deoxy-D-galactitol), L-rhamnite (6-deoxy-L-mannitol), β-rhamnohexite (7-deoxy-L-*glycero*-L-*talo*-heptitol), and α-rhamnohexite (7-deoxy-L-*glycero*-L-*galacto*-heptitol) by Bertrand's sorbose bacterium (*A. xylinum*) did not proceed seemed to imply that the Bertrand rule could not be applied for the dehydrogenation of ω-deoxyalditols. Further work using *G. oxydans*, surveyed by Janke (1960), led to the conclusion that only the ω-deoxyalditols of the following configuration were dehydrogenated:

$$\begin{array}{c} | \\ H-C-OH \\ | \\ H-C-OH \\ | \\ HO-C-H \\ | \\ CH_3 \end{array} \quad \text{or} \quad \begin{array}{c} | \\ H-C-OH \\ | \\ H-C-OH \\ | \\ H-C-OH \\ | \\ CH_3 \end{array}$$

This specificity may be explained with the use of the Bertrand–Hudson rule in such a way that the terminal methyl group is taken to be equal to hydrogen, i.e., the terminal two-carbon group CH_3—CHOH is considered to be a broadened form of the group CH_2OH. Two D-*cis*-hydroxyls are then in the terminal group and the alditol is dehydrogenated in accordance with the Bertrand–Hudson rule, in this case at the third carbon from the terminus (Richtmayer et al., 1950). The terminal primary alcoholic group may also be substituted by other functions (—OCH_3, —SC_2H_5, CH_3COO—) (Hough et al., 1959).

By the action of G. oxydans cells, 4-deoxy-L-*glycero*-petulose was obtained from 2-deoxy-D-*erythro*-pentitol (Linek et al., 1979), 5-deoxy-5-fluoro-L-sorbose from 2-deoxy-2-fluoro-D-glucitol (Kulhánek et al., 1977), and 4-deoxy-4-fluoro-D-fructose from 3-deoxy-3-D-mannitol (Buděšínský et al., 1984). By these reactions, the validity of the Bertrand–Hudson rule was extended to other deoxyalditols and deoxyhalogen-alditols. Another dehydrogenation of D-fructose to yield D-*threo*-2,5-hexodiulose is also in accordance with this rule.

The Bertrand–Hudson rule, as well as the gluconate rule cited below, are valid exclusively for dehydrogenations performed by whole cells, i.e., in bacterial cultures or by the action of cultivated and isolated cells. This was the protocol of Bertrand's and Hudson's experiments, on the basis of which the rules were derived. The rules represent conclusions from the predominant action of the particulate DHs located in cytoplasmic membrane. Cytosol DHs with different specificities may eventually be manifested, but much more slowly, under conditions of prolonged fermentation, by the production of small amounts of side metabolites (Kulhánek and Tadra, 1973).

B. Aldoses

As a rule, microbial dehydrogenations of aldoses give rise to 1,4- or 1,5-lactones of aldonic acids initially, without furanose or pyranose rings being opened. Subsequently, the lactone is enzymically or nonenzymically hydrolyzed to the acid. Thus, the reaction is microbial dehydrogenation, but not a typical ketofermentation. Ketoaldonic acids are more often produced from aldoses, which are more readily available than aldonic acids. Here, the fact is that the microorganisms dehydrogenating aldonic acids are usually equipped with an enzyme system for oxidation of the corresponding aldose.

Data on the microbial dehydrogenations of aldoses to aldonic acids are covered in the survey papers cited at the beginning of this review. When summarizing the experimentally derived relationships between configuration and bacterial dehydrogenation of aldoses in table form (Kulhánek, 1980), it was found that only limited literature data were available, which were contradictory to each other in some cases (Kulhánek, 1984). The reviewed results suggest that no simple dependence on configuration can be found so far for the dehydrogenations of aldoses.

C. Aldonic Acids

After Bertrand's work, it was supposed that the relationship discovered by him for the dehydrogenation of alditols was also automatically valid for the dehydrogenation of acids. The same conclusion was reached even much later (Rainbow, 1961; Anderson and Magasnik, 1971). However, in 1935, Bernhauer and Görlich demonstrated the production of D-arabino-2-hexulosonic acid by the action of *Acetobacter gluconicum* on glucose, which had hydroxyls in the position *trans* at the second and third carbons. At that time, this finding was considered to be an exception to the validity of the Bertrand rule. Results of papers elucidating this issue are surveyed in this section.

In aldonic acids, bacterial dehydrogenations proceed similarly to alditols, only at the second and next to last carbons. However, there exist strains of bacteria performing dehydrogenations of aldonic acids exclusively (*P. aeruginosa*) or predominantly (*G. oxydans*) at one of the positions, i.e., either at C-2 or the next to last carbon. Distinguishing between these strains is based on their behavior towards D-gluconate.

For the dehydrogenation of gluconate at C-2, suitable strains dehydrogenate D-gluconate to D-*arabino*-2-hexulosonate with a high effectiveness, i.e., quickly, completely and in high concentrations if possible. This is the case for some selected strains of *P. aeruginosa*. For the dehydrogenation of aldonate at the next to last carbon, suitable strains of *G. oxydans* dehydrogenate D-gluconate to D-*xylo*-5-hexulosonate with a high effectiveness. These strains are also usually highly effective when used for sorbose and related fermentations.

A series of aldonic acids was tested to derive the relationship between the configuration of aldonic acid and its bacterial dehydrogenation. This was later designated the gluconate rule (Kulhánek, 1953b, 1980; Kulhánek and Tadra, 1975).

At C-2, the dehydrogenation proceeds only with L-arabinonate, D-xylonate, D-galactonate, D-gluconate, L-idonate, and D-*glycero*-D-*galacto*-heptonate, which are of the L-*threo*- configuration at C-2 and C-3. The dehydrogenation of all other tested gluconates did not occur, i.e., of D-arabinonate, D-*glycero*-D-*ido*-heptonate, having a D-*threo*-

configuration at C-2 and C-3; of D-ribonate, D-gulonate, L-mannonate, D-*glycero*-L-*manno*-heptonate, and D-*glycero*-D-*gulo*-heptonate, having a D-*erythro*- configuration at C-2 and C-3; and finally of D-mannonate, having a L-*erythro*- configuration at C-2 and C-3.

At the next to last carbon, the dehydrogenation occurred only for D-arabinonate, D-ribonate, D-mannonate, D-gluconate, D-*glycero*-D-*galacto*-heptonate, D-*glycero*-D-*gulo*-heptonate, and D-*glycero*-D-*ido*-heptonate, which have a D-*erythro*- configuration at the next to last and adjacent asymmetric carbons. The dehydrogenation did not occur with L-arabonate or L-mannonate, where the noted positions are of the L-*erythro*- configuration; with D-xylonate, D-galactonate, D-gulonate, or D-*glycero*-L-*mano*-heptonate, which are of the D-*threo*- configuration; nor with L-idonate, with an L-*threo*- configuration.

All of these results lead to the following conclusions: (1) the L-*threo*-configuration is necessary for dehydrogenation at C-2, and (2) the D-*erythro*- configuration is essential for the dehydrogenation at the next to last carbon:

$$
\begin{array}{c}
COO^- \\
| \\
H-C-OH \\
| \\
HO-C-H \\
| \\
H-C-OH \\
| \\
H-C-OH \\
| \\
CH_2OH
\end{array}
\xrightarrow{-2H}
\begin{array}{c}
COO^- \\
| \\
CO \\
| \\
HO-C-H \\
| \\
H-C-OH \\
| \\
CO \\
| \\
CH_2OH
\end{array}
$$

(with a $-2H$ step shown between the middle portion as well)

This relationship may also be stated by saying that an aldonate is dehydrogenated by bacteria only at a position similar to the site of the dehydrogenation of D-gluconate and only if the aldonate has the same configuration as D-gluconate. This relationship is designated the gluconate rule.

As examples of the application of the gluconate rule, dehydrogenations of some rare aldoses may be presented. Through the action of both *P. aeruginosa* and *G. oxydans*, D-xylose, L-mannose, and D-talose yielded only nonreducing sugar metabolites of acidic nature. This is in accordance with the fact that none of these aldoses has the configuration required by the gluconate rule for the dehydrogenation to ketogluconic acids. Therefore, aldonic acids must be the metabolites. Another example is represented by the dehydrogenation of D-fucose that yields a reducing sugar of acidic nature, the 2-keto-derivative, according to the gluconate rule. To be able to predict the structure of a metabolite helps to elucidate and confirm its structure.

The production of D-*threo*-2,5-hexodiulosonate from D-gluconate or D-glucose, D-*arabino*-2-hexulosonate being an intermediate, is also in accordance with the gluconate rule.

D. OTHER STRUCTURE–DEHYDROGENATION RELATIONSHIPS

1. D-*Glucarate*

The action of a strain of *P. aeruginosa* that transformed D-gluconate to D-*arabino*-2-hexulosonate converted D-glucarate to D-*arabino*-hexulosarate. This was the first ketoaldarate reported (Kulhánek et al., 1983). Strains of *G. oxydans* dehydrogenating D-gluconate to D-*xylo*-hexulosonate produced only small quantities of the same ketoderivative. It was suggested that in agreement with the Bertrand–Hudson rules, enzyme inhibition was caused by the presence of a carboxyl group at C-6, and a lower activity of the bacteria was obtained as compared to the simultaneously occurring dehydrogenation of D-gluconate at C-2.

2. D-*Glucuronate*

In the presence of both groups of oxidizing bacteria, D-glucuronate remained unchanged in solution. In this case, the carboxyl at C-6 functioned to sterically hinder the dehydrogenation at C-1, C-2, and C-5, which occurred with D-glucose.

V. Microorganisms and Fermentation Technique

Dehydrogenations of nonphosphorylated monosaccharides have been known for a long time to occur in acetic bacteria (Brown, 1886, 1887). Later, they were also detected in bacteria of the genus *Pseudomonas* (Pervozvanskii, 1939; Stubbs et al., 1940) and, occasionally, in *Serratia* (Misenheimer et al., 1965), *Enterobacter* (Ishizaki et al., 1973), and *Klebsiella* (Shinagawa et al., 1978a). Characteristic of these "oxidizing bacteria" is the presence of D-gluconate DH, which does not occur in other aerobic bacteria (Shinagawa et al., 1978a).

The capacity of fungi, especially of black aspergilli, to rapidly oxidize D-glucose to D-gluconate has been known for years (Molliard, 1924).

A series of systems and names has been proposed for the classification of acetic bacteria (De Ley, 1960; Rainbow, 1966; Asai, 1968). However, it has been found that there exist no features that would identify the numerous individual species reliably and definitively. Distinct production of a brown dye passing into the soils, so characteristic of the species G. *melanogenus*, can be lost spontaneously with a shift in production of D-*threo*-2,5-hexodiulosonate to D-*xylo*-5-hexolusonate (De Ley and Stouthamer, 1959). The recent classification of acetic bacteria includes two genera, *Acetobacter* and *Gluconobacter*. The *Gluconobacter* genus includes a single species, G. *oxydans*, with other species originally separated now classified as subspecies. Both genera are included in the section of Gram-negative aerobic rods and cocci. In this section, the genus *Pseudomonas* is classified in the family Pseudomonadaceae, the genera *Acetobacter* and *Gluconobacter* in the family Acetobacteraceae (Krieg and Holt, 1984). The metabolism of sugars is different in the two genera. Production of acetic acid from ethanol is high in bacteria of the genus *Acetobacter* and low or negligible in the genus *Gluconobacter*. Oxidation of D-glucose and ketogenic activity is low or absent in the genus *Acetobacter* and high in the genus *Gluconobacter* (Adachi et al., 1978a).

In the genus *Pseudomonas*, the classification of individual species is also rather problematic due to their variability. In early work, the name *P. fluorescens* was used and later this was replaced by *P. aeruginosa*. Other examples of reclassification of species used for preparative purposes can be found in the literature (Pfeifer et al., 1958). In our experiments, practically all typical strains of *P. aeruginosa*, i.e., the strains releasing pyocyanin into the soil, have rapidly converted D-gluconate in the first passage, giving high yields of D-*arabino*-2-hexulosonate.

A. Dehydrogenations of Individual Groups of Monosaccharides

For the dehydrogenation of aldoses to aldonic acids, *A. niger* is suitable, as it is devoid of an enzyme system for further dehydrogenation of the formed aldonic acid to ketoaldonic acid. *Aspergillus niger* has the narrow specificity of glucose oxidase, and so the use of *A. niger* is limited to preparations of D-gluconic acid and 2-deoxy-D-*arabino*-hexonic acid. For the same reasons, *P. aeruginosa* is suitable for the dehydrogenation of 2-deoxyaldoses to 2-deoxyaldonic acids. This bacterium dehydrogenates aldonic acids of appropriate configuration to 2-ketoaldonic acids only. This is not the case with 2-deoxyaldoses that can yield only aldonic acids (Kulhánek et al., 1975, 1979). Similarly, this bacterium might be used to dehydrogenate aldoses with incorrect configuration at C-2, according to the gluconate rule. To prepare 2-ketoaldonic acids, P. aeruginosa is most suitable. To prepare (ω-1)-ketoaldonic acids, strains of G. oxydans are used that can usually carry out rapid dehydrogenations of alditols as well. Steric requirements for the dehydrogenation of aldonic acids are dictated by the gluconate rule and for the dehydrogenation of alditols by the Bertrand–Hudson rule. To prepare calcium D-threo-2,5-hexodiulosonate and D-threo-2,5-hexodiulose, selected strains of G. oxydans subsp. melanogenus are used (Kulhánek, 1966).

B. Fermentation Technique

The following cases may occur during fermentation dehydrogenations of monosaccharides: (1) Monosaccharide makes a suitable source of carbon for the growth of the microorganism and is dehydrogenated; (2) monosaccharide is a suitable source of carbon but dehydrogenation does not proceed to any degree because of steric hindrance; (3) monosaccharide is not a suitable source of carbon for the growth of the microorganism but it is dehydrogenated by the grown cells; (4) monosaccharide is not a suitable source of carbon and is not dehydrogenated by cells grown on another suitable saccharide; (5) monosaccharide is not a suitable source of carbon and its presence is inhibitory to the growth of cells on another assimilatable, "physiological" saccharide present simultaneously. Cases 1 and 2 are the simplest, and studies of these types have been carried out since the "classical" times (Brown, 1886, 1887; Bertrand, 1898). Work that can be included in groups 3 and 4 above are more recent. The problem of the growth of cells for the dehydrogenation of a monosaccharide that is not a suitable source of carbon has been solved by Bernhauer by using supplementation of the fermentation medium with a small amount of an assimilatable monosaccharide that makes growth possible. Such a monosaccharide has been designated the growth inducer (Bernhauer, 1950). Another possibility for types 3 and 4 is the utilization of washed cells grown on another suitable medium. Aseptic treatment of such procedures can be rather difficult and therefore 1% toluene (Harold, 1970) or, even better, 0.002–0.01% D-chloramphenicol (Brook, 1961) is added as a short-term preservative. In case 5, these latter approaches represent possibilities for initiating a dehydrogenation study.

C. Enzyme Inhibitors in the Fermentations

In the 1950s, King and Cheldelin (1952a,b, 1953) and Fewster (1957, 1958a,b) first used enzyme inhibitors when studying direct oxidations of sugars. Using manometric experiments, they found that during the oxidation of D-glucitol by resting cells of G. *oxydans* in the presence of 2,4-dinitrophenol at a concentration of 1×10^{-4} mol/liter, the oxygen consumption stopped at 0.5 mol/mol D-glucitol, which suggested dehydrogenation to L-sorbose. In a control experiment, the oxygen consumption was 4 mol/mol D-glucitol. Growth of the cells on the medium with D-glucitol was not influenced by this inhibitor up to a concentration of 4×10^{-4} mol/liter. The 5% D-glucitol medium supplemented with dinitrophenol contained higher amounts of reducing sugars than the medium devoid of dinitrophenol (King and Cheldelin, 1952a,b). Later, the authors reported that a concentration of 1.6×10^{-4} mol/liter dinitrophenol inhibited the oxidation of dihydroxyacetone (the oxidation of glycerol was stopped at the stage of dihydroxyacetone) and the subsequent oxidation of L-sorbose by the resting cells, however, this concentration did not influence the growth on glycerol and D-glucitol (King and Cheldelin, 1953). In the presence of 1×10^{-3} mol/liter dinitrophenol or 2×10^{-4} mol/liter phenylmercuric acetate, the yield of ketogluconates from D-glucose or D-gluconate was higher as well (Fewster, 1958a,b). Dried P. *aeruginosa* cells oxidized D-glucose with the consumption of 1 mol oxygen per mol D-glucose and D-gluconate with the consumption of 0.5 mol oxygen per mol D-gluconate to form 2-keto-D-gluconate, which was not oxidized further. This oxidation was not inhibited either by sodium fluoride at the concentration of 0.056 mol/liter or iodoacetate at the concentration of 0.005 mol/liter (Stokes and Campbell, 1951).

It has been proven for preparative fermentations that an optimal inhibitor concentration in the medium can be found, i.e., the concentration at which the fermentation is not being decelerated and the product yield is increased. The optimal addition of sodium fluoride must be determined empirically for various raw materials and tap water, due to the formation of insoluble calcium fluoride.

The influence of addition of an inhibitor may be illustrated by the well-known fermentative dehydrogenation of D-arabinitol to D-xylulose, which was purposely allowed to continue after reaching the maximal content. A total volume of 200 ml of the medium containing 6 g of D-arabinitol, filtered autolysate of 2.5 g baker's yeast, and 100 ml of tap water was fermented with G. *oxydans* CCM 2370 in the presence of fluoride. Substantial consequences of the inhibitor on the fermenta-

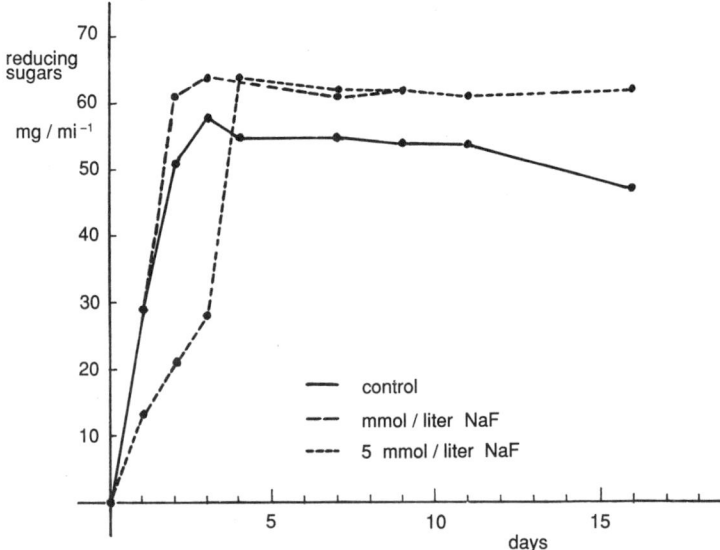

FIG. 3. Influence of fluorides on the fermentative dehydrogenation of D-arabinitol by *Gluconobacter oxydans*. The starting concentration was 6 g of arabinitol in 100 ml of soil. Enrichment of the soil in the course of sterilization and fermentation in a shaking apparatus was not corrected. Checking by chromatographic analysis confirmed the formation of a single reducing metabolite, D-*threo*-pentulose (D-xylulose).

tion are shown in Fig. 3. (1) In all experiments, the maximal content of D-xylulose was reached after 3–4 days of fermentation; in a control experiment, it was 9–10% lower than in the experiments with inhibitors. (2) Another 14 days of fermentation resulted in a 3% decrease of the D-xylulose content in the experiments with inhibitors and a 19% decrease in the control experiment as compared with the maximum reached in the same experiment. After 16 days of fermentation, the total D-xylulose content in the control was about 75% of the content in the medium with inhibitors. (3) Fluoride added at a concentration of 5×10^{-3} mol/liter slowed the fermentation by approximately 1 day, but the maximal xylulose content was practically the same as in the experiment with a lower fluoride content. (4) In the control experiment, the dehydrogenation did not proceed quantitatively. By chromatographic analysis, traces of the starting D-arabinitol were detected that were not found in the experiments with inhibitors, analyzed simultaneously (Kulhánek, 1980).

In the case of sorbose fermentation using G. oxydans strains, the addition to common media of sodium fluoride in concentrations of 1×10^{-3} to 5×10^{-3} mol/liter increased the yield of isolated sorbose by up to 10%, the addition of 2,4-dinitrophenol in concentrations of 1.6×10^{-4} to 3.2×10^{-4} mol/liter by up to 9%, diethyldithiocarbamate in a concentration of 1×10^{-4} mol/liter by up to 9% and monoiodoacetic acid in a concentration of 2.5×10^{-3} mol/liter by up to 3%. One effect caused by inhibitor addition and detected by analytical monitoring of the experiment was a deceleration of D-threo-2,5-hexodiulose formation, i.e., deceleration of the subsequent dehydrogenation ("overoxidation") of L-sorbose (Kulhánek, 1966, 1983). In addition, our later work demonstrated that in strains that did not show the formation of D-threo-2,5-hexodiulose using current procedures for sorbose production, the addition of sodium fluoride resulted in an enhanced yield of isolated sorbose of about 2% against the control.

For the production of D-fructose and calcium D-gluconate, mother liquors remaining after the crystallization of D-glucose from invert sugar were processed by fermentative dehydrogenation of the residual D-glucose to D-gluconate, while the remaining D-fructose was not substantially transformed. To decrease any dissimilation of D-fructose, sodium azide addition has proved effective in concentrations of approximately 1×10^{-4} mol/liter (Kulhánek and Tadra, 1983).

Clearly, enzyme inhibitor additions to microbial dehydrogenations of sugars on the preparative scale afforded a favorable effect on ketoderivative yields in many cases. This yield increase, as compared with the control, seems to be at the expense of the part of the starting substance. The cytosol enzymes are partially inhibited and a greater amount of the starting substance is available for dehydrogenation by the action of the membrane enzymes. The mechanism of such activity is evidently dictated by the nature of individual enzyme inhibitors. For example, with fluorides it may be the inhibition of kinases that catalyzes the first phosphorylative reactions of the total degradation by removal of magnesium ions from the medium (Hackett, 1960).

VI. Preparative and Industrial Applications

A. Dehydrogenations of Alditols

Early literature has been surveyed in previous review articles (Cheldelin et al., 1962; Janke, 1960; Asai, 1968; Spencer and Gorin, 1968).

1. Glycerol

Since the 1960s the use of dihydroxyacetone in cosmetics for skin tanning without sunburn has stimulated its production. Using a selected strain of G. oxydans, a medium containing 11% glycerol is fermented. The mixture is shaken, and after 72 hours, dihydroxyacetone was obtained at a concentration of 90 mg/ml. The optimal pH is 5.5 and addition of corn steep liquor is favorable (Green, 1961; Flickinger and Perlman, 1977). By gradual additions of glycerol during the growth stage, the final concentration can be enhanced up to 175 g/liter (Friedland et al., 1961).

More recently, G. oxydans ATCC 621 (from the American Type Culture Collection) has been used as an active inoculum in the logarithmic phase of growth under extremely aerobic conditions for the total fermentation of a medium containing 20% glycerol, 3% corn steep liquor, 0.3% ammonium fumarate, and 1% calcium carbonate during a period of 52–56 hours (Yamada et al., 1979a,b). According to another paper, media containing 15% glycerol have been fermented by a selected strain quantitatively (98–100%) in 48 hours. The cells can be separated and reused (Mironova et al., 1981). A three-reactor system for the continuous fermentation of glycerol has been described as well (Izuo et al., 1980). Additionally, resting cells can be used (Pomortseva et al., 1974), as can cells immobilized with polyacrylamide (Nabe et al., 1979; Makhotkina et al., 1981) or alginate (Adlercreutz et al., 1985) gels. Satisfactory isolation of crystalline dihydroxyacetone proceeds only in the absence of any residual glycerol. Deionization of the filtered medium is carried out on a basic anion exchanger. The final product is crystallized from alcohol (Ermolayev et al., 1979).

2. Erythritol

The dehydrogenation of meso-erythritol to L-erythrulose by the action of the "sorbose bacterium" has been known since Bertrand's classic work. Subsequently, this reaction was performed in 4.5% solution by the action of G. oxydans are a course of 7 days (Fulmer and Underkofler, 1947). Altro-heptulose and L-threitol have been detected as side products (Hu et al., 1965).

3. D-Arabinitol

The dehydrogenation of D-arabinitol to D-xylulose (D-threo-pentulose) by the action of the sorbose bacterium was described by Bertrand. He assumed a similar reaction, L-arabinitol → L-xylulose, to proceed as well. The issue of the dehydrogenation of alditols with the L-erythro- configuration was clarified later (Kulhánek and Tadra, 1973). An ancient recipe for the preparative utilization of Acetobacter xylinum claims a yield of 42% D-arabinitol (Hough and Theobald, 1962). The reaction using G. oxydans has been described by Hann et al. (1938). Of the later preparative procedures (Moses and Ferrier, 1962; Onishi and Suzuki, 1969), ours, mentioned in the preceding section, is the most productive. Product purification on a cellulose column is advantageous (Stankovič et al., 1982a). The reaction has also been studied with the use of immobilized cells (Ohmoto et al., 1983).

4. Ribitol

Recently, preparative procedures have been described utilizing G. oxydans (Moses and Ferrier, 1962; Stankovič et al., 1982b; Simpson 1966).

5. Allitol

The dehydrogenation of allitol by the action of A. xylinum yielded L-psikose (L-allulose) (Steiger and Reichstein, 1935). Later, G. oxydans was used for this reaction (Carr et al., 1968).

6. D-Glucitol

L-Sorbose generated by the bacterial dehydrogenation of D-glucitol was first isolated by Pelouze (1852) from spontaneously fermented rowan squash that had been left standing for a period longer than 1 year. However, in the following decades, attempts to reproduce this process were rare (for details, see Bernhauer, 1938, 1950). It was Bertrand (1896a,b) who prepared an enrichment culture designated as the "sorbose bacterium" and used it for dehydrogenations of some other alditols. He suggested that the sorbose bacterium might have been identical to the acetic bacterium *Bacterium xylinum* (now *Acetobacter pasteurianus*) described by Brown (1886, 1887). Later, this identity was proven (Emmerling, 1899). The constitution and structure of L-sorbose were also determined (Bertrand, 1904). More recently, other acetic bacteria have been used for the preparation of L-sorbose—*Acetobacter melanogenus, Bacterium (Acetobacter) suboxydans,* and *Bacterium gluconicum*, which are classified as the species G. oxydans at present. Stationary fermentations were replaced by submerged fermentations in aerated Kluyver flasks (Bernhauer, 1938), later by rotating metal drums (Wells et al., 1937), and recently, in the era of antibiotics, in modern, aerated, and stirred fermentors (Sherwood, 1947). Side metabolites of the sorbose fermentation have been identified—D-fructose and D-threo-2,5-hexodiulose (Kulhánek and Ševčíková, 1965), and conceivably also traces of 2-hexulosonic acid (Wakisaka et al., 1965a,b). L-Sorbose was produced in growing quantities as an intermediate during the synthesis of vitamin C according to Reichstein and Grüssner, with G. oxydans being used exclusively. The sorbose fermentation has been a subject of many studies, which have been recently reviewed by Kulhánek (1970) and Crawford and Crawford (1980). The most detailed description of the industrial process has been given by Shnaydman (1973).

Zolotarev et al. (1970) have found that iron content at 0.004–0.1/liter in the medium has practically no influence on the fermentation. But at Ni concentrations over 0.06 g/liter in the medium, the growth was markedly supressed. For a sufficient reduction of the Ni content in the technical-grade glucitol filtration is suitable. An addition to the medium of 1 g/liter of ammonium nitrate has also proved favorable.

A great deal of attention has been paid to oxygen transfer and its level in the medium during experiments in a shaking apparatus (Sukharevich and Razumovskaya, 1968). Currently used cotton stoppers may limit oxygen transfer. Caps made of a cotton layer placed between gauze layers seem more suitable (Gaden, 1962; Schultz, 1964; Yamada et al., 1978). A fermentation has been proposed proceeding in a circulating, closed atmosphere of oxygen, with carbon dioxide being captured in an absorber (Glushkov et al., 1978). Oxygen aeration allows 98% utilization of oxygen as compared with its 25% utilization when air flow is used. Thus, a very low oxygen flow can satisfy the culture demands (Lessing and Watson, 1982). However, Oosterhuis et al. (1985) have pointed out an adverse effect on the growth and product formation of oxygen at concentrations above 60%.

L-Sorbose can also be obtained from mixtures of glucitol containing up to 5% D-mannitol, in which case the yield of isolated L-sorbose is unaffected by the D-fructose present (Pomortseva and Krasilnikova, 1978). The presence of certain enzyme inhibitors at concentrations that do not slow the fermentation has improved yields (Kulhánek and Tadra, 1979; Kulhánek, 1983). The most significant enhancement of yield has occurred in strains tending to "over-oxidize," i.e., D-threo-2,5-hexodiulose is formed. The effect of vitamins on the growth of G. oxydans and the dehydrogenation of D-glucitol has been studied extensively (Solov'eva and Pomortseva, 1980). Resting G. oxydans cells may be recycled for the sorbose fermentation (Pomortseva et al., 1983), as can cells immobilized on polyacrylamide gel (Pomortseva and

Krasilnikova, 1983). Conversion with cells of G. oxydans immobilized on alginate has been markedly enhanced by stirring under an overpressure of oxygen (Asahi Chem. Ind., 1984).

A number of papers dealing with continuous sorbose fermentations were cited by Shnaydman in his monograph (1973). A low-capacity plant has been put into operation. The processed solution that fermented 75–90% was transferred from the top of a plate tower to the first or second finishing fermentor in turn. Here, the procedure was finished discontinuously, reaching 95% conversion (Zarak et al., 1975). Recently, conversions of 88–92% have been reached in a plate tower at higher dilution rates; however, additional oxidation of 94–96% was necessary in a special fermentor. At lower dilution rates, 94–96% conversion was already reached in the fermentation tower (Nikol'skaya et al., 1984). Dependence of the growth rate on the concentration of D-glucitol and L-sorbose has also been studied in the continuous fermentation (Pomortseva et al., 1980). Recirculation of cells using a special filter substantially increased the rate of sorbose formation (Bull and Young, 1981).

7. Heptitols and Octitols

Results obtained for dehydrogenations of higher alditols were surveyed by Cheldelin et al. (1962) and Asai (1968). From other studies, L-gluco-heptulose obtained by the action of G. oxydans on meso-glycero-gulo-heptitol with a yield of 88% should be mentioned (Maclay et al., 1942). In a similar way, D-glycero-D-altro-heptitol yielded L-guloheptulose of a syrupy consistency, which was in turn transformed by the action of acids to the crystalline 2,7-anhydro-β-L-gulo-heptulopyranose (Stewart et al., 1952). D-glycero-D-ido-Heptitol yielded D-idoheptulose in the form of 2,7-anhydride (Pratt et al., 1952), meso-glycero-allo-heptitol yielded L-allo-heptulose, and D-glycero-D-altroheptitol yielded D-talo-heptulose (Pratt and Richtmayer, 1955).

8. Deoxyalditols

Early literature concerning dehydrogenations of deoxyalditols, especially 1- or ω-deoxyalditols, was surveyed by the authors mentioned in the preceding paragraph. More recently, 6-deoxy-L-sorbose was prepared from 1-deoxy-D-glucitol, 6-deoxy-L-psikose from 6-deoxy-L-allitol (Kaufmann and Reichstein, 1967), 4-deoxy-L-glycero-pentulose from 2-deoxy-D-erythro-pentitol (Linek et al., 1979), and 5-deoxy-D-threo-hexulose from 2-deoxy-D-arabino-hexitol using immobilized G. oxydans cells (Tiwary et al., 1986).

9. Deoxyaminoalditols

Using stationary cultivation of G. oxydans, 2-acetamido-2-deoxy-D-glucitol was converted to 5-acetamido-5-deoxy-L-xylo-hexulose (Jones et al., 1961a), 1-deoxy-1-N-methylacetamido-D-glucitol to syrupy 6-deoxy-6-N-methylacetoamido-L-xylo-hexulose (Jones et al., 1961b), 2-acetamido-1,2-dideoxy-D-glucitol to 5-acetamido-5,6-dideoxy-L-xylo-hexulose (Jones et al., 1961b), 1-acetamido-1-deoxy-D-arabinitol to 5-acetamido-5-deoxy-D-threo-pentulose (Jones et al., 1962), and 1-acetamido-1-deoxy-D-ribitol to 5-acetamido-5-deoxy-L-erythro-pentulose (Jones et al., 1962). 6-amino-6-deoxyl-L-Sorbose was prepared from 1-amino-1-deoxy-D-glucitol (Ninast and Schedel, 1980a,b).

10. Deoxyhalogenalditols

By dehydrogenation of 2-deoxy-2-fluoro-D-glucitol, 5-deoxy-5-fluoro-L-sorbose was prepared (Kulhánek et al., 1977). 4-deoxy-4-fluoro-D-Fructose was prepared from 3-deoxy-3-fluoro-D-mannitol (Buděšínský et al., 1984).

11. Deoxy-S-alditols

Various terminally substituted pentitols and hexitols with L-xylo- or D-ribo- configuration at three carbon atoms adjacent to the terminal carbon were oxidized to ketoses by G. suboxydans. The terminal hydroxyl group of alditol was replaced by —H, —Me, —SEt, or —OAc groups. 6-deoxy-6-S-ethyl-L-Sorbose was prepared by dehydrogenation of 1-deoxy-1-S-ethyl-D-glucitol. 6-deoxy-6-S-ethyl-L-Galactol and 1-O-acetyl-DL-galactitol appeared to yield the corresponding 3-hexuloses (Hough et al., 1959). 5-deoxy-5-S-ethyl-D-threo-Pentulose was prepared from 1-deoxy-1-S-ethyl-D-arabinitol (Jones and Mitchell, 1959).

12. D-Mannitol

The fermentative dehydrogenation of D-mannitol to D-fructose was first described by Brown in 1886 and 1887. Early literature data about this procedure were summarized by Bernhauer (1938). In both surface (Fulmer et al., 1939) and submerged (Fulmer and Underkofler, 1947; Peterson et al., 1956) cultures of G. oxydans, it was possible to reach yields of up to 85% at 25 g/liter^{-1} mannitol in the medium. Abbott Laboratories (North Chicago, Illinois) used this procedure for the industrial production of fructose (Cushing and Davis, 1958). However, due to the high price of mannitol compared to that of invert sugar, this otherwise elegant method cannot be profitable (Kulhánek and Tadra, 1972).

Studies with labeled mannitol proved that the two ends of the molecule are not equivalent for dehydrogenation. Isbell and Karabinos (1952) have found that dehydrogenation of 1-C^{14}-D-mannitol depended on cultivation conditions. In a further study with 2-C^{14}-D-mannitol, the dehydrogenation at C-2 has been found to proceed somewhat more slowly than at C-5 (Frush and Tregoning, 1958). For preparative purposes, it is necessary to select suitable strains, as most strains dehydrogenate the primarily formed fructose further to D-*threo*-2,5-hexodiulose.

13. D-Glucitol → D-Fructose

This unusual reaction was observed in an ultraviolet mutant of *Pseudomonas coronafaciens* (Isono et al., 1963). *Bacillus megatherium*, *Bacillus fructosus* nov. sp., *Pseudomonas fluorescens*, *Pseudomonas boreopolis*, and *B. fructosus* gave yields of up to 85% in the media containing up to 15% D-glucitol during the course of 2–10 days, which has been considered to be sufficient for industrial purposes (Ueda, 1965; Ueda and Higashi, 1964; Higashi, 1966). In a screening study, strains of *B. megatherium* produced the largest quantities of fructose. One strain of the genus *Bacillus* akin to *B. megatherium* has been given the name *B. fructosus* (Ueda et al., 1967). *Kurthia zorphii* produced lower yields—3.7% fructose per liter, starting from a 10% solution of D-glucitol (Miashiro et al., 1976). Our study with *B. fructosus* ATCC 15451 has proven its narrow specificity. The bacterium does not attack other saccharides of similar structure, e.g., galactitol or lactitol. To my knowledge, its enzymological evaluation is still to be performed. Up to now, it is clear that this reaction is catalyzed by other than the Bertrand–Hudson enzyme. It seems to be located in cytoplasmic membrane as well.

B. Dehydrogenations of Ketoses

The production of D-*threo*-2,5-hexodiulose (5-keto-D-fructose) by further dehydrogenation of D-fructose may serve as an example of ketose dehydrogenation. The history of its discovery and demonstration of its structure are of interest (Kulhánek, 1980). For some time, it was supposed to be 6-aldo-D-fructose. The capacity to dehydrogenate D-fructose to its 5-keto derivative is widely distributed in acetic bacteria (Terada et al., 1961; Carr et al., 1963). It was identified as a side or subsequent product of D-mannitol dehydrogenation and a side product of D-glucitol dehydrogenation, accumulation during prolonged sorbose fermentation (Kulhánek and Ševčíková, 1962, 1965; Wakisaka et al., 1965a,b). The optimal pH for the accumulation of D-threo-2,5-hexodiulose is 4.0. During the exponential growth phase of at neutral pH, it was not accumulated due to fast reutilization by reduction to fructose (Mowshowitz et al., 1974). D-Glucitol or L-sorbose were the most suitable sources of carbon for the preparation of D-threo-2,5-hexodiulose by both growing and resting cells of the strain G. oxydans. D-Fructose yielded only negligible amounts of this product. After 10 hours only L-sorbose was present in the medium containing 4% D-glucitol. After 25–30 hours, the maximal content of D-threo-2,5-hexodiulose was reached, i.e., approximately 3% (Sato et al., 1969). In conditions of a favorable ratio C:N in the medium and a starting level of 10% D-glucitol, a yield about 90% has been reached after 72 hours (Asano et al., 1970).

C. Dehydrogenations of Aldoses

1. D-Glucose

D-Gluconic acid, particularly in the form of calcium or sodium salts or in the lactone form, is used in the fodder and foodstuff industry and in other industrial areas. On an industrial scale, it is produced in fermentors by A. niger by oxidation of up to 30% D-glucose solutions. Calcium gluconate is obtained by fermentation in the presence of calcium carbonate. Its crystallization during the fermentation, due to its limited solubility, is avoided by borate ion addition. During the production of alkaline gluconate, the produced acid is continuously neutralized by automatically added hydroxide solution (Peppler, 1967; Rehm, 1967; Underkofler and Hickey, 1954; Prescott and Dunn, 1959). Subsequently, this procedure has been elaborated on in more detail (Mahmond et al., 1977; Nyeste et al., 1980; Ghosh and Ghose, 1978). The use of molasses as a raw material was also tested (Grigorov et al., 1979), and the use of immobilized glucose oxidase was described in other papers (Richter and Heinecker, 1979; Coppens, 1980; Kirstein et al., 1980; Ishimori et al., 1981; Hartmeier, 1981). The preparation of labeled D-gluconic acid, using a similar procedure, was described (Zemek et al., 1981). In similar conditions, 1-thio-D-glucose was oxidized to the corresponding disulphide (Oda et al., 1974). A survey of technologies and literature was published by Wilt (1972), and a literature review on the utilization of various fungi was published by Mandal and Chatterjee (1985). Studies continued with the use of immobilized glucose oxidase (Tsukamoto et al., 1982; Chang et al., 1984). More favorable seems to be the use of immobilized cells (Iwata, 1983). Mycelia permeabilized with 2-propanol have proven better than the originally used glucose oxidase–catalase preparations (Doeppner and Hartmeier, 1984). The classic fermentation is also being improved. When the content of dissolved oxygen is kept at 10–40 ppm at 36°C, the 20% medium was fermented in the course of 12 hours; the admixture of 1–40 ppm iron prevented any lowering of the activity of the recycled microorganism (Daicel Chem. Ind., 1984). Addition of MnO_2 0.02 g/liter to the medium, favoring the degradation of hydrogen peroxide, accelerated the fermentation. In contrast, other studies indicate a fermentation time of 36 hours with the starting 20% concentration (Jaksa et al., 1983).

For the isolation of D-fructose from mixtures with D-glucose A. niger (Tsukamoto, 1947; Holstein and Holsing, 1962), G. oxydans (Kett, 1963), glucose oxidase complexed with catalase from A. niger (Holstein and Holsing, 1965), or a complex invertase–glucose oxidase–catalase was used (D'Souza and Nadkarni, 1980). It should be pointed out here that, in general, the possibility of repeatedly using mycelium isolated by flotation or other suitably treated whole cells should be considered since the utilization of enzyme preparations appears to be too expensive. In our work using A. niger, D-fructose and D-gluconate were obtained from the mixture resulting from the process of glucose crystallization from invert sugar, with fructose increased to 55–60% in the dry matter. Addition of sodium azide was favorable to decrease the dissimilation of fructose (Kulhánek and Tadra, 1981).

2. Other Aldoses

Data concerning direct oxidation of aldoses were surveyed by Bernhauer (1950), De Ley (1960), Janke (1960), and Müller (1960). More recently, Weimberg (1961) has found that the resting as well as growing cells of *Pseudomonas fragi* dehydrogenate D-arabinose, L-arabinose, D-xylose, and D-ribose to the 1,4-lactone of the corresponding aldonic acids, which are accumulated as such. The next step is delactonization, which has been, however, proven to be an enzymic reaction only in the case of D-arabino-1,4-lactone. In a study by De Ley (1961), a series of acetic bacteria species oxidized mannose, galactose, L-arabinose, xylose, and ribose to aldonic acids. In the case of xylose, mannose, and galactose, oxygen consumption evidenced their dehydrogenation to ketoaldonic acids. Ohsugi et al. (1970) used *Micrococcus* for the preparation of D-xylonic acid by oxidation of up to 15% solutions of D-xylose. This paper also included a survey of the literature concerning

D-xylonic acid production by microorganisms. A nonspecific *Pseudomonas* bacterium dehydrogenated D-fucose to D-fucono-1,5-lactone, which was in turn spontaneously hydrolyzed to D-fuconate (Dahms and Anderson, 1972). The bacterium *Enterobacter cloacae*, isolated from soil, converted D-xylose to D-xylonate in solutions of up to 20% xylose (Ishizaki et al., 1973).

3. *Deoxyaldoses*

The oxidation of 2-deoxy-D-glucose to 2-deoxy-D-gluconic acid was observed by Eagon (1971) in manometric experiments with *P. aeruginosa*. Later, this reaction was used to prepare 2-deoxy-D-gluconic acid (Kulhánek et al., 1975). In a similar way, 2-deoxy-D-*lyxo*-hexonic acid was prepared from 2-deoxy-D-*lyxo*-hexose (Kulhánek et al., 1979).

D. Dehydrogenations of Aldonates at C-2

1. L-*Arabonate*

According to our experience, the dehydrogenation of L-arabonate with growing *P. aeruginosa* IHE 5/40 cultures proceeded quickly. The product, a reducing saccharide of acidic nature, yielded by paper chromatography the color of a 2-ketoaldonic acid. Although L-*erythro*-2-pentulosonic acid has been prepared chemically (Gakhokidze, 1941), isolation of the ketoderivative was not possible due to its low stability. The low stability of 2-pentulosonic acid was noted by Price and Reichstein (1937).

2. D-*Xylonate*

The preparation of D-*threo*-2-pentulosonic acid with *Pseudomonas mildenbergii* is protected by Japanese patents (Yasuda, 1968, 1969).

3. D-*Galactonate*

Using *G. oxydans*, Ettel et al. (1952) transformed D-galactonate to D-*xylo*-2-hexulonate with a yield of 73%. The reaction also proceeded with *Pseudomonas fluorescens* (Yokazawa, 1952; Asai et al., 1952). For preparative purposes, *P. aeruginosa* has been used (Kulhánek, 1953a). Also, D-galactose can be used as the starting substance even in 5% solution (Mastropietro-Cancellieri and Tiecco, 1961).

4. D-*Gluconate*

D-*arabino*-2-Hexulosonic acid (2-keto-D-gluconic acid) was used as an intermediate, yielding isoascorbic acid by enolization and lactonization. Isoascorbic acid was used as an antioxidant or oxygen-removing compound in the food industry, since it is less expensive than ascorbic acid. D-*arabino*-2-Hexulosonic acid is more readily available than L-*xylo*-2-hexulosonic acid (2-keto-L-gulonic acid), similarly yielding L-ascorbic acid (Kulhánek, 1970).

Also, the development of large-scale vitamin C production, and a subsequent price decrease, resulted in wide use of L-ascorbic acid for these purposes. Early papers concerning the preparation of D-*arabino*-2-hexulosonic acid have been reviewed (Underkofler and Hickey, 1954; Prescott and Dunn, 1959; Janke, 1960). *Pseudomonas fluorescens* or *P. aeruginosa* were used for the dehydrogenation of D-glucose or D-gluconate. In more recent studies, a saccharified starch solution was used directly for the fermentation. D-*arabino*-2-Hexulosonic acid was usually isolated as the calcium salt, the free acid being known only in a syrupy form. Its isolation in crystalline form has been described only recently (Faubl and Kolb, 1978).

5. L-*Idonate*

L-*xylo*-2-Hexulosonic acid (2-keto-L-gulonic acid) is an intermediate in the production of L-ascorbic acid. Although for large-scale production the chemical synthesis of vitamin C by the Reichstein and Grüssner process is used exclusively, efforts to improve its biochemical preparation continue. Older studies were described in a preceding survey (Kulhánek, 1970).

The best available method is the long-known fermentation of a mixture of D-gluconate with L-idonate obtained by the catalytic hydrogenation of D-*xylo*-5-hexulosonate with bacteria of the genus *Pseudomonas*. The synthesized D-*arabino*-2-hexulosonate D-gluconate is metabolized quantitatively and crystalline L-*xylo*-2-hexulosonic acid was isolated from the solution (Kulhánek, 1953a). The work continued, utilizing *A. melanogenus*, *A. suboxydans* (Mochizuki et al., 1969), and *P. fluorescens* (Fujisawa et al., 1963; Kita et al., 1978).

Other methods of biochemical preparation of L-*xylo*-2-hexulosonic acid deviate in their chemistry from the theme of this publication and will be mentioned only briefly. The preparations from D-glucitol (Takeda Chem. Ind., 1975) or L-sorbose (Yan et al., 1981) using artificial mutants gave yields of about 40% from 5–10% solutions in the course of 150 hours. The preparation of L-*xylo*-2-hexulosonate by means of the partial and stereospecific hydrogenation of D-*threo*-2,5-hexodiulosonate are mentioned here in the appropriate section.

E. Dehydrogenations of Aldonates at C-(ω-1)

1. D-Gluconate

The production of D-*xylo*-5-hexulosonic acid (5-keto-D-gluconic acid) from D-glucose solution by acetic bacteria in the presence of calcium carbonate is the oldest known keto fermentation along with the D-mannitol dehydrogenation referred to above.

D-*xylo*-5-Hexulosonic acid was probably first obtained by Boutroux (1886) by the fermentation of glucose or calcium gluconate with micrococci, later classified as acetic bacteria. The author assumed that a compound analogous to D-glucuronic acid, but with reversed configuration (L-guluronic acid), was involved. Later, he confirmed its correct structure as proposed by Bertrand. It was also prepared by means of the sorbose bacterium (Bertrand, 1904), chemically and, later, by means of G. oxydans. Other literature data are mentioned in reviews (Underkofler and Hickey, 1954; Janke, 1960; Asai, 1968; Kulhánek, 1970). The yield was 90% of the calcium salt from 10% glucose, when the pH was kept at about 4.0, and the simultaneous degradation of D-*xylo*-hexulosonate was very low (Yamazaki, 1954). The yields of the crystallized product were about 90–95% of the fermented sugar. Sugar cane molasses was also used as a starting raw material. It is advantageous to use glucoamylase starch hydrolysate. This procedure circumvents the isolation of crystalline glucose. Particularly useful is the very low solubility of sythesized calcium D-*xylo*-5-hexulosonate in water, allowing for the spontaneous isolation from the fermented solution. Its crystallization during the fermentation is not slowed down by impurities present in the crude, filtered glucoamylase starch hydrolysate. Phosphate ions and protein degradation products serve as nutrients for the acetic acid bacteria (Kulhànek and Tadra, 1981).

D-*xylo*-5-Hexulosonic acid is produced almost exclusively by acetic bacteria. Its production by unidentified pseudomonads has been also mentioned, but the classification of these bacteria may be problematic due to the morphological similarity of some *Acetobacter* and *Pseudomonas* species (Stewart, 1959). Small amounts of this compound were also detected during the citrate fermentation with *A. niger* (Martin and Steel, 1955). D-*xylo*-5-Hexulosonic acid served as an intermediate in several newer "after-Reichstein" syntheses of vitamin C (Kulhánek, 1970).

In 1940, a patent concerning the preparation of D-tartaric acid by chemical oxidation of D-xylo-5-hexulosonic acid or by oxidation of its salts by oxygen in the presence of vanadium pentoxide or manganese dioxide, etc., was approved (Pasternak and Brown, 1940). Shortly after that, another patent (Kamlet, 1943) for the preparation of D-tartaric acid by the fermentation of 10% glucose solution with G. oxydans ATCC 621 in the presence of corn-steep liquor or yeast autolysate was published. The fermentation lasted 8–24 hours and the product was isolated in the form of potassium hydrogen tartrate. Addition of vanadium oxide to the fermentation medium should have accelerated the fermentation and inhibited the formation of oxalate. In spite of the fact that the patent did not inspire confidence, it was tested in our laboratory during the post-war shortage of the foodstuff organic acids using available Gluconobacter strains. No production of tartaric acid could be demonstrated. Similar results were also obtained in more recent studies (Krumphanzl et al., 1968). In spite of this, yields of 3–6.3 g of tartaric acid per liter of fermentation medium were described in studies with G. oxydans mutant (Kotera et al., 1972). However, it appears that this procedure cannot compete with the fermentation production of citric acid, and, as far as the isolation of tartaric acid is concerned, with natural viticulture sources (Kopřiva and Hlaváček, 1971).

2. *2-Deoxy-D-gluconate*

Fewster (1958a,b) inferred that oxygen consumption during the oxidation of 2-deoxy-D-glucose by G. oxydans cells suggested formation of a ketoderivative and accumulation of a "reducing acid." Preparation of 2-deoxy-D-threo-5-hexulosonic acid was realized later by Berman and Magasnik (1966).

3. *D-Mannonate*

The reducing sugar detected analytically during the growth of acetic bacteria in the solution of calcium D-mannonate (Bernhauer, 1938; Bernhauer and Knobloch, 1939–1940; Knobloch and Tietze, 1941) or D-mannose (Frateur et al., 1954) was assumed to be identical to D-arabino-2-hexulosonic acid (2-keto-D-gluconic, 2-keto-D-mannonic acid). According to the consumption of oxygen in manometric experiments, resting cells of G. suboxydans and Gluconobacter liquefaciens oxidized D-mannose slightly beyond the stage of aldonic acid (De Ley, 1961). Our study demonstrated that D-mannonic acid was mainly attacked by strains capable of dehydrogenation of aldonic acids at the penultimate carbon (Kulhánek, 1953a,b). Later, the resulting keto acid was proven to be D-*lyxo*-5-hexulosonic acid (Kulhánek and Tadra, 1977).

F. DEHYDROGENATIONS OF ALDONATES AT BOTH C-2 AND C-(ω-1)

D-*threo*-2,5-Hexodiulosonic acid (2,5-diketo-D-gluconic acid) is of increasing significance as a possible intermediate in the synthesis of vitamin C. By means of partial and stereospecific reduction, it is possible to directly synthesize L-*xylo*-2-hexulosonic acid (2-keto-L-gulonic acid) where the industrial synthesis requires four steps (Kulhánek, 1970). It follows from the literature data that D-*threo*-2,5-hexodiulosonic acid was probably obtained by Takahashi and Asai (1931, 1932) from glucose by fermentation with *Bacterium hoshigaki* var. *glucuronicum*. After the separation of D-*xylo*-5-hexulosonate by filtration and of gluconate by crystallization, they precipitated the filtrate with methanol and obtained a reducing calcium salt, which they considered to be L-glucuronate. A mixture of D-*arabino*-2-hexulosonate with D-*threo*-2,5-hexodiulosonate, which had not been prepared biochemically at that time, was probably involved. The presence of a diketoderivative was indicated by the described reduction of Fehling solution in the cold.

Bernhauer and Knobloch (1938) obtained a similar metabolite in a fermentation with *B. gluconicum*. The authors assumed that 6-aldehydegluconic acid (L-guluronic acid) was involved. Later, they found that the compound was produced particularly by melanogenic strains (Riedl-Tumová and Bernhauer, 1950). D-*threo*-2,5-hexodiulosonic acid was identified by Katznelson *et al.* (1953). It was found to be produced from glucose via gluconate and D-*arabino*-2-hexulosonate, by the action of the cytoplasmic membrane enzymes. This was later demonstrated with *G. liquefaciens* (Stouthamer, 1961). The compound is not stable, especially at pH higher than 4.5 and decomposes readily to brown products. In early work, its isolation was only possible by freeze-drying the calcium salt, but later by chromatography on an ion-exchange resin (Stroshane and Perlman, 1977) or by crystallization of the free acid (Oga *et al.*, 1972). In other studies, *A. melanogenus* was used (Datta and Katznelson, 1956; Stroshane and Perlman, 1977), followed by *Pseudomonas albosezamae* nov. sp. (Wakisaka, 1964), *G. liquefaciens* (Aida *et al.*, 1957; Stouthamer, 1961), *Acetobacter fragum* nov. sp. (Oga *et al.*, 1972), *Acetobacter cerinus* (Kita and Fenton, 1981; Kita and Hall, 1979, 1981), and recently various strains of the genus *Erwinia* (Sonoyama *et al.*, 1982b, 1986).

Significant effort was exerted to prepare L-*xylo*-2-hexulosonic acid from D-*threo*-2,5-hexodiulosonic acid. The best result from chemical reduction by hydridoborates is a mixture of both 2-hexulosonic acids (Andrews et al., 1979). The degradation of D-*arabino*-2-hexulosonate in this mixture (Kita et al., 1978) resembled the isolation of L-*xylo*-2-hexulosonate from the mixture of D-gluconate with L-idonate by the fermentation with P. *aeruginosa* (Kulhánek, 1953a). The biochemical hydrogenation was the subject of a series of patents, according to which the product was obtained in concentrations of tenths of 1%. For the fermentation of D-*threo*-2,5-hexodiulosonate solutions, bacteria of the genera *Corynebacterium*, *Brevibacterium*, or *Citrobacter* were used. Alternatively, a similar two-phase procedure may be applied starting with D-glucose (Kita and Hall, 1980).

The whole problem appeared to be solved by a procedure in which D-*threo*-2,5-hexodiulosonate was obtained from glucose ta a high concentration, with a yield of 94.5% by conversion with an *Erwinia* mutant. The number of viable cells in the fermentation solution was decreased by sodium dodecyl sulfate. The medium containing glucose as a hydrogen donor was added to a culture of a *Corynebacterium* mutant for 50 hours. In the 10 m^3 fermentors, 10.6% solutions of calcium L-*xylo*-2-hexulosonate were obtained with a yield of 84.6%, glucose serving as the starting sugar (Sonoyama et al., 1982a). Thus, these results seem to indicate that the simplest synthesis of vitamin C, D-glucose → D-*threo*-2,5-hexodiulosonic acid → L-*xylo*-2-hexulosonic acid → L-ascorbic acid, might be performed on an industrial scale in the near future.

More recently, a possibility was reported of a two-stage process for vitamin C production, utilizing genetically modified bacteria. A gene encoding the corresponding 2,5-diketo-D-gluconate reductase was isolated from the species *Corynebacterium*. By its incorporation into the bacterium *Erwinia herbicola*, which fermented D-glucose to D-*threo*-2,5-hexodiulosonate itself, a new bacterium was obtained, fermenting D-glucose to L-*xylo*-2-hexulosonate in a single step (Estell et al., 1983; Anderson et al., 1985; Anonymous, 1985).

REFERENCES

Adachi, O. (1979). *J. Agric. Chem. Soc. Jpn.* **53**, R77–R86.
Adachi, O., Tayama, K., Shinagawa, E., Matsushita, K., and Ameyama, M. (1978a). *Agric. Biol. Chem.* **42**, 2045–2056.
Adachi, O., Miyagawa, E, Shinagawa, E., Matsushita, K., and Ameyama, M. (1978b). *Agric. Biol. Chem.* **42**, 2331–2340.
Adachi, O., Matsushita, K., Shinagawa, E., and Ameyama, M. (1979). *J. Biochem. (Tokyo)* **86**, 699–709.

Adachi, O., Matsushita, K., Shinagawa, E., and Ameyama, M. (1980a). *Agric. Biol. Chem.* **44**, 155–164.
Adachi, O., Tayama, K., Shinagawa, E., Matsushita, K., and Ameyama, M. (1980b). *Agric. Biol. Chem.* **44**, 503–515.
Adachi, O., Shinagawa, E., Matsushita, K., and Ameyama, M. (1982). *Agric. Biol. Chem.* **46**, 2859–2863.
Adlercreutz, P., Holst, O., and Mattiasson, B. (1985). *Appl. Microbiol. Biotechnol.* **22**, 1–7.
Aida, K., Fujii, M., and Asai, T. (1957). *Bull. Agric. Chem. Soc. Jpn.* **21**, 30.
Ameyama, M., Tayama, K., Shinagawa, E., Matsushita, K., and Adachi, O. (1978). *Agric. Biol. Chem.* **42**, 2347–2354.
Ameyama, M., Shinagawa, E., Matsushita, K., and Adachi, O. (1981a). *J. Bacteriol.* **145**, 814–823.
Ameyama, M., Shinagawa, E., Matsushita, K., and Adachi, O. (1981b). *Agric. Biol. Chem.* **45**, 851–861.
Anderson, R. L., and Dahms, A. S. (1975). In "Methods in Enzymology" (W. A. Wood, ed.), Vol. 41, pp. 147–150. Academic Press, New York.
Anderson, S., Marks, C. B., Lazarus, R., Miller, J., Stafford, K., Seymour, J., Light, D., Rastetler, W., and Estell, D. (1985). *Science* **230**, 144–149.
Anderson, W. A., and Magasnik, B. (1971). *J. Biol. Chem.* **246**, 5653–5661.
Andrews, G., Bacoa, B., Crawford, T., and Breitenbach, R. (1979). *J. Chem. Soc., Chem. Commun.* **17**, 740–741.
Anonymous (1985). *Chem. Eng. News* **63**, 6.
Arcus, A. C., and Edson, N. L. (1956). *Biochem. J.* **64**, 385–394.
Asai, T. (1968). "Acetic Acid Bacteria." Univ. of Tokyo Press, Tokyo.
Asai, T., Aida, K., and Ueno, Y. (1952). *J. Agric. Chem. Soc. Jpn.* **25**, 625–630.
Asahi Chem. Ind. (1984). Japanese Patent 84/02691.
Asano, K., Imada, K., Oga, S., and Sato, K. (1970). Japanese Patent 70/24391.
Batzing, B. L., and Claus, G. W. (1973). *J. Bacteriol.* **113**, 1455.
Berman, T., and Magasnik, B. (1966). *J. Biol. Chem.* **241**, 807.
Bernhauer, K. (1938). *Ergeb. Enzymforsch.* **7**, 246–280.
Bernhauer, K. (1950). *Ergeb. Enzymforsch.* **11**, 151–496.
Bernhauer, K., and Görlich, B. (1935). *Biochem. Z.* **280**, 367–374.
Bernhauer, K., and Knobloch, H. (1938). *Naturwissenschaften* **26**, 819.
Bernhauer, K., and Knobloch, H. (1939–1940). *Biochem. Z.* **303**, 308.
Bertrand, G. (1896a). *Bull. Soc. Chim. Fr.* [3] **15**, 627.
Bertrand, G. (1896b). *C. R. Hebd. Seances Acad. Sci.* **122**, 900–903.
Bertrand, G. (1898). *Bull. Soc. Chim. Fr.* [3] **19**, 347–349.
Bertrand, G. (1904). *Ann. Chim. Phys.* **8**, 181–288.
Birdsell, D. C., and Cota-Robles, E. H. (1970). *Biochim. Biophys. Acta* **216**, 250–261.
Boutroux, L. (1886). *C. R. Hebd. Seances Acad. Sci.* **102**, 924–927.
Brock. T. D. (1961). *Bacteriol. Rev.* **25**, 32–48.
Brown, A. J. (1886). *J. Chem. Soc.* **49**, 172–187.
Brown, A. J. (1887). *J. Chem. Soc.* **51**, 638–642.
Buděšíský, M., Černý, M., Doležalová, J., Kulhánek, M., Pacák, J., and Tadra, M. (1984). *Collect. Czech. Chem. Commun.* **49**, 267–274.
Bull, D. N., and Young, M. D. (1981). *Biotechnol. Bioeng.* **23**, 373–389.
Bygrave, F. L., and Shaw, D. R. D. (1961). *Proc. Univ. Otago Med. Sch.* **39**, 15–16.
Carr, J. G., Coggins, R. A., and Whiting, G. C. (1963). *Chem. Ind. (London)* p. 1279.
Carr, J. G., Hough, L., and Coggins, R. A. (1968). *Phytochemistry* **7**, 1.
Chang, H. N., Joo, I. S., and Ghim, Y. S. (1984). *Biotechnol. Lett.* **6**, 487–492.

Cheldelin, V. H. (1961). "Metabolic Pathways in Microorganisms." Wiley, New York.
Cheldelin, V. H., Wang, C. H., and King, T. E. (1962). In "Comparative Biochemistry" (M. Florkin and H. S. Mason, eds.), Vol. 3, Part A, pp. 427–502. Academic Press, New York.
Claus, G. W., Batzing, B. L., Baker, C. A., and Goebel, E. M. (1975). J. Bacteriol. **123**, 1169–1183.
Cline, A. L., and Hü, L. S. A. (1965). J. Biol. Chem. **240**, 4488–4493, 4498–4503.
Coppens, G. (1980). European Patent Appl. 14,011.
Crawford, T. C., and Crawford, S. A. (1980). Adv. Carbohydr. Chem. Biochem. **37**, 79–155.
Cummins, J. T., Cheldelin, V. H., and King, T. E. (1957). J. Biol. Chem. **226**, 301–306.
Cushing, I. B., and Davis, R. V. (1958). J. Am. Pharm. Assoc., Sci. Ed. **47**, 765–768.
Dahms, A. S., and Anderson, R. L. (1972). J. Biol. Chem. **247**, 2222–2227.
Daicel Chem. Ind. (1984). Japanese Patent 84/55190 84/55191.
Datta, A. G., and Katznelson, H. (1956). Arch. Biochem. Biophys. **65**, 576.
De Ley, J. (1960). J. Appl. Bacteriol. **23**, 400–401.
De Ley, J. (1961). J. Gen Microbiol. **24**, 31.
De Ley, J., and Dochy, R. (1960a). Biochim. Biophys. Acta **40**, 277–279.
De Ley, J., and Dochy, R. (1960b). Biochim. Biophys. Acta **42**, 538–541.
De Ley, J., and Stouthamer, J. (1959). Biochim. Biophys. Acta **34**, 171–183.
Dickerson, R. E., and Timkovich, R. (1975). In "The Enzymes" (P. D. Boyer, ed.), 3rd ed., Vol. 11, pp. 397–547. Academic Press, New York.
Doeppner, T., and Hartmeier, W. (1984). Starch/Staerke **36**, 283–287.
Dolin, M. J. (1961). In "The Bacteria" (I. C. Gunzalus and R. Y. Stanier, eds.), Vol. 2, p. 319. Academic Press, New York.
D'Souza, S. F., and Nadkarni, G. B. (1980). Biotechnol. Bioeng. **22**, 2179–2189.
Eagon, R. G. (1958). Can. J. Microbiol. **4**, 1–7.
Eagon, R. G. (1963). Biochem. Biophys. Res. Commun. **12**, 274–279.
Eagon, R. G. (1971). Can. J. Biochem. **49**, 606–613.
Edson, N. L. (1953). Rep. Meet. Aust. N. Z. Assoc. Adv. Sci. **29**, 281–299.
Emmerling, O. (1899). Ber. Dtsch. Chem. Ges. **32**, 541–542.
Ermolayev, E. D., Zemturis, M., Dagys, A., Viesture, Z., Gutmanis, A., Malei, S. M., Shnitko, M. R., and Khagendorf, A. A. (1979). USSR Patent 694,533.
Estell, D. A., Light, D. R., Rasteter, W. H., Lazarus, R. A., and Miller, J. V. (1983). European Patent 132308.
Ettel, V., Liebster, J., and Tadra, M. (1952). Chem. Listy **46**, 45–48.
Faubl, H., and Kolb, C. A. (1978). Carbohydr. Res. **63**, 315–317.
Fewster, J. A. (1957). Biochem. J. **66**, 9P.
Fewster, J. A. (1958a). Biochem. J. **68**, 19P.
Fewster, J. A. (1958b). Biochem. J. **69**, 582.
Flickinger, M. C., and Perlman, D. (1977). Appl. Environ. Microbiol. **33**, 706–712.
Forrest, W. W., and Walker, D. J. (1971). Adv. Microb. Physiol. **5**, 213–217.
Frateur, J., Simonart, P., and Coulon, T. (1954). Antonie van Leeuwenhoek **20**, 111.
Friedland, W. C., Alm, A. G., Barlett, M. C., and Hansen, C. J. (1961). Abstr. Pap. 140th Meet. Am. Chem. Soc. p. 10P.
Frush, H. L., and Tregoning, L. J. (1958). Science **128**, 597.
Fujisawa, T., Yamazaki, M., Nishiyama, K., Shimizum, K., and Sawai, H. (1963). Japanese Patent 63/7725.
Fulmer, E. I., and Underkofler, L. A. (1947). Iowa State Coll. J. Sci. **21**, 251–270.
Fulmer, E. I., Dunning, J. W., and Underkofler, L. A. (1939). Iowa State Coll. J. Sci. **13**, 279–281.

Gaden, E. L. (1962). Biotechnol. Bioeng. **4**, 99–103.
Gakhokidze, A. M. (1941). Zh. Obshch. Chim. **11**, 109–116.
Ghosh, P. and Ghose, T. K. (1978). J. Ferment. Technol. **56**, 139–144.
Glushkov, G. K., Zolotarev, M. D., Moskvin, M. D., Tchorevskaya, Z. G., Khafizova, A. D., and Poponina, R. P. (1978). USSR Patent 594,170.
Green, S. R. (1961). J. Biochem. Microbiol. Technol. Eng. **3**, 351–356.
Grigorov, I., Dzherova, A., and Aleksieva, P. (1979). Priroda (Sofia) **28**, 67–70.
Hackett, D. P. (1960). Handb. Pflanzenphysiol. **12**(2), 23–41.
Hann, R. M., Tilden, E. B., and Hudson, C. R. (1938). J. Am. Chem. Soc. **60**, 1201–1203.
Harold, F. M. (1970). Adv. Microb. Physiol. **4**, 45–10.
Harold, F. M. (1972). Bacteriol. Rev. **36**, 172–230.
Hartmeier, W. (1981). Starch/Staerke **33**, 97–102.
Hauge, J. G. (1966). In "Methods Enzymology" (W. A. Wood, ed.), Vol. 9, pp. 92–98. Academic Press, New York.
Higashi, S. (1966). Dempunto Gijutsu Kenkyu Kaiho **33**, 1–8.
Holstein, A. G., and Holsing, G. C. (1962). U.S. Patent 3,050,444.
Holstein, A. G., and Holsing, G. C. (1965). British Patent 1,006,903.
Hough, L., and Theobald, R. S. (1962). Methods Carbohydr. Chem. **1**, 94.
Hough, L., Jones, J. K. N., and Mitchell, D. L. (1959). Can. J. Chem. **37**, 725–730.
Hu, C. L., McComb, E. A., and Rending, V. V. (1965). Arch. Biochem. Biophys. **110**, 350.
Imanoga, Y., Hirato-Sewatake, Y., Arita-Hashimoto, Y., Itou-Shibouta, Y., and Katoh-Sembra, R. (1979). Proc. Jpn. Acad. Ser. B **55**, 264–269.
Isbell, H. S., and Karabinos, J. V. (1952). Natl. Bur. Stand. (U.S.) **48**, 438–440.
Ishimori, Y., Karube, Y., and Suzuki, S. (1981). Biotechnol. Bioeng. **23**, 2601–2608.
Ishizaki, H., Ihara, T., Yoshitaka, J., Shimamura, M., and Imai, T. (1973). J. Agric. Chem. Soc. Jpn. **47**, 755–761.
Isono, M., Nakanishi, T., and Sasajima, K. (1963). Japanese Patent 66/357.
Iwata, K. K. (1983). Japanese Patent 83/76097.
Izuo, N. Nabe, K., Yamada, S., and Chibate, I. (1980). J. Ferment. Technol. **58**, 221–226.
Jaksa, I., Gyuran, J., Nemeth, S., Trischler, F., and Udvardy Nagy, I. (1983). Hungarian Patent 28770.
Janke, A. (1960). Handb. Pflanzenphysiol. **12**(1), 670–746.
Janke, A. (1962). Arch. Mikrobiol. **41**, 79–114.
Jones, J. K. N., and Mitchell, D. L. (1959). Can. J. Chem. **37**, 1561–1566.
Jones, J. K. N., Perry, M. B., and Turner, J. C. (1961a). Can. J. Chem. **39**, 965–972.
Jones, J. K. N., Perry, M. B., and Turner, J. C. (1961b). Can. J. Chem. **39**, 2400–2410.
Jones, J. K. N., Perry, M. B., and Turner, J. C. (1962). Can. J. Chem. **40**, 503–510.
Kamlet, J. (1943). U.S. Patent 2,314,831.
Katznelson, H., Tanenbaum, S. W., and Tatum, E. L. (1953). J. Biol. Chem. **204**, 43–59.
Kaufmann, H., and Reichstein, T. (1967). Helv. Chim. Acta **50**, 2280–2287.
Kersters, K., Wood, W. A., and De Ley, J. (1965). J. Biol. Chem. **240**, 965–974.
Kett, G. (1963). British Patent 923,858.
King, T. E. (1966). In "Methods in Enzymology" (W. A. Wood, ed.), Vol. 9, pp. 98–103. Academic Press, New York.
King, T. E., and Cheldelin, V. H. (1952a). Science **115**, 14–15.
King, T. E., and Cheldelin, V. H. (1952b). J. Biol. Chem. **198**, 135–141.
King, T. E., and Cheldelin, V. H. (1953). J. Bacteriol. **66**, 581.
King, T. E., and Cheldelin, V. H. (1957). J. Biol. Chem. **224**, 579–590.
King, T. E., and Cheldelin, V. H. (1958). Biochem. J. **68**, 31P–32P.
Kirstein, D., Besserdich, H., and Kahring, J. (1980). Chem. Tech. (Leipzig) **32**, 466–468.
Kita, D. A., and Fenton, D. M. (1981). BRD Patent 3,036,413.

Kita, D. A., and Hall, K. E. (1979). BRD Patent 2,849,393.
Kita, D. A., and Hall, K. E. (1980). U.S. Patent 4,245,049.
Kita, D. A., and Hall, K. E. (1981). U.S. Patent 4,263,402.
Kita, D. A., Gagne, J. W., and Fenton, D. M. (1978). U.S. Patent 4,202,942.
Klungsöyr, L., King, T. E., and Cheldelin, V. H. (1957). *J. Biol. Chem.* **227,** 135-149.
Knobloch, H., and Tietze, H. (1941). *Biochem. Z.* **309,** 399-414.
Kopřiva, B., and Hlaváček, V. (1971). *Sb. Vys. Sk. Chem.-Technol. Praze, Potraviny* **E32,** 23-52.
Kotera, U., Umehara, Kodama, T., Minoda, Y., and Yamada, K. (1972). *Agric. Biol. Chem.* **36,** 1307-1313.
Krieg, N. R., and Holt, J. G. (1984). "Bergey's Manual of Systematic Bacteriology," Vol. 1. Williams & Wilkins, Baltimore, Maryland.
Krumphanzl, V., Dyr, J., Honzová, H., and Pardon, J. (1968). *Sb. Vys. Sk. Chem.-Technol. Praze, Potraviny* **E21,** 19-24.
Kulhánek, M. (1953a). *Chem. Listy* **47,** 1071-1074.
Kulhánek, M. (1953b). *Chem. Listy* **47,** 1081-1085; *Chem. Abstr.* **48,** 4044.
Kulhánek, M. (1966). Thesis, VŠCHT Praha.
Kulhánek, M. (1970). *Adv. Appl. Microbiol.* **12,** 11-33.
Kulhánek, M. (1980). Thesis, VŠCHT Praha.
Kulhánek, M. (1983). Patent ČSSR 209,251.
Kulhánek, M. (1984). In "Modern Biotechnology" (V. Krumphanzl and Z. Řeháček, eds.), Vol. 2, pp. 614-767. Inst. Microbiol., Prague.
Kulhánek, M., and Ševčíková, Z. (1962). *Folia Microbiol. (Prague)* **7,** 288-297.
Kulhánek, M., and Ševčíková, Z. (1965). *Folia Microbiol. (Prague)* **10,** 362-364.
Kulhánek, M., and Tadra, M. (1972). *LC, Listy Cukrov.* **88,** 31-35.
Kulhánek, M., and Tadra, M. (1973). *Zentralbl. Bakteriol., Parasitenk., Infektionskr. Hyg., Abt. 2, Naturwiss.: Allg., Landwirtsch. Tech. Mikrobiol.* **128,** 25-30.
Kulhánek, M., and Tadra, M. (1975). *Folia Microbiol. (Prague)* **20,** 74.
Kulhánek, M., and Tadra, M. (1977). *Folia Microbiol. (Prague)* **22,** 373-375.
Kulhánek, M., and Tadra, M. (1979). ČSSR Patent 180,669.
Kulhánek, M., and Tadra, M. (1981). ČSSR Patent 188,604.
Kulhánek, M., and Tadra, M. (1983). ČSSR Patent 209,255.
Kulhánek, M., Tadra, M., Linek, K., and Kučár, Š. (1975). *Folia Microbiol. (Prague)* **20,** 409-411.
Kulhánek, M., Tadra, M., Pacák, J., Trejbalová, H., and Černý, M. (1977). *Folia Microbiol. (Prague)* **22,** 295-297.
Kulhánek, M., Tadra, M., Linek, K., and Kučár, Š. (1979). *Folia Microbiol. (Prague)* **24,** 185-187.
Kulhánek, M., Tadra, M., Tuma, J., Buděšínský, M., and Černý, M. (1983). ČSSR Patent 222,577.
Lessing, L., and Watson, T. G. (1982). *S. Afr. Food Rev.* **9,** S107-S108.
Linek, K., Sandtnerová, R., Sticzay, T., Kováčík, V., Kulhánek, M., and Tadra, M. (1979). *Carbohydr. Res.* **76,** 290-294.
Maclay, W. D., Hahn, R. M., and Hudson, C. S. (1942). *J. Am. Chem. Soc.* **64,** 1606.
Mahmond, S. A. Z., El-Sawy, and Nour El-Din, O. O. (1977). *Egypt. J. Food Sci.* **5,** 9-20, 21-29.
Makhotkina, T. A., Pomortseva, N. V., Lomova, I. E:, and Nikolaev, P. I. (1981). *Prikl. Biokhim. Mikrobiol.* **17,** 102-106.
Mandal, S. K., and Chatterjee, S. P. (1985). *Folia Microbiol. (Prague)* **30,** 414-419.
Martin, S. M., and Steel, R. (1955). *Can. J. Microbiol.* **1,** 470-472.

Mastropietro-Cancellieri, M. F., and Tiecco, G. F. (1961). *Rend. Ist. Super Sanita (Ital. Ed.)* **24**, 754.
Matsushita, K., Shinagawa, E., Adachi, O., and Ameyama, M. (1979a). *J. Biochem. (Tokyo)* **85**, 1173–1181.
Matsushita, K., Shinagawa, E., Adachi, O., and Ameyama, M. (1979b). *J. Biochem. (Tokyo)* **86**, 249–256.
Matsushita, K., Ohno, Y., Shinagawa, E., Adachi, O., and Ameyama, M. (1980a). *Agric. Biol. Chem.* **44**, 1505–1512.
Matsushita, K., Yamada, M., Shinagawa, E., Adachi, O., and Ameyama, M. (1980b). *J. Bacteriol.* **141**, 389–392.
Matsushita, K., Shinagawa, E., Adachi, O., and Ameyama, M. (1982a). *FEBS Lett.* **1982**, 255–258.
Matsushita, K., Shinagawa, E., Adachi, O., and Ameyama, M. (1982b). *J. Biochem. (Tokyo)* **92**, 1607–1613.
Matsushita, K., Ohno, Y., Shinagawa, E., Adachi, O., and Ameyama, M. (1982c). *Agric. Biol. Chem.* **46**, 1007–1011.
Miashiro, S., Nakamura, J., and Hirose, Y. (1976). Japanese Patent 76/82,787.
Mironova, T. N., Petukhova, N. I., Maksimenko, V. N., and Nikolaev, P. I. (1981). USSR Patent 857,264.
Misenheimer, T. J., Anderson, R. F., Lagoda, A. A., and Tyler, D. D. (1965). *Appl. Microbiol.* **13**, 393–396.
Mochizuki, K., Kanzaki, T., Okazaki, H., and Doi, M. (1969). Japanese Patent 69/08,074.
Molliard, M. (1924). *C. R. Hebd. Seances Acad. Sci.* **178**, 41–45.
Moses, V., and Ferrier, R. J. (1962). *Biochem. J.* **83**, 8–14.
Mowshowitz, S., Awigad, G., and England, S. (1974). *J. Bacteriol.* **118**, 1051–1058.
Müller, D. (1960). *Handb. Pflanzenphysiol.* **12**(1), 635–669.
Nabe, K., Izuo, N., Yamada, S., and Chibata, I. (1979). *Appl. Environ. Microbiol.* **38**, 1056–1060.
Nakamura, T., and Ogura, Y. (1962). *J. Biochem. (Tokyo)* **52**, 214–220.
Nikol'skaya, N. A., Belyakova, M. S., Statkevich, B. D., Pomortseva, N. V., Abramova, N. I., and Zarutskii, V. V. (1984). *Khim.-Farm. Zh.* **18**, 879–882.
Ninast, G., and Schedel, M. (1980a). Ger. Offen. 2,834,122.
Ninast, G., and Schedel, M. (1980b). European Patent Appl. 8031.
Nyeste, L., Sevella, L., Szigeti, L., Szöke, A., and Holló, J. (1980). *Eur. J. Appl. Microbiol. Biotechnol.* **10**, 87–94.
Oda, Y., Ninjoji, A., Ikuta, J., Kawabata, D., and Honda, K. (1974). *Agric. Biol. Chem.* **38**, 2161–2165.
Oga, S., Sato, K., Imada, K., and Asano, K. (1972). U.S. Patent 3,790,444.
Ohmoto, S., Tozawa, Y., and Ueda, K. (1983). *J. Ferment. Technol.* **61**, 373–378.
Ohsugi, M., Tochikura, T., and Ogata, K. (1970). *Agric. Biol. Chem.* **34**, 357–363.
O'Malley, J. J., and Weaver, J. L. (1972). *Biochemistry* **11**, 3527–3532.
Onishi, H., and Suzuki, T. (1969). *Appl. Microbiol.* **18**, 1031–1035.
Oosterhuis, N. M. G., Groesbeek, N. M., Kossen, N. W. F., and Schenk, E. S. (1985). *Appl. Microbiol. Biotechnol.* **21**, 42–49.
Pasternak, R., and Brown, E. V. (1940). U.S. Patent 2,194,921.
Peppler, H. J. (1967). "Microbial Technology." Reinhold, New York.
Pelouze, J. (1852). *Ann. Chim. Phys.* **35**, 222.
Pervozvanskii, W. W. (1939). *Mikrobiologiya* **8**, 149.
Peterson, M. H., Friedland, W. C., Denison, F. W., and Sylvester, J. C. (1956). *Appl. Microbiol.* **4**, 316–322.

Pfeifer, V. T., Vojnovich, C., Heger, E. N., Nelson, G. E. N., and Haynes, W. C. (1958). Ind. Eng. Chem. **50,** 1009–1012.
Pomortseva, N. V., and Krasilnikova, T. N. (1978). Khim.-Farm. Zh. **12,** 88–91.
Pomortseva, N. V., and Krasilnikova, T. N. (1983). Khim.-Farm. Zh. **27,** 721–725.
Pomortseva, N. V., Krasilnikova, T. N., Palejeva, M. A., and Nikolajev, P. I. (1974). Prikl. Biokhim. Mikrobiol. **10,** 59–63.
Pomortseva, N. V., Ignatov, Yu. L., and Yanchevskii, S. V. (1980). Prikl. Biokhim. Mikrobiol. **16,** 890–894.
Pomortseva, N. V., Solov'eva, K.-A., Krasilnokova, T. N., and Suvorova, E. E. (1983). Prikl. Biokhim. Mikrobiol. **19,** 250–255.
Pratt, J. W., and Richtmayer, N. K. (1955). J. Am. Chem. Soc. **77,** 6326–6328.
Pratt, J. W., Richtmayer, N. K., and Hudson, C. S. (1952). J. Am. Chem. Soc. **74,** 2210–2214.
Prescott, S. C., and Dunn, C. C. (1959). "Industrielle Mikrobiologie." Deutscher Verlag, Wissenschaften, Berlin.
Price, R., and Reichstein, T. (1937). Helv. Chim. Acta **20,** 101–109.
Rainbow, C. (1961). Prog. Ind. Microbiol. **3,** 43–70.
Rainbow, C. (1966). Wallerstein Lab. Commun. **29,** 5–15.
Ramakrishnan, T., and Campbell, J. J. R. (1955). Biochim. Biophys. Acta **17,** 122–127.
Rao, M. R. R. (1957). Annu. Rev. Microbiol. **11,** 317–338.
Razumovskaya, Z. G. (1956). Uch. Zap. Leningr. Gos. Univ. im. A. A. Zhdanova, Ser. Biol. Nauk **41,** 5.
Rehm, H. J. (1967). "Industrielle Mikrobiologie." Springer, Berlin.
Richter, G., and Heinecker, H. (1979). Starch/Staerke **31,** 418–422.
Richtmayer, N. K., Stewart, L. C., and Hudson, C. S. (1950). J. Am. Chem. Soc. **72,** 4934–4937.
Riedl-Tumová, E., and Bernhauer, K. (1950). Biochem. Z. **320,** 472.
Sasajima, K., and Isono, M. (1968). Agric. Biol. Chem. **32,** 161–169.
Sato, K., Yamada, Y., Aida, K., and Uemura, T. (1969). Agric. Biol. Chem. **33,** 1612–1618.
Schultz, J. S. (1964). Appl. Microbiol. **12,** 305–310.
Shaw, D. R. D. (1956). Biochem. J. **64,** 394–405.
Sherwood, I. R. (1947). Aust. Chem. Inst. J. Proc. **14,** 221–232.
Shinagawe, E., Chiyonobu, T., Adachi, O., and Ameyame, M. (1976). Agric. Biol. Chem. **40,** 475–483.
Shinagawa, E., Chiyonobu, T., Matsushita, K., Adachi, O., and Ameyama, M. (1978a). Agric. Biol. Chem. **42,** 1055–1057.
Shinagawa, E., Matsushita, K., Adachi, O., and Ameyama, M. (1978b). Agric. Biol. Chem. **42,** 2355–2361.
Shinagawa, E., Matsushita, K., Adachi, O., and Ameyama, M. (1981). Agric. Biol. Chem. **45,** 1079–1085.
Shinagawa, E., Matsushita, K., Adachi, O., and Ameyama, M. (1982). Agric. Biol. Chem. **46,** 135–141.
Shnaydman, L. O. (1973). "Proizvodstvo Vitaminov," Pishchevaya Promyshlennost, Moskva.
Simpson, F. J. (1966). In "Methods in Enzymology" (W. A. Wood, ed.), Vol. 9, pp. 11–46. Academic Press, New York.
Smith, L. (1968). In "Biological Oxidations" (T. P. Singer, ed.), pp. 55–122. (Wiley Interscience), New York.
Solov'eva, K. A., and Pomortseva, N. V. (1980). Khim.-Farm. Zh. **14,** 67–70.

Sonoyama, T., Tani, H., Matsuda, K., Kageyama, B., Tanimoto, M., Kobayashi, K., Yagi, S., Kyotani, H., and Mitsushima, K. (1982a). *Appl. Environ. Microbiol.* **43,** 1064–1069.
Sonoyama, T., Yagi, S., Kageyama, B., and Tanimoto, M. (1982b). European Patent 46,282.
Sonoyama, T., Yagi, S., Kageyama, B., and Tanimoto, M. (1986). Japanese Patent 86/63278.
Spencer, J. F. T., and Gorin, P. A. G. (1968). *Prog. Ind. Microbiol.* **7,** 177.
Stankovič, L., Linek, K., and Kulhánek, M. (1982a). Czech. Authorship Certificate 197527.
Stankovič, L., Linek, K., and Kulhánek, M. (1982b). Czech. Authorship certificate 197528.
Steiger, M., and Reichstein, T. (1935). *Helv. Chim. Acta* **18,** 790–799.
Stewart, D. J. (1959). *Nature (London)* **183,** 1133–1134.
Stewart, L. C., Richtmayer, N. K., and Hudson, C. S. (1952). *J. Am. Chem. Soc.* **74,** 2206–2210.
Stokes, F. N., and Campbell, J. J. R. (1951). *Arch. Biochem. Biophys.* **30,** 121–125.
Stouthamer, A. H. (1961). *Biochim. Biophys. Acta* **48,** 484–500.
Stouthamer, A. H. (1962). *Biochim. Biophys. Acta* **56,** 19–32.
Stroshane, R. M., and Terlman, D. (1977). *Biotechnol. Bioeng.* **19,** 459–465.
Stubbs, J. J., Lockwood, L. B., Roe, E. T., Tabenkin, B., and Ward, G. E. (1940). *Ind. Eng. Chem.* **32,** 1626–1631.
Sukharevich, B. I., and Razumovskaya, Z. G. (1968). *Mikrobiologiya* **37,** 832–836.
Swoboda, B. E. P., and Massey, V. (1965). *J. Biol. Chem.* **240,** 2209–2215.
Takahashi, T., and Asai, T. (1931). *J. Agric. Chem. Soc. Jpn.* **7,** 1.
Takahashi, T., and Asai, T. (1932). *J. Agric. Chem. Soc. Jpn.* **8,** 703.
Takeda Chem. Ind. (1975). Japanese Patent J75022-113.
Terada, O., Suzuki, S., and Kinishita, S. (1961). *Nippon Nogei Kagaku Kaishi* **35,** 178–182.
Tiwary, K. N., Dhawale, M. R., Szarek, W. A., Hay, g. W., and Kropinski, A. M. B. (1986). *Carbohydr. Res.* **156,** 19–24.
Touster, O., and Shaw, D. R. D. (1962). *Physiol. Rev.* **42,** 181–225.
Trutko, S. M., Akimenko, V. K., and Lozinov, A. B. (1977). *Izv. Akad. Nauk SSR, Ser. Biol.* pp. 422–428.
Tsukamoto, M. (1947). *J. Ferment. Technol.* **25,** 142.
Tsukamoto, T., Morita, S., and Okada, J. (1982). *Chem. Pharm. Bull.* **30,** 1539–1549.
Ueda, K. M. (1965). French Patent 1,399,027.
Ueda, K. M., and Higashi, S. (1964). *Kogyo Kagaku Zasshi* **67,** 926–931.
Ueda, K. M., Higashi, S., and Origuchi, K. (1967). *J. Ferment. Technol.* **45,** 541–549.
Underkofler, L. A., and Hickey, R. J. (1954). "Industrial Fermentations," Vol. 1. Chem. Publ., New York.
Uspenskaya, S. N., and Loitsyanskaya, M. S. (1979). *Mikrobiologiya* **48,** 400–405.
Van Dijken, J. P., and Veenhuis, M. (1980). *Eur. J. Appl. Microbiol. Biotechnol.* **9,** 75–83.
Votoĉek, E., Valentin, F., and Rac, F. (1930). *Collect. Czech. Chem. Commun.* **2,** 402–413.
Wakisaka, Y. (1964). *Agric. Biol. Chem.* **28,** 819.
Wakisaka, Y., Ishida, L., Kubota, F., Kyotani, H., Kowachi, N., Sangen, S., Fukui, K., and Kimura, T. (1965a). *Annu. Rep. Shionogi Res. Lab.* **15,** 69–76.
Wakisaka, Y., Ishida, H., and Kubota, F. (1965b). *Annu. Rep. Shionogi Res. Lab.* **15,** 77–83.
Weibull, C. (1953). *J. Bacteriol.* **66,** 697–702.

Weimberg, R. (1961). *J. Biol. Chem.* **236**, 629–635.
Wells, P. A., Stubbs, J. J., Lockwood, L. B., and Roe, E. T. (1937). *Ind. Eng. Chem.* **29**, 1385–1388.
White, S. A., and Claus, G. W. (1982). *J. Bacteriol.* **150**, 934–943.
Whiting, P. H., Midgley, M., and Cawes, E. A. (1976). *J. Gen. Microbiol.* **92**, 304–310.
Widmer, C., King, T. E., and Cheldelin, V. H. (1956). *J. Bacteriol.* **71**, 737.
Wilt, H. G. J. (1972). *Ind. Eng. Chem. Prod. Res. Dev.* **11**, 370–373.
Wolf, J. (1960). *Handb. Pflazenphysiol.* **12**(1), LXIII–CCLXXX.
Wood, W. A., and Schwerdt, R. F. (1953). *J. Biol. Chem.* **201**, 501–511.
Yamada, S., Wada, M., and Chibata, I. (1978). *J. Ferment. Technol.* **56**, 20–28, 29–34.
Yamada, S., Nabe, K., Izuo, N., Wada, M., and Chibata, I. (1979a). *J. Ferment. Technol.* **57**, 215–220.
Yamada, S., Nabe, K., Izuo, N., and Chibata, I. (1979b). *J. Ferment. Technol.* **57**, 221–226.
Yamazaki, M. (1954). *J. Agric. Chem. Soc. Jpn.* **28**, 748–751.
Yan, Z. Z., Tao, Z. X., Yu, L. H., Yin, G. L., Ning, W. Z., Wang, C. H., Wang, S. D., Jiang, H. F., Yu, J. F. et al. (1981). *Wei Sheng Wu Hsueh Pau* **21**, 185–191.
Yasuda, N. (1968). Japanese Patent 68/28,950.
Yasuda, N. (1969). Japanese Patent 69/13,140.
Yokozawa, K. (1952). *J. Agric. Chem. Soc. Jpn.* **26**, 415.
Zarak, V. A., Aganova, E. V., Beljakova, M. S., Nikolskaya, L. A., Pogak, E. M., Chudoshena, Z. A., Barablina, N. V., Obushenkov, G. J., Kogan, M. I., Savost'yanov, G. I., Anshakov, A. F., Beygelman, I. A., Bogatyreva, E. D., Shchekochikhin, G. F., and Khrinachev, I. E. (1975). USSR Patent 461,942.
Zemek, J., Kučár, S., Augustin, J., and Kolina, J. (1981). USSR Patent 188,626.
Zolotarev, N. S., Tkhorevskaya, Z. G., Moskvin, M. D., Vasilenko, K. D., and Barsegyan, I. V. (1970). *Khim.-Farm. Zh.* **4**, 28–30.

Antitumor and Antiviral Substances from Fungi

SHUNG-CHANG JONG* AND RICHARD DONOVICK†

*American Type Culture Collection
Rockville, Maryland 20852

†Developmental Therapeutics Program
Division of Cancer Treatment
National Cancer Institute
Frederick Cancer Research Facility
Frederick, Maryland 21701

I. Introduction
II. Historical Background
III. Screening of Fungi
IV. Antibiotics Produced by Fungi
V. Antitumor–Antiviral Substances Produced by Fungi
   A. Fungi Imperfecti
   B. Basidiomycetes
   C. Ascomycetes
   D. Phycomycetes
References

I. Introduction

The United States National Cancer Institute (NCI) has recently intensified its emphasis upon natural products as a source of new drug discovery. The initial impetus for these efforts stemmed both from the abundant precedents for the discovery of a wide variety of valuable medicinal agents across essentially all major pharmaceutical classes and from a comprehensive restructuring and expansion of scope of the NCI drug-screening programs. For example, a major new anticancer drug-screening program deemphasizes the use of typically insensitive nonspecific in vivo animal tumor models, and instead emphasizes the use of panels of very sensitive in vitro "disease-oriented" human tumor cell line-based models for the initial stage of screening (for reviews and other background, see Boyd, 1986; Boyd et al., 1988a,b). The NCI has also more recently assumed responsibilities for the implementation of a national acquired immunodeficiency syndrome (AIDS) antiviral drug-screening and drug development program in parallel and in complement to the anticancer drug program (for a review, see Boyd, 1988).

Natural product areas receiving special emphasis in the NCI programs include plants, marine organisms, and selected classes of mircoorganisms. Among the microbial sources, fungi are currently of major interest. The precedents for discovery of novel substances with anticancer and/or antiviral activities from fungi are extensive, yet the more systematic and successful exploration of fungi for potentially useful new anticancer and/or antiviral drugs has undoubtedly been constrained by limitations in available screening systems. With the increased emphasis on development and application of sensitive and specific new anticancer and antiviral [e.g., anti-Human Immunodeficient Virus (HIV)] screens by NCI, the industry, and elsewhere, it is timely to review comprehensively the existing literature concerning antitumor and antiviral substances from fungi. It is hoped that this compendium will facilitate the expeditious and efficient further exploration and/or reexploration of fungi for new medicinal products of urgent priority.

II. Historical Background

The discovery of penicillin from fungi marked a milestone in the development of antibiotics. This discovery led to a revolution in the medical, veterinary, and agriculture sciences, and indeed in all of the biological sciences during the last four decades. The knowledge and techniques acquired from the development of penicillins for human chemotherapy have proved to be basic, and are applicable to the entire field of antibiotics and to the field of industrial microbiology.

Of the ≈4000 known antibiotics of microbial origin, ≈70% are elaborated by Actinomycetes (especially species of *Streptomyces*), 20% by fungi, and 10% by eubacteria (Porter, 1976; Perlman, 1979). Most, but by no means all, of the antibiotic-producing organisms have been isolated from soils in programs designed to discover new antibiotics. S. A. Waksman, a Nobel Prize winner in 1952 because of his discovery of streptomycin, was the first to use soils as the source of microorganisms for a large-scale screening program for new antibiotics. Although in the past, actinomycetes were found to predominate in soils, the flora of microorganisms has shifted as civilization has expanded due to the changes in habitats of plants, insects, and animals, and repeated applications of agriculture chemicals (e.g., fertilizers, insecticides, fungicides, herbicides, etc.).

The fungi produce many antibiotic substances besides penicillin, some of which have been reported to show antiviral and/or antitumor activity. In light of the continued search for such substances, a systematic exploration of these microorganisms as potential producers of antiviral and/or antitumor substances is merited.

The metabolic pathways of fungi can be divided into two broad areas: the primary metabolism that provides the fungi with energy, synthetic intermediates and key macromolecules, such as protein and DNA; and the secondary metabolism that yields a range of species-specific compounds possessing no obvious function in cell growth (Rose, 1979). In view of the different character of the synthesis in the successive growth phases, the term "balanced growth" (Borrow et al., 1961) or "trophophase" (Bu'Lock, 1965) has been used to refer to the rapid growth stage in which synthesis of primary metabolites takes place, and "storage and maintenance phases" (Borrow et al., 1961) or "idiophase" (Bu'Lock, 1965) is used to describe the stationary stage of growth in which the secondary metabolites are synthesized. Antibiotics produced by fungi are secondary metabolites, since they appear to play no essential role in cell metabolism and are synthesized only after the organism has passed through part or all of the period of balanced growth (trophophase).

A systematic survey of the secondary metabolites of fungi was initiated in 1922 by Raistrick and collaborators (Rose, 1979). Their work dealt primarily with the isolation, identification, and synthesis of chemical substances produced by fungi and can be considered to be a study of the organic chemistry of fungus products rather than of fungus metabolism. In the 1940s and 1950s, scientists emphasized potentially therapeutic metabolites toxic to pathogenic microorganisms. More recently, investigators have shifted to those secondary substances having specific effects on tissues of plants or animals, in the search for compounds showing possible utility as antitumor and/or antiviral agents.

III. Screening of Fungi

Fungi (molds, yeasts, and mushrooms) are widely distributed in nature and are especially abundant in soils and in spoilage of foodstuffs. They are almost always associated with the decay of organic debris of dead plants and animals, and are regularly isolated from fresh and marine water, and less frequently from oil. Nearly all fungi thrive on plant materials in their natural habitats. However, screening programs devoted to the search for new antibiotics usually used soils as the source of fungi. As a consequence, the screening tests have been limited to certain groups of fungi, particularly the groups that predominate in the soil, so that the real potentiality of other groups of fungi to produce antibiotics has not been determined.

Identification of a newly isolated antibiotic requires an enormous amount of work and money. If the substance identified turns out to be already known, this effort is wasted. Little, if any, effort has been made to identify the selected organisms below the genus level in the preliminary studies and a great deal of duplication inevitably has occurred. This obviously has led to the undesirable duplication of the pursuit of identical compounds as well. Furthermore, an area of information which is sadly lacking in the enormous accumulation of screens that have been carried out since the earliest antibiotic screening programs is that of negative results, i.e., those genera that have been found to produce no substances active in the test methods employed in a particular screen. This leaves large gaps in our knowledge of those genera of the greatest or least interest.

Therefore, a collection of fungi representing different taxonomic groups from many natural habitats is certainly a valuable source in developing a screening program to search for new antibiotics. A systematic exploration of identified cultures could be more efficient and economical than screening programs utilizing raw substrates (such as soils) as the starting material. For example, the American Type Culture Collection (ATCC) maintains over 24,000 strains of living fungi representing 5000 species. These consist of 50% Fungi Imperfecti, 30% Ascomycetes, 10% Phycomycetes, 9% Basidiomycetes, and 1% Myxomycetes. They provide the tools for experimental investigation in practically all of the biological sciences, including medicine, veterinary science, and agriculture. The fact that only 518 strains of ATCC fungi have been reported to produce 207 antibiotics (Table I) only emphasizes the potential of these organisms as sources of therapeutic agents, despite the relatively narrow spectra of fungi examined. The reported antibiotic-producing fungi listed in Table I consist of 233 Fungi Imperfecti, 31 Ascomycetes, 26 Basidiomycetes, and 25 Phycomycetes (Jong and Gantt, 1987).

IV. Antibiotics Produced by Fungi

In 1951, Brian reviewed the reports of antibiotic production in various taxonomic groups of fungi. Of the 96 antibiotics produced by fungi, 57 were well characterized. Forty-eight were elaborated from Fungi Imperfecti, eight from Basidiomycetes, and one from Ascomycetes. Phycomycetes were almost completely neglected. In 1966, Broadbent summarized the work in this area since 1951. From the additional results available to him, he answered the following two questions posed by Brian: (1) Is the capacity to produce antibiotics spread evenly over all taxonomic groups of the fungi? (2) Is there a

TABLE I

ATCC Antibiotic-Producing Fungi

Antibiotic	ATCC No.	Fungus	Type of antibiotic
Acetylene dicarboxylic acid	28535	*Fusarium merismoides*	Antimicrobial
Acronycine	8688a, 9245	*Cunninghamella echinulata* var. *elegans*	Antineoplastic
Acrostalic acid	36937	*Verticillium* sp.	Antimicrobial
Acrostalidic acid	36937	*Verticillium* sp.	Antimicrobial
6-Aminopenicillanic acid	13742	*Cunninghamella echinulata* var. *elegans*	Antimicrobial
	20378	*Leptosphaerulina australis*	
	1216a	*Mucor circinelloides* f. *lusitanicus*	
	20377	*Phialomyces macrosporus*	
	20375	*Pseudofusarium fusarioideum*	
	20379	*Robillarda* sp.	
Antiamoebin	16541	*Cephalosporium pimprina*	Antimicrobial
	16411	*Emericellopsis poonensis*	
	16540	*Emericellopsis synnematicola*	
Anticarcinogenic polysaccharide	20603	*Isaria atypicola*	Antineoplastic
Antifungal agents (unnamed)	34441–34442	*Aspergillus caespitosus*	Antimicrobial
	46829	*Fusarium solani*	
Antimicrobial polypeptide	22064	*Rhizomucor miehei*	Antimicrobial
Antiprotozoal agent (unnamed)	20119	*Trametes albida*	Antimicrobial
Aphidicolin	28300	*Cephalosporium aphidicola*	Antiviral
	26952	*Nigrospora oryzae*	
Aranotins	34649	*Amauroascus aureus*	Antiviral
Asperetin	42568	*Eurotium repens*	Antimicrobial
Aspergillic acid	9170	*Aspergillus flavus*	Antimicrobial
	24741	*Aspergillus flavus*	

(continued)

TABLE I (continued)

Antibiotic	ATCC No.	Fungus	Type of antibiotic
Asperphenamate	11013	Aspergillus flavipes	Antimicrobial
	9056	Penicillium brevicompactum	
Asposterol (=aspochalasine complex)	32452	Aspergillus microcysticus	Antimicrobial
Astechrome	10020	Aspergillus terreus	Antimicrobial
Asterriquinone	10020	Aspergillus terreus	Antineoplastic
	46560	Aspergillus terreus var. africanus	
Aszonapyrone A	16867	Aspergillus zonatus	Antimicrobial
Atrovenetin	13351–13352	Penicillium melinii	Antimicrobial
Avenaciolide	16861	Aspergillus avenaceus	Antimicrobial
15-Azasterol	28804	Scytalidium flavo-brunneum	Antimicrobial
A-22082	58396	Aspergillus nidulans	Antimicrobial
	58398	Emericella rugulosa	
A-25822	28804	Scytalidium flavo-brunneum	Antimicrobial
A-26771B	28797	Penicillium turbatum	Antimicrobial
A-30912	58397	Aspergillus nidulans var. roseus	Antimicrobial
	58398	Aspergillus rugulosus	
A-32390A	36938	Pyrenochaeta sp.	Antimicrobial
N-Benzoyl-l-phenylalaninol	11013	Aspergillus flavipes	Antimicrobial
Bikaverin	28708	Fusarium oxysporum	Antimicrobial, antineoplastic
Bleomycinic acid	20351, 20352, 20355	Fusarium anguioides Fusarium roseum	Antineoplastic
	20353, 20354	Helminthosporium zonatum	
Bostrycin	22255	Alternaria eichhorniae	Antimicrobial
Botrylactone	44534	Botrytis cinerea	Antimicrobial
Brefeldins A and C	58665	Eupenicillium brefeldianum	Antimicrobial, antiviral

188

Calvatic acid	20424	Calvatia craniformis	Antimicrobial
	52736–52737	Lycoperdon pyriforme	Antineoplastic
Candidulin	13686	Aspergillus candidus	Antimicrobial
Cephalosporin	20427	Acremonium chrysogenum	Antimicrobial
	46117	Acremonium chrysogenum	
Cephalosporin P	14615	Acremonium chrysogenum	Antimicrobial
Cephalosporin C	14553	Acremonium chrysogenum	Antimicrobial
	20425–20426	Acremonium chrysogenum	
	20427–20428	Acremonium chrysogenum	
	36225	Acremonium chrysogenum	
	48272	Acremonium chrysogenum	
	20359–20360	Cephalosporium polyaleurum	
	42450	Cephalosporium sp.	
	14645	Emericellopsis minima	
	34702	Paecilomyces persicinus	
Chaetiacandin	60324	Monochaetia dimorphospora	Antimicrobial
Chaetocin	26417	Chaetomium minutum	Antimicrobial, antineoplastic, antiviral
Chaetomin	10195	Chaetomium cochliodes	Antimicrobial, antineoplastic, antiviral
3′Chloro-5:2′dihydroxy-3:78-trimethoxyflavone	20022–20023	Aspergillus candidus	Antimicrobial
Chloroflavonin	20022	Aspergillus candidus	Antimicrobial
Chloromycorrhizin A	36554	Mycorrhizal fungus	Antimicrobial
Coriolin	20305	Coriolus consors	Antimicrobial, antineoplastic
Cortalcerone	36306	Pulcherricium caeruleum	Antimicrobial

(continued)

TABLE I (continued)

Antibiotic	ATCC No.	Fungus	Type of antibiotic
Cyathin	38343	*Cyathus earlei*	Antimicrobial
	28392, 28396	*Cyathus helenae*	
Cyclodepsipeptide	36781	*Fusarium acuminatum*	Antimicrobial
Cyclopaldic acid	18232	*Penicillium verrucosum* var. *alum*	Antimicrobial
Cyclosporins A and B (=antibiotics S7481/F-1 and F-2)	56272	*Cylindrocarpon lucidum*	Antimicrobial
Cyclosporins A, B, C, D, E, and G	34921	*Trichoderma inflatum*	
	34921	*Beauveria nivea*	Antimicrobial
Deacetoxycephalosporin C	11550, 14553	*Acremonium chrysogenum*	Antimicrobial
	20416	*Acremonium chrysogenum*	
	20420	*Arachnomyces minimus*	
	20359–20360	*Cephalosporium polyaleurum*	
	20415	*Cephalosporium polyaleurum*	
	20370	*Cephalosporium* sp.	
	20422	*Emericellopsis minima*	
	20417	*Paecilomyces carneus*	
	20418	*Paecilomyces persicinus*	
	34702	*Paecilomyces persicinus*	
	20421	*Spiroidium fuscum*	
Deacetylcephalosporin C	20371	*Acremonium chrysogenum*	Antimicrobial
	34702	*Paecilomyces persicinus*	
Deflectins	16807	*Aspergillus deflectus*	Antimicrobial, antineoplastic
12-α-Deoxytetracyline	13807	*Thielavia terricola*	Antimicrobial
Dermadin	36387	*Trichoderma viride*	Antineoplastic
Desferritriacetylfusigen	16807	*Aspergillus deflectus*	Antimicrobial

Diacetoxyscirpenol	20227	*Fusarium lateritium*	Antineoplastic
	26533	*Fusarium sporotrichioides*	
Dihydrocompactin	24631	*Fusarium tricinctum*	Antimicrobial
Dihydrocyclosporin G	20606	*Penicillium citrinum*	Antimicrobial
Diketopiperazines (active against helminths)	34921	*Beauveria nivea*	Antiparasitic
1',2'-Dihydrorotenone	32079	*Penicillium italicum*	Antineoplastic
Diketocoriolin B	8688a	*Cunninghamella echinulata* var. *elegans*	Antimicrobial,
	20305	*Coriolus consors*	antineoplastic
Duclauxin	10500	*Talaromyces stipitatus*	Antimicrobial, antineoplastic
E	32332	*Phoma exigua* var. *linicola*	Antimicrobial
	32811	*Phoma exigua* var. *sambuci-nigrae*	
	32812	*Phoma hedericola*	
Emericin complex	32460	*Emericellopsis minima*	Antimicrobial
Epipolythiopiperazinedione	28797	*Penicillium turbatum*	Antiviral
Epirodin	36925, 56595	*Epicoccum nigrum*	Antimicrobial
Equisetin	28805	*Fusarium equisetin*	Antimicrobial, antineoplastic
Ergosterol peroxide	36292	*Alternaria dianthicola*	Antineoplastic
	32320	*Rhizoctonia repens*	
Ethylene oxide-α/β dicarboxylic acids	10894	*Aspergillus fumigatus*	Antimicrobial
Falcarindiol	44022	*Myocentrospora acerina*	Antimicrobial
Flavipucine (=glutamicine)	26499	*Aspergillus flavus*	Antimicrobial
Frequetin	11080	*Penicillium glabrum*	Antimicrobial
Fumigatin	10894	*Aspergillus fumigatus*	Antimicrobial
Funicin	16846	*Aspergillus funiculosus*	Antimicrobial
Fusidic acid	42458	*Calcarisporium antibioticum*	Antimicrobial
FWH-775	16508	*Aspergillus niger*	Antiviral
(+)Geodin	24839	*Aspergillus terreus*	Antimicrobial

(*continued*)

TABLE I (continued)

Antibiotic	ATCC No.	Fungus	Type of antibiotic
Giotoxin	9197	Aspergillus fumigatus	Antimicrobial
	10894	Aspergillus fumigatus	Antineoplastic
	34961	Aspergillus ruber	Antiviral
	38361	Aspergillus sp.	
	9438	Penicillium spinulosum	
	36199	Trichoderma viride	
	9275	Trichoderma viride	
	34650	Trichoderma viride	
Gladiolic acid	9437	Penicillium gladiolii	Antimicrobial
Gliovirin	52045	Gliocladium virens	Antimicrobial
Grahamimycins A and B	20502	Cytospora sp.	Antimicrobial
Griseofulvin	28132	Khuskia oryzae	Antimicrobial
	28133	Nigrospora oryzae	
	58614	Penicillium concentricum	
	11885	Penicillium griseofulvum	
	24632	Penicillium griseofulvum	
	58620	Penicillium griseofulvum	
	9439	Penicillium janczewskii	
	58633	Penicillium viridicatum	
Hadicidin	14973	Penicillium glabrum	Antineoplastic
	14974	Penicillium thomii	
Helenine	18318	Penicillium funiculosum	Antiviral
Helvolic acid	9197	Aspergillus fumigatus	Antimicrobial
	10894	Aspergillus fumigatus	
Hemigossypal	26289	Verticillium dahliae	Antimicrobial
Humicolin	11079	Aspergillus asperescens	Antimicrobial
Hyalodendrin	28250	Hyalodendron sp.	Antimicrobial, antineoplastic, antiviral

9-Hydroxyacronycine	9142	*Aspergillus niger*	Antineoplastic
	9144, 36190	*Cunninghamella echinulata* var. *echinulata*	Antimicrobial
8- and 9-Hydroxyellipticines	10060	*Aspergillus alliaceus*	Antineoplastic
Hydroxygriseofulvin	28243	*Cephalosporium curtipes*	Antimicrobial
3'-Hydroxyrotenone	8688a	*Cunninghamella echinulata* var. *elegans*	Antineoplastic
3'-Hydroxyverrucarin A	11145	*Rhizopus oryzae*	Antineoplastic
16-Hydroxyverrucarins A and B	11145	*Rhizopus oryzae*	Antineoplastic
Hypothemycin	44392	*Hypomyces trichothecoides*	Antimicrobial
Islandicin	10127	*Penicillium islandicum*	Antimicrobial
Iso-acrostalidic acid	36937	*Verticillium* sp.	Antimicrobial
4-Isoavenaciolide	16861	*Aspergillus avenaceus*	Antimicrobial
Isocyclosporin G	34921	*Beauveria nivea*	Antimicrobial
Isopenicillin N	26818	*Penicillium chrysogenum*	Antimicrobial
Isororidin A	24571	*Myrothecium verrucaria*	Antimicrobial
Isororidin E	24571	*Myrothecium verrucaria*	Antimicrobial
Javanicin	48021	*Fusarium solani*	Antimicrobial
KA-6606	20501	*Saccharopolyspora hirsuta*	Antimicrobial
Leucinostatin	26839	*Paecilomyces lilacinus*	Antimicrobial, antineoplastic
Leucomycin derivatives	7159	*Beauveria bassiana*	Antimicrobial
	36190	*Cunninghamella echinulata* var. *echinulata*	
	8983, 9245	*Cunninghamella echinulata* var. *elegans*	
	46515	*Dermaloma* sp.	
	46553	*Lentodontium squamulosum*	
	11715	*Mortierella ramanniana* var. *angulispora*	
	10114	*Paecilomyces lilacinum*	
	10002	*Penicillium chrysogenum*	
	10435	*Penicillium daleae*	
	10455	*Penicillium janthinellum*	
	46513	*Penicillium* sp.	
	11816	*Pestalotia microspora*	

(continued)

TABLE I (continued)

Antibiotic	ATCC No.	Fungus	Type of antibiotic
	8992	*Thamnostylum piriforme*	
	8678	*Trichoderma viride*	
Linomycin sulfoxides	8740	*Aspergillus carbonarius*	Antimicrobial
	9029	*Aspergillus niger*	
LL-S88α	1012	*Aspergillus terreus*	Antiviral
	10029, 10690	*Aspergillus terreus*	
	12238, 34648	*Aspergillus terreus*	
Lycopersin	28708	*Fusarium oxysporum*	Antimicrobial
L-Lysine α-oxidase	20536	*Trichoderma viride*	Antineoplastic
Magnamycin derivatives	20538	*Trichoderma viride*	Antimicrobial
	7159	*Beauveria bassiana*	
	36190	*Cunninghamella echinulata* var. *echinulata*	
	8983, 9245	*Cunninghamella echinulata* var. *elegans*	
	46515	*Dermaloma* sp.	
	46553	*Lentodium squamulosum*	
	11715	*Mortierella ramanniana* var. *anguilispora*	
	10114	*Paecilomyces lilacinum*	
	10002	*Penicillium chrysogenum*	
	10435	*Penicillium daleae*	
	10455	*Penicillium janthinellum*	
	46513	*Penicillium* sp.	
	11816	*Pestalotia microspora*	
	8992	*Thamnostylum piriforme*	
	8678	*Trichoderma viride*	
	32459	*Verticillium lateritium*	
Melinacidin complex			Antimicrobial, antineoplastic

6-MFA	28706	Aspergillus ochraceus	Antiviral
Microline	36719	Gilmaniella humicola	Antimicrobial
Mieheim	16457	Rhizomucor miehei	Antimicrobial
	46343–46346	Rhizomucor miehei	
Monocerin	24641	Drechslera monoceras	Antimicrobial
Mycelianamide	11885	Penicillium griseofulvum	Antimicrobial
	36199	Penicillium spinulosum	
Mycomycin	22338	Odontia bicolor	Antimicrobial
Mycophenolic acid	9056	Penicillium brevi-compactum	Antineoplastic
	10111	Penicillium brevi-compactum	
	16024	Penicillium brevi-compactum	
	46514	Penicillium brevi-compactum	
	58606	Penicillium brevi-compactum	
	34905	Penicillium roqueforti	
	46836, 46837	Penicillium roqueforti	
	58626–58628	Penicillium roqueforti	
Mycorrhizin A	36554	Mycorrhizal fungus	Antimicrobial
Mycosporin 2	38321	Gnomonia leptosyla	Antimicrobial
M48541 and II	44497	Chaetomella raphigera	Antimicrobial
Namotin and nemotinic acid	14681	Basidiomycete	Antimicrobial
	14710	Perenniporia subacida	
	7159	Beauveria bassiana	
Niddamycin derivatives	36190	Cunninghamella echinulata var. echinulata	Antimicrobial
	8983, 9245	Cunninghamella echinulata var. elegans	
	46515	Dermaloma sp.	
	46553	Lentodium squamulosum	
	11715	Mortierella ramanniana var. anguilispora	

(continued)

TABLE I (continued)

Antibiotic	ATCC No.	Fungus	Type of antibiotic
	10114	*Paecilomyces lilacinus*	
	10002	*Penicillium chrysogenum*	
	10435	*Penicillium daleae*	
	10455	*Penicillium janthinellum*	
	46513	*Penicillium sp.*	
	11816	*Pestalotia microspora*	
	8992	*Thamnostylum piriforme*	
	8678	*Trichoderma viride*	
Nidulin, nornidulin	24528	*Aspergillus nidulans*	Antimicrobial
Nitrogen-containing antineoplastic agents	10032	*Emericella unguis*	Antineoplastic
	20565	*Coriolus consors*	
	20562	*Coriolus pargamenus*	
	20561	*Trametes hirsuta*	
	20545	*Trametes versicolor*	
β-Nitropropionic acid	11500	*Aspergillus flavus*	Antimicrobial
	7252, 12892	*Aspergillus oryzae*	
	13351, 13352	*Penicillium melinii*	
Novalichin	24148	*Paecilomyces fusisporus*	Antimicrobial
Noxiversin	11923	*Penicillium griseoroseum*	Antiviral
Ophiocordin	36865	*Cordyceps ophioglossoides*	Antimicrobial
Orsellinic acid	24528	*Aspergillus nidulans*	Antimicrobial
	18223	*Penicillium madriti*	
Ovicide (active against helminths)	52620	*Acremonium butyi*	Antiparasitic
	52621	*Cunninghamella elegans var. elegans*	
	52622	*Gliocladium catenualtum*	
	52623	*Paecilomyces lilacinus*	

Paracelsin	52624	Paecilomyces marquandii	Antimicrobial
Patulin (=clavacin, clavatin, claviformin)	52625	Verticillium bulbilosum	Antimicrobial
	26921	Trichoderma reesei	
	1007	Aspergillus clavatus	
	9192, 9198	Aspergillus clavatus	
	9599–9601	Aspergillus clavatus	
	36841	Byssochlamys fulva	
	36614	Byssochlamys nivea	
	42746	Byssochlamys nivea	
	56268	Byssochlamys nivea	
	42708	Penicillium crustosum	
	28876–28898	Penicillium expansum	
	28907	Penicillium expansum	
	36200	Penicillium expansum	
	42709–42710	Penicillium expansum	
	58619	Penicillium expansum	
	9260	Penicillium griseofulvum	
	10120	Penicillium griseofulvum	
	18172	Penicillium griseofulvum	
	24550	Penicillium griseofulvum	
	24632	Penicillium griseofulvum	
	28131	Penicillium griseofulvum	
	46037	Penicillium griseofulvum	
	58620	Penicillium griseofulvum	
Pecilocin (=variotin)	13435	Paecilomyces varoti var. antibiotics	Antimicrobial
Penicillin	14553	Acremonium chysogenum	
	28901–28905	Aspergillus nidulans	

(continued)

TABLE I (continued)

Antibiotic	ATCC No.	Fungus	Type of antibiotic
	32353	Aspergillus nidulans	
	10002	Penicillium chrysogenum	
	10003	Penicillium chrysogenum	
	10134	Penicillium chrysogenum	
	10238	Penicillium chrysogenum	
	11709–11710	Penicillium chrysogenum	
	12687–12690	Penicillium chrysogenum	
	20444	Penicillium chrysogenum	
	26818	Penicillium chrysogenum	
	28089	Penicillium chrysogenum	
	28593	Penicillium chrysogenum	
	58611	Penicillium chrysogenum	
	9178	Penicillium notatum	
	9179	Penicillium notatum	
	9479, 11625	Penicillium notatum	
Penicillin N	36411	Aspergillus aculeatus	Antimicrobial
	14553	Acremonium chrysogenum	
	14615	Acremonium chrysogenum	
	36225	Acremonium chrysogenum	
	11661	Emericellopsis minima	
	14645	Emericellopsis minima	
	34702	Paecilomyces persicinus	
Penicillin X	10135	Penicillium chrysogenum	Antimicrobial
Protein-bound polysaccharides	20545	Trametes versicolor	Antineoplastic
	20547–20560	Trametes versicolor	

Pterin deaminase	20413	Aspergillus sp.	Antineoplastic
	20414	Aspergillus sp.	
	20410	Backusella lamprosporus	
	20411, 20412	Penicillium sp.	
Puberulic acid	20409	Rhizopus oryzae	Antimicrobial
Pyrenolides A, B, and C	8732	Penicillium puberulum	Antimicrobial
Quadrone	44532	Pyrenophora teres	Antineoplastic
Regulin	20516	Aspergillus terreus	Antineoplastic
Restrictocin	34506	Aspergillus restrictus	Antineoplastic
Rhizoxin	34475	Aspergillus fumigatus	Antimicrobial, antineoplastic
	20577	Rhizopus sp.	
RIT-2214	20389	Acremonium chrysogenum	Antimicrobial
Roridin A	24571	Myrothecium verrucaria	Antimicrobial
Roridin E	24571	Myrothecium verrucaria	Antineoplastic
Roridin E-2	20540	Myrothecium verrucaria	Antineoplastic
Roridin J	24571	Myrothecium verrucaria	Antimicrobial, antineoplastic
Roridin L-2	20605	Myrothecium roridum	
Rubrofusarin	8740	Aspergillus carbonarius	Antineoplastic
Scytalidin	16675	Scytalidium album	Antimicrobial
Scytalone	16675	Sctalidium album	Antimicrobial
SC-28762 and SC-28763	28806	Paecilomyces varioti	Antimicrobial
SC-30532	28806	Paecilomyces varioti	Antimicrobial
Sillucin	16458, 16459	Rhizomucor pusillus	Antimicrobial
	46341, 46342	Rhizomucor pusillus	
Sirodesmins A, B, C, and G	22064	Rhizomucor miehei	Antiviral
	36539	Sirodesmium diversum	

(continued)

199

TABLE I (continued)

Antibiotic	ATCC No.	Fungus	Type of antibiotic
SL 1846	34920	Pseudeurotium ovalis	Antineoplastic
SL 2266	36386	Sordaria araneosa	Antineoplastic
SL 3364	36385	Sepedonium chrysospermum	Antineoplastic
SL 3440	36384	Diheterospora chlamydosporia	Antineoplastic
Sordarin	36386	Sordaria araneosa	Antimicrobial
Spheropsidin	34505	Phoma sp.	Antimicrobial
Statalon	14586	Penicillium brevicompactum	Antiviral
Stemlon	20183	Stemphylium botryosum	Antiviral
Terrecyclic acid A	20516	Aspergillus terreus	Antineoplastic
Terreic acid	12238	Aspergillus terreus	Antimicrobial
Tetronic acids	16797	Aspergillus panamensis	Antineoplastic
Thermozymocidin	20349	Mycelia sterilia	Antimicrobial
Trichorzianine A IIIc	20672	Trichoderma harzianum	Antimicrobial
Tris–dechloronornidulin	24528	Aspergillus nidulans	Antimicrobial
T-2636 C (monoacyl derivative)	20245	Aspergillus sojae	Antimicrobial
U-53,946	56273	Paecilomyces abruptus	Antimicrobial
Unnamed	34441, 34442	Aspergillus caesiptosus	Antimicrobial
Unnamed	28221	Byssochlamys nivea	Antimicrobial
Unnamed	28799	Byssochlamys fulva	Antiviral
Unnamed	56822	Candida guilliermondii	Antimicrobial
Unnamed	11574	Coriolus consors	Antineoplastic
Unnamed	46829	Fusarium solani	Antineoplastic
Unnamed	11081	Gliocladium sp.	Antimicrobial

Unnamed	8715	*Gloeophyllum trabeum*	Antineoplastic
Unnamed	11453	*Lentinus lepideus*	Antineoplastic
Unnamed	9419	*Lenzites saepiaria*	Antineoplastic
Unnamed	24148	*Paecilomyces fusisporus*	Antimicrobial
Unnamed	9480	*Penicillium chrysogenum*	Antiviral
Unnamed	10002, 10003	*Penicillium chrysogenum*	Antiviral
Unnamed	10238, 11707	*Penicillium chrysogenum*	Antiviral
Unnamed	13799	*Penicillium chrysogenum*	Antiviral
Unnamed	20399	*Penicillium rugulosum*	Antimicrobial
Unnamed	20398	*Penicillium variable*	Antimicrobial
Unnamed	28144	*Stilbella erythrocephala*	Antimicrobial
Unnamed	60977–60981	*Torulaspora hansenii*	Antimicrobial
Unnamed	36936	*Trichoderma todica*	Antiviral
Unnamed	10938	*Xylobolus subpileatus*	Antineoplastic
Vergosin	26289	*Verticillium dahliae*	Antimicrobial
Vermicillin	26015	*Talaromyces flavus var. flavus*	Antineoplastic
Vermiculine	26015	*Talaromyces flavus var. flavus*	Antimicrobial
	60775	*Talaromyces wortmannii*	
Vermistatin	26015	*Talaromyces flavus var. flavus*	Antineoplastic
Vioxanthin (semi-)	42743	*Penicillium citreonigrum*	Antimicrobial
WF-1360	20577	*Rhizopus sp.*	Antineoplastic
ZN-6	14700	*Fusidium coccineum*	Antimicrobial
1233	24941	*Cephalosporium sp.*	Antimicrobial

tendency for related fungi to produce similar antibiotics and for distinct taxonomic groups to produce distinct antibiotics? His answer to the first was "no" and to the second, "yes." Broadbent listed 12 active species of Phycomycetes, but only one compound—an antibacterial antibiotic ramycin—was isolated from *Mucor ramannianus*. Ramycin later appeared to be identical to fusidic acid produced by an imperfect fungus *Fusidium coccineum* (Vanderheghe et al., 1965).

The antifungal agents produced by fungi are shown in Table II. There are 106 antifungal antibiotics produced by 146 fungi, which include 113 Fungi Imperfecti, 22 Ascomycetes, 10 Basidiomycetes, and 1 Phycomycete. Among these antibiotics, only griseofulvin is currently in clinical use. Griseofulvin is active *in vitro* and *in vivo* against dermatophytes such as *Epidermophyton*, *Microsporum*, and *Trichophyton*. It is elaborated by *Cephalosporium curtipes*, *Khuskia oryzae*, *Nigrospora oryzae*, *Penicillium griseofulvum*, *Penicillium janczewski*, *Penicillium nigricans*, and *Penicillium patulum*.

During the last two decades, the search for antiviral and antitumor substances has been intensified, as has that for antibiotics effective against bacteria and fungi. Although many more antibiotics have been discovered from fungi, Broadbent's answers to Brian's questions remain the same. More antibiotics have been discovered from Fungi Imperfecti and Basidiomycetes than from other major groups of fungi. Again, the systematic screening of soil microorganisms, where these groups abound, has been the method most commonly used in the search for new antibiotics. Whether the numbers of antibiotics so far obtained from each of the major taxonomic groups of fungi reflect the real potentialities of these groups is therefore very difficult to ascertain. The fact that very few negative results are published further complicates this issue. Nevertheless, it can be ascertained that in screening for antibiotics, different groups of fungi have been very unevenly studied.

V. Antitumor–Antiviral Substances Produced by Fungi

As the great threat to human life by both neoplastic diseases and AIDS continues to increase, the pursuit of antitumor–antiviral antibiotics takes on a compelling urgency. According to Hata (1977), over 300 antitumor substances had been isolated from microbial origin. Of these, 76% were derived from Actinomycetes, 13% from fungi, and 11% from bacteria. Among the 43 antitumor substances from fungi, 23 were obtained from Fungi Imperfecti, 15 from Basidiomyctes, and 5 from Ascomycetes. Phycomycetes were not mentioned.

However, our recent survey indicates that at least 220 antitumor

TABLE II
ANTIFUNGAL SUBSTANCES PRODUCED BY FUNGI

Antibiotic	Fungus
Aculeacins	Aspergillus aculeatus
Agrocybin	Agrocybe dura
Alamethicin	Trichoderma viride
Alibidin	Penicillium albidum
Alternaric acid	Alternaria solani
Antibiotic	
A-2	Myrothecium verrucaria
A-22082	Aspergillus nidulans
A25822A	Geotrichum flave-brienneum
A-25822B	Geotrichum flave-brienneum
A-25822D	Geotrichum flave-brienneum
A-25822H	Geotrichum flave-brienneum
A-25822L	Geotrichum flave-brienneum
A-25822M	Geotrichum flave-brienneum
A-25822N	Geotrichum flave-brienneum
A-26771B	Penicillium turbatum
A-30641	Aspergillus tamarii
A-30912	Aspergillus rugulosus
A-32204	Aspergillus echinulatus
A-32390A	Pyrenochaeta
AB-22	Pichia polymorpha
Antibiotic OM-1	Oudemansiella mucida
Antibiotic S-7481/F 1	Cylindrocarpon lucidum
Antibiotic SL7810	Aspergillus rugulosus
Antibiotic YO-7396	Colephoma
Antibiotic 417A	Penicillium claviforme
	Aspergillus flaschentraegeri, Fusarium acuminatum, Scytalidium aurantiacum
Ascochitin	Ascochyta pisi
Athelstatin	Aspergillus niger
Aurantiogliocladin gliorosein	Gliocladium roseum
Avenaciolide	Aspergillus avenaceous
Botric acid	Botrytis sp.
Canadensolide	Penicillium canadense
Candidulin	Aspergillus candidus
Canescin	Penicillium canescens
Cerulenin	Sartorya fumigata
Chloroflavonin	Aspergillus candidus
Chloromycorrhizin A	Mycorrhizal fungus
Cladosporin	Cladosporium cladosporioides
Cochliodinol	Chaetomium cochiodes, Chaetomium globosum

(continued)

TABLE II (continued)

Antibiotic	Fungus
Cortinellin	*Lentinus edodes*
Crotocin (antibiotic T)	*Cephalosporium crotocinigenum*
Cyanein (brefeldin A)	*Penicillium cyaneum*, *Penicillium brefeldianum*, *Phyllosticta* (CBS 47,963)
Cyathin	*Cyathus helenae*
Cyclopaldic Acid	*Penicillium cyclopium*
Cylindrocladin	*Cylindrocladium ilicicola*
Desferritriacetyl-fusigenin	*Aspergillus deflectus* (CBS 109.55)
5,6-Dihydro-5(R)-acetoxy-6(S)-1′,2-trans-epoxypropyl-2H-pyran-2-one	*Aspergillus* sp. (NRRL 5769)
Fumigachlorin	*Sartorya fumigata* var. *spinosa*
Funiculosin	*Penicillium funiculosum*
Gladiolic acid	*Penicillium gladioli*
Gliotoxin	*Aspergillus fumigatus*, *Aspergillus* sp., *Gliocladium virens*, *Penicillium terlikowskii*, *Penicillium viride*
Glutinosin	*Myrothecium verrucaria*
Griseofulvin (hydroxy)	*Cephalosporium curtipes*, *Khuskia oryzae*, *Nigrospora oryzae*, *Penicillium griseofulvum*, *Penicillium janczewski*, *Penicillium nigricans*, *Penicillium patulum*
Helicocerin	*Helicoceras oryzae*
Helmintin	*Helminthosporium* sp.
Humicolin	*Aspergillus asperescens*
Hyalodendrin	*Hyalodendron* sp.
Isotentoxine	*Alternaria mali*
Itabashillin	*Penicillium oxalicum* var. *itabashikum*
Javanicin (solanione)	*Fusarium javanicum*
Kojic acid	*Aspergillus flavus*, *Aspergillus oryzae*, *Aspergillus oryzae* var. *viride*, *Aspergillus tamarii*, *Penicillium kojigenum*

(continued)

TABLE II (continued)

Antibiotic	Fungus
Lambertellin	*Lambertella hicoriae*
Lenzitin	*Lenzites saepiaria*
Leucinostatin	*Paecilomyces lilacinum*
Lilacinin	*Penicillium lilacinum*
LL-N313	*Sporormia affinis*
LL-Z127	*Acrostalagmus sp.*
LL-1271	*Acrostalagmus sp.*
Lunatoic acid A	*Cochliobolus lunata*
Lunatoic acid B	*Cochliobolus lunata*
Mollisin	*Mollisia caesia*
Monocerin	*Bipolaris monoceras*
Mycobacillin	*Candida albicans*
Mycomycin	*Odontia bicolor*
Mycophenolic acid	*Penicillium brevi-compactum,* *Penicillium stoloiferum*
Nidulin, nornidulin	*Aspergillus nidulans,* *Aspergillus unguis*
Novalichin	*Paecilomyces fusisporus*
Oosporein (chaetomidin)	*Phlebia mellea,* *Serpula lacrimans*
Ophiobolin (cochliobolin)	*Cochliobolus miyabeanus,* *Cochliobolus heterostrophus*
Ophicocordin	*Cordyceps ophioglossoides*
Papulacandin A, B, C, and D	*Papularia sphaerosperma*
Patulin (clavicin, clavatin, claviformin, expansin, mycoin, penicidin	*Aspergillus clavatus,* *Penicillium claviforme,* *Penicillium expansum,* *Penicillium griseofulvum,* *Penicillium patulum*
Pearlmycin	*Botryotrichum piluliferman* (FERM-P 1647)
Phomalactone	*Phoma minispora*
Puloilloric acid	*Penicillium puloillorum*
Pyrenophorin	*Pyrenophora avenae*
Quadrilineatin	*Aspergillus quadrileneatus*
Radicinin	*Stemyphylium radicinum*
Rhizopchin	*Rhizopus chinensis*
Rubrogliocladin	*Gliocladium roseum*
Scytalidin	*Scytalidium aurantiacum*
Sellenin (cellulenin)	*Cephalosporium cellulens*
Siccanin	*Helminthosporium siccans*
Sphaeropsidin	*Phoma sp.*
Striatin A, B, and C	*Cyathus striatus*
Stysadin	*Stysanus medius*

(continued)

TABLE II (continued)

Antibiotic	Fungus
Talaron	*Talaromyces flavus* var. *flavus*
Tardin	*Penicillium tardum*
Terric acid	*Aspergillus terreus*
Thermozymocidin	*Mycelia sterilia*
Trichodermin	*Trichoderma viride*
Trichothecin	*Trichothecium roseum*
Trichoviridin	*Trichoderma viride*
Usmic acid	*Cetraria islandica*, *Cladonia alpestris*, *Cladonia cristallela*, *Cladonia leptoclada*, *Cladonia sylvatica*, *Ramalina reticulata*, *Usnea barbata*
Variotin	*Paecilomyces varioti* var. *antibiotics*
Versicolin	*Aspergillus versicolor*
X-Viridin	*Gliocladium virens*
β-Viridin	*Trichoderma viride*
Zizanin	*Helminthosporium zizaniae*

agents and 42 antiviral agents have been reported to be produced by the fungi (Table III and IV). The antitumor agents were elaborated by 208 Fungi Imperfecti, 109 Basidiomycetes, 34 Ascomycetes, and 13 Phycomycetes. The antiviral agents were isolated from 66 Fungi Imperfecti, 12 Ascomycetes, 7 Basidiomycetes, and 1 Phycomycetes.

A. FUNGI IMPERFECTI

1. Mycoviruses

The substances helenine, statolon, noxiversion, and 6-MFA are not directly viricidal. Their antiviral activities are known to reside in mycoviruses that act as inducers of endogenous interferon in the host body to protect against lethal doses of infectious mammalian viruses (Maheshwari and Gupta, 1973; Cooke and Stevenson, 1965; Kleinschmidt et al., 1964). However, commercial exploitation of these substances is not promising, due to difficulties in isolation and concentration (Kleinschmidt and Ellis, 1975). Helenine is isolated from *Penicil-*

TABLE III
Antitumor–Antiviral Substances Produced by Fungi

Antibiotic	Reference	Antibiotic	Reference
Acetylaranotin	Miller et al. (1968)	Bleomycinic acid	Umezawa et al. (1974)
Aflatoxin B_1	Lemarinier and Jaquet (1975)	Botryodiplodin	McCurry and Abe (1973)
α-Sarcin	Endo and Wool (1982)	Bredinin	Okamoto et al. (1978)
Altenuene	Magan et al. (1984)	Brefeldin A	Hutchinson (1983)
Altenario	Magan et al. (1984)	Brefeldins A and C	Sunagawa et al. (1983)
Altenuisol	Pero et al. (1973a)		
Altertoxins, etc.	Pero et al. (1973b)		
3'-Amino-3'-deoxyade	Guarino and Kredien (1963)	Calonectrin	Gardner et al. (1972)
Anguidine	Liao et al. (1976)	Calvacin	Roland et al. (1960)
Anguidine, ovalicin	Hartmann et al. (1978)	Calvatic acid	Okuda and Fugiwara (1982)
Antitumor	Espenshade and Griffith (1966)	Catenarin	Anke et al. (1981)
Antiviral	Kandefer-Szerszen et al. (1980)	Chaetocin	Hauser et al. (1972)
		Chaetoglobosins A, B, C, and F	Umeda et al. (1975)
Antiviral	Takehara et al. (1984)		
Antiviral	Sokoloff and Toda (1967b)	Chaetoglobosin C	Springer et al. (1976)
Antiviral	Villard et al. (1980)	Chaetoglobosins D and E	Notori (1977)
Aphidicolin	Starratt and Loschiavo (1974)	Chaetoglobosin K	Springer et al. (1980)
Apoaranotin	Trown et al. (1968)	Chaetoglobosins A–J	Sekita et al. (1982)
Aranoflavins	Mizuno et al. (1970)		
		Chaetomin	McInnes et al. (1976)
Aranotin	Neuss et al. (1968)	Chlamydocin	Close and Huguenin (1974)
Ascochlorin	Aldridge et al. (1972)	Cordycepin	Y. Ito et al. (1981)
Ascoforanose	Glasby (1976)	Coriolan (polysaccharide)	Ito et al. (1979)
L-Asperaginase	Broome (1965)		
Asperlicin	Goetz et al. (1985)	Coriolin	Nishimura et al. (1980b)
Asperlin	Lesage and Perlin (1978a,b)		
		Crinipellin	Kupka et al. (1979)
Aspterric acid	Tsuda et al. (1978)	Crotocin	Glaz et al. (1966)
Asterriquinone	Yamamoto et al. (1976)	Cyanein	Betina et al. (1964)
		Cyclopin	Naficy (1965)
(+)-R-Avellaneol	Mair et al. (1982)	Cylindrochlorin	Kato et al. (1969a)
		Cytochalasin	Cuppoletti et al. (1981)
Baccharin	Jarvis et al. (1981)		
Basidalin	Iinuma et al. (1983)	Cytochalasin A	Cunningham et al. (1979)
Bikaverin	Iijima et al. (1979)		

(continued)

TABLE III (continued)

Antibiotic	Reference	Antibiotic	Reference
Cytochalasins A and B	Wells et al. (1981)	Epipolythiopiperazinedione	Michel et al. (1974)
Cytochalasin B	Shearer and Moore (1981)	Epoformin	Glasby (1976)
Cytochalasins B and D	Graf et al. (1974)	Equisetin	Vesonder et al. (1979)
Cytochalasin C	Carter (1967)	Ergosterol peroxide	Arditti et al. (1972)
Cytochalasins C and D	Aldridge and Turner (1969)		Brown and Jacobs (1975)
Cytochalasin D	Tabuchi et al. (1981)	Fatty acids	Fuska et al. (1974a)
Cytochalasin E	Glinsukon and Suvannapur (1982)	Fermentation product	Gregory et al. (1966)
		Fermentation product	Riondel et al. (1981)
Cytochalasins K–M	Fex (1981)	Flammulin	Komatsu et al. (1963)
Cytochalasin M	Albertsson et al. (1981)	Fomecin A	Anchel et al. (1952)
Cytochalasins	Aldridge et al. (1973)	Frequentin	Fuska et al. (1972)
Cytochalasins	Low et al. (1979)	Fungal extract	Villa and Agostoni (1972)
Dactylarin	Miko et al. (1979)	Funiculosin	Ando et al. (1969)
Decumbin	Singleton and Bahonos (1958)	Fumagillin	Eble and Hansen (1951)
Deflectins A and B	Anke et al. (1981)	Funginon	
Dehydro-α-lapachone	Rosazza (1979)	Fusarenone	Uneo et al. (1967)
		Fusarin C	Farber and Sanders (1986)
Dermadine	Coats et al. (1971)		
Diacetoxyscirpenol	von Stahelin et al. (1968)	Fusariocin C	T. Ito et al. (1981)
		FWH-775	Joshi (1974)
1',2'-Dihydrorotenone	Sariasleni and Rosazza (1983)	Ganoderic acids	Toth et al. (1983)
1',2'-Dihydrorotenone (1',2'-dikydro-)	Sariasleni and Rosazza (1983)	Gliocladic acid	Itoh et al. (1982)
		Gliotoxin	Bu'Lock and Leich (1975)
Dithiadiketopiperazine	Brannon et al. (1971)	β-D-glucan	Kato et al. (1983)
		β-D-glucan, H11	Kanayama et al. (1983)
Double-stranded nucleic acid	Sutherland et al. (1971)	β-D-glucans	Misaki et al. (1981)
Duclauxin	Kuhr et al. (1973)	β-D-glucans	Usui et al. (1983)
		β-D-glucan, EA3, EA5	Ikekawa et al. (1982)
Emitanin	Sakakida and Ikekawa (1980)	Glucomannan	Oka et al. (1969)
Emodin	H. Anke et al. (1980a)	Glycoprotein	Saltarelli (1980)
		Gregatins A, B, D, etc.	T. Anke et al. (1980)
Endopeptidase	Roth et al. (1980)		

(continued)

TABLE III (continued)

Antibiotic	Reference	Antibiotic	Reference
Hadacidin	Dulaney and Gray (1962)	Melinacidin	Reusser (1968)
Helenine	Shope (1966)	Melinacidins	Argoudelis (1972)
Helminthosporal	Mayo et al. (1962)	Merulinic acids	Giannetti et al. (1978)
Heteroglycan	Ukai et al. (1983)	6-MFA	Maheshwari et al. (1978)
Hyalodendrin A	Strunz (1975)		
3'-Hydroxyverrucarins A and B	Pavanasasivam and Jarvis (1983)	Monoglycerides	Tamura et al. (1968c)
3'-Hydroxyacromycine	Betts et al. (1974)	Monordene	W.R. Grace & Co. (1980)
8 and 9-Hydroxyellipticines	Chien and Rosazza (1979)	Mycophenolic acid	Suzuki et al. (1976)
		Mycotoxins	Ueno (1985)
3'-Hydroxyrotenone	Sariasleni and Rosazza (1983)	Naematolon	Backens et al. (1984)
3'-Hydroxyverrucarin A	Pavanasasivan and Jarvis (1983)	Nebularine	Lofgren and Luning (1953)
14-α-Hydroxywithaferin A	Rosazza et al. (1978)	Neoschizophyllan	Kikumoto et al. (1978)
		Nitrogen-containing antineoplastic agents	Ueno et al. (1980)
Illudins	Anchel et al. (1950)		
Jawaharene	Mitra and Roy (1963)	Nivalenone	Tatsuno (1968)
		Noxiversin	Cooke and Stevenson (1965)
KS-2-B	Kirin Brewery Co. Ltd. (1980)	Oosporein	Vining et al. (1962)
		Oudemansin	Anke et al. (1979)
L-Lysine	Fukamura and Kato (1980)	Penicillic acid	Susuki et al. (1970)
L-Lysine oxidase	Khaduev et al. (1987)	Peptide K582M	Kondo (1980)
Lectin	Lin and Chou (1984)	Phenylalanine ammonia-lyase	Abell et al. (1973)
		Phamamide	Ferezou et al. (1979)
Lentinan	Sasaki and Takasuka (1976)	Pleuromutilin	Kavanagh et al. (1951)
Leucinostatin	Ishiguro and Arai (1976)	Poine	Elpidina (1959)
Leucinostatins A and B	Fukushima (1983)	Polysaccharide AS-1	Nanba and Kuroda (1987a)
LL-S88a	Trown et al. (1968)	Polysaccharide ATSO	Naruse and Takeda (1974)
Lomatiol, glucoside	Otten and Rosazza (1981)	Polysaccharide D-11	Sigiura et al. (1980)
D-Mannan	Matsumoto et al. (1980)	Polysaccharide GL-I	Miyazaki and Nishijima (1981)

(continued)

TABLE III (continued)

Antibiotic	Reference	Antibiotic	Reference
Polysaccharide H11	Kannayama et al. (1983)	Rhizoxin	Kiyoto (1986)
Polysaccharide PSK	Naruse and Takeda (1974)	Roridin E-2	Freckman et al. (1984)
Polysaccharide	Sasaki and Uchida (1987)	Roridin J	Jarvis et al. (1980)
Polysaccharide	Nanba and Kuroda (1987a)	Roridin L-2	R. J. Bloem et al. (1983b)
Polysaccharide enzyme	Hayashida and Watanabe (1983)	Roridin, verrucarin	Jarvis et al. (1981)
Polysaccharides EA3, EA5, EA501	Ikekawa et al. (1982)	Roseotoxin	Engstrom et al. (1975)
Polysaccharides	Kamasuka et al. (1968)	Rubratoxin B	Lemarinier and Jaquet (1975)
Polysaccharides	Komnatsu et al. (1973)	Rubrofusarin	Galmarini et al. (1974)
Polysaccharides	Mizuno and Hazama (1986)	Rugulosin	Nakamura et al. (1974)
Polysaccharides	Mizuno et al. (1986)	Satratoxins F and G	Eppley et al. (1980)
Polysaccharides	Ohno et al. (1985a)	Schizophyllan	Komatsu et al. (1969)
Polysaccharides	Sasaki et al. (1985)	Scleroglucan	Singh et al. (1974)
Polysaccharides	Nakajima et al. (1983)	Scorodonin	T. Anke et al. (1980)
Polysaccharides	Ukai et al. (1983)	Sirodesmin PL	Ferezou et al. (1980)
Polysaccharides	Yoshioka et al. (1975)	SL 1846	Sigg and Stoll (1969)
Polysaccharopeptide	Yang et al. (1987)	Statolon	Kleinschmidt and Ellis (1975)
Poricin	Schillings and Ruelius (1968)	Stemlon	Cole and Planterose (1973)
Protein-bound polysaccharides	Hotta et al. (1981)	Sterigmatocystins	Bradner et al. (1975)
Pterin deaminase	Kusakabe et al. (1979c)	Swainsonine	Hino et al. (1985)
Purine	Bieselle et al. (1955)	Tenuazonic acid	Gatenbeck and Sierakewicz (1973)
Quadrone	Calton et al. (1979)	Terrecyclic acid A	Beale et al. (1984)
Quinone	Gupta et al. (1984)	Terreic acid	Takahashi et al. (1961)
Regulin	Olson and Goerner (1966)	Tetronic acids	Anke et al. (1980a)
Restrictocin	Olson et al. (1963)	Trichodermin	Weindling and Emerson (1936)
Rhizopin	Das and Pal (1974)	Trichothecenes	Jarvis et al. (1983)

(continued)

TABLE III (continued)

Antibiotic	Reference	Antibiotic	Reference
Trichothecenes	Mohr et al. (1984)	Verrucarin	ElKady and Moubash (1982)
Trichothecenes	Smitka et al. (1984)		
Trichothecin	Sorenson et al. (1975)	Verrucarins, roridins	Achimi et al. (1974)
Trihydroxytetralones	Fujimoto and Satoh (1986)	Verrucarol	Jarvis et al. (1984)
Triornicin	Frederick et al. (1981b)	Verticillin A	Katagiri et al. (1970)
		Verticillins A–C	Minato et al. (1973)
Triterpene acids	Valisolalao et al. (1980)	Vertisporin	Minato et al. (1975)
		Vesiculogen	Suzuki et al. (1982)
Tumor inhibitor	Saltarelli (1980)	Viocristin	H. Anke et al. (1980b)
Vermicillin	Fuska et al. (1979a)		
Vermiculine	Jones et al. (1984)	Zygosporins	Minato et al. (1972)
Vermistatin	Fuska et al. (1979b)	Zymosan	Bradner et al. (1958)

lium funiculosum, statolon from Penicillium stoloniferum, noxiversion from Penicillium cyaneo-fulvum, and 6-MFA from Aspergillus ochraceus.

An antiviral substance, cyclopin, from a strain of Penicillium cyclopium effectively prevents multiplication and cytopathic effects of arbovirus groups A and B in human and chick cell cultures. Its mode of action and certain of its physical and chemical activities resemble interferon. However, human cell cultures are more sensitive to the activity of cyclopin than are chick embryo cells (Naficy, 1965).

2. Mycotoxins

Mycotoxins are fungal secondary metabolites produced in foodstuffs which are known to be casual agents of mycotoxicoses in animals and humans. Depending on the level of selective toxicity and dosage, some mycotoxins are reported as antitumor and/or antiviral agents.

Aflatoxin B_1 produced by Aspergillus flavus and Aspergillus parasiticus is the most potential carcinogen of the aflatoxin group, but has weak antitumor effects (Green, 1968; Lemarinier and Jaquet, 1975). Sterigmatocystins are produced primarily by Aspergillus versicolor and have carcinogenic properties similar to the structurally related aflatoxins. However, sterigmatocystins have much less toxic potential than aflatoxins, and have been reported to have significant inhibitory effects on the transplanted mouse leukemias P-388 and L-1210.

TABLE IV

FUNGI KNOWN TO PRODUCE ANTITUMOR–ANTIVIRAL SUBSTANCES

Organism	Compound	Type of activity	Reference
Fungi imperfecti			
Acrostalagmus cinnabarinus var. *melinacidicus*	Melinacidins I–IV (dithiodiketopiperazine)	Antitumor	Argoudellis (1972)
Alternaria alternata	Altenuene	Antitumor	Magan et al. (1984)
Alternaria alternata	Altenuisol	Antitumor	Pero et al. (1973a)
Alternaria alternata	Alternariol	Antitumor	Magan et al. (1984)
Alternaria alternata	Altertoxins	Antitumor	Pero et al. (1973b)
Alternaria alternata	Tenuazonic acid	Antitumor, antiviral	Gatenbeck and Sierakewicz (1973)
Alternaria carthami	Brefeldin A	Antiviral	Hutchinson (1983)
Alternaria dianthicola	Ergosterol peroxide	Antitumor	Brown and Jacobs (1975)
Alternaria tenuissima	Tenuazonic acid	Antitumor, antiviral	Griffin and Chu (1983)
Alternaria imperfecta	Ascotoxine (brefeldin A)	Antitumor	Susuki et al. (1970)
Ascochyta viciae	Ascochlorin (terpene)	Antitumor	Tamura et al. (1968b)
Ascochyta viciae	Ascofuranone	Antitumor, antiviral	Sasaki et al. (1971)
Aspergillus alliaceus	Asperlicin	Antitumor	Goetz et al. (1985)
Aspergillus amstelodami	Sterigmatocystins	Antitumor	Bradner et al. (1975)
Aspergillus carbonarius	Rubrofusarin	Antitumor	Galmarini et al. (1974)
Aspergillus chevalieri	Gliotoxin	Antitumor, antiviral	Reilly et al. (1953)
Aspergillus chevalieri	Sterigmatocystins	Antitumor	Bradner et al. (1975)
Aspergillus clavatus	Cytochalasins A–E, K–M	Antitumor	Katagiri and Matsuura (1971)
Aspergillus cristatus	Catenarin, emodin, vincristine	Antitumor	H. Anke et al. (1980a)
Aspergillus deflectus	Deflectins A and B	Antitumor	Anke et al. (1981)
Aspergillus flavus	Aflatoxin B_1	Antitumor	Green (1968)
Aspergillus flavus	Rubratoxin B	Antitumor	Lemarinier and Jaquet (1975)
Aspergillus flavus	Sterigmatocystins	Antitumor	Bradner et al. (1975)
Aspergillus fumigatus	Fumagillin	Antitumor	Eble and Hanson (1951)
Aspergillus fumigatus	Funginon	Antitumor	Omura (1965)
Aspergillus fumigatus	Gliotoxin	Antitumor, antiviral	Weindling and Emerson (1936)

Species	Compound	Activity	Reference
Aspergillus funiculosum	Funiculosin (anthraquinone)	Antitumor	Ando et al. (1969)
Aspergillus giganteus	α-Sarcin (peptide)	Antitumor	Olson (1963), Olson and Goerner (1965), Olson et al. (1965)
Aspergillus gymnosardae	Pterin deaminase	Antitumor	Kusakabe et al. (1976)
Aspergillus japonicus	Polysaccharides	Antitumor	Sakai et al. (1968)
Aspergillus melleus	Penicillic acid	Antitumor, antiviral	Suzuki et al. (1971)
Aspergillus nidulans	Asperline (lactone)	Antitumor	Tahave et al. (1971)
Aspergillus nidulans	Brefeldin A	Antitumor, antiviral	Owen and Bruyan (1965)
Aspergillus nidulans	Cordycepin (3'-deoxypurine nucleoside)	Antitumor	Kodama et al. (1979)
Aspergillus nidulans	Sterigmatocystins	Antitumor	Bradner et al. (1975)
Aspergillus nidulans	Asperlin	Antitumor	Lesage and Perli (1978a)
Aspergillus niger	Polysaccharides	Antitumor	Sakai et al. (1968)
Aspergillus niger	FWH-775	Antiviral	Joshi (1974)
Aspergillus niger	Jawaharene	Antitumor	Mitra and Roy (1963)
Aspergillus niger	Endopeptidase	Antiviral	Roth et al. (1980)
Aspergillus niger	9-Hydroxyacronycine	Antitumor	Betts et al. (1974)
Aspergillus ochraceus	Penicillic acid	Antitumor, antiviral	Suzuki et al. (1971)
Aspergillus ochraceus	6-MFA (mycovirus, interferon)	Antiviral	Maheshwari et al. (1978)
Aspergillus oryzae	Oryzachlorin (dithiodiketopiperazine)	Antitumor	Kato et al. (1969b)
Aspergillus oryzae	Polysaccharide	Antitumor	Sasaki and Uchida (1987)
Aspergillus oryzae	Pterin deaminase	Antitumor	Kusakabe et al. (1976)
Aspergillus ostianus	Penicillic acid	Antitumor, antiviral	Suzuki et al. (1971)
Aspergillus panamensis	Tetronic acid	Antitumor	Anke et al. (1980a)
Aspergillus panamensis	Gregatins A, B, D, etc.	Antitumor	Anke et al. (1980a)
Aspergillus parasiticus	Aflatoxin B_1	Antitumor	Lemarinier and Jaquet (1975)
Aspergillus parasiticus	Sterigmatocystins	Antitumor	Bradner et al. (1975)
Aspergillus restrictus	Restrictocin	Antitumor	Olson and Goerner (1966)
Aspergillus restrictus	Ragulin	Antitumor	Olson and Goerner (1966)
Aspergillus ruber	Gliotoxin	Antitumor, antiviral	Reilly et al. (1953)
Aspergillus ruber	Sterigmatocystins	Antitumor	Bradner et al. (1975)

(continued)

TABLE IV (continued)

Organism	Compound	Type of activity	Reference
Aspergillus rugulosus	Sterigmatocystins	Antitumor	Bradner et al. (1975)
Aspergillus sulphureus	Penicillic acid	Antitumor, antiviral	Suzuki et al. (1971)
Aspergillus tamarii	Pterin deaminase	Antitumor	Kusakabe et al. (1976)
Aspergillus terreus	Acetylaranotin (LL-S88)	Antiviral	Miller et al. (1968)
Aspergillus terreus	Aspterric acid	Antitumor	Tsuda et al. (1978)
Aspergillus terreus	Asteriquinone	Antitumor	Yamamoto et al. (1976)
Aspergillus terreus	Gliotoxin	Antitumor, antiviral	Bu'Lock and Leich (1975)
Aspergillus terreus	Penicillic acid	Antitumor, antiviral	Kawasaki (1972)
Aspergillus terreus	Quadrone	Antitumor	Calton et al. (1978)
Aspergillus terreus	Terreic acid (benzoquinone)	Antitumor	Takahashi et al. (1961)
Aspergillus ustus	Sterigmatocystins	Antitumor	Bradner et al. (1975)
Aspergillus versicolor	Sterigmatocystins	Antitumor	Bradner et al. (1975)
Aspergillus sp. (ATCC 20413 and 20414)	Pterin deaminase	Antitumor	Kusakabe et al. (1976)
Aspergillus sp.	Sterigmatocystins	Antitumor	Bradner et al. (1975)
Beauveria bassiana	Oosporein	Antitumor	Vining et al. (1962)
Bipolaris sorokiniana	Sterigmatocystins	Antitumor	Bradner et al. (1975)
Botryodiplodia theobromae	Botryodiplodin (tetrahydrofuron)	Antitumor	McCurry and Abe (1973)
Byssochlamys fulva	Unnamed	Antiviral	Sokoloff and Toda (1967a)
Candida albicans	Glycoprotein	Antitumor	Saltarelli (1980)
Candida albicans	Mannan	Antitumor	Kumano et al. (1973)
Candida albicans	Tumor inhibitor	Antitumor	Saltarelli (1980)
Candida guilliermondii	Polysaccharides hydroglucan	Antitumor	Mankowski et al. (1957)
Candida tropicalis	Polysaccharides	Antiviral	Kovalenko et al. (1978)
Candida utilis	Glucomannan	Antitumor	Oka et al. (1969)
Cephalosporium aphidicola	Aphidicolin (tetracyclic diterpoid)	Antiviral	Bucknall et al. (1973)
Cephalosporium crotocinigenum	Crotocin (trichothecene)	Antitumor	Glaz et al. (1966)
Cephalosporium diospyri	Fatty acids	Antitumor	Ando et al. (1968)
Chrysosporium merdarium	Anthraquinones	Antitumor	Slater et al. (1971)

Cercospora oryzae	Monoglycerides	Antitumor	Tamura et al. (1968c)
Chaetomium globosum	Gliocladic acid	Antitumor	Itoh et al. (1982)
Cladosporium fulvum	Polysaccharides	Antitumor	Komatsu et al. (1973)
Curvularia lunata	Cytochalasins A–H, J	Antitumor	Katagiri and Matsuura (1971)
Curvularia lunata	Dehydrolapachone	Antitumor	Rosazza (1979)
Cylindrocladium ilicicola	Ascochlorin (terpens)	Antitumor	Aldridge et al. (1972)
Cylindrocladium ilicola	Cylindrochladin	Antitumor	Matsushima (1970)
Cylindrocladium sp.	Cylindrochlorin	Antiviral	Kato et al. (1970)
Dactylaria lutea	Dactylarin	Antitumor	Horakova et al. (1974)
Diheterospora chlamydospora	Chlamydocin (tetrapeptide)	Antitumor	Close and Huguenin (1974)
Diheterospora chalydospora	Monordene	Antitumor	McCapra et al. (1964)
Drechslera cynodontis	Anthraquinones (cynodontin)	Antitumor	White and Johnson (1971)
Epicoccum purpurascens	Triornicin	Antitumor	Frederick (1981a)
Fusarium anguioides	Anguidine (trichothecene)	Antitumor	Loeffler et al. (1964)
Fusarium anguioides	Bleomycinic acid	Antitumor	Umezawa et al. (1974)
Fusarium aquaeductum	Fusarenone	Antitumor	Ueno et al. (1975)
Fusarium aquaeductum	Fusarenon X	Antitumor	
Fusarium avenaceus	Anguidine	Antitumor	Stahelm et al. (1968)
Fusarium concolor	Anguidine	Antitumor	Stahelm et al. (1968)
Fusarium culmorum	Anguidine	Antitumor	Stahelm et al. (1968)
Fusarium episphaeria	Fusarenone	Antitumor	Ueno et al. (1975)
Fusarium episphaeria	Nivalenol	Antitumor	Tatsuno (1968)
Fusarium equiseti	Anguidine	Antitumor	Stahelm et al. (1968)
Fusarium equiseti	Equisetin	Antitumor	Vesonder et al. (1979)
Fusarium lycopersici	Bikaverin (benzoxanthone)	Antitumor	Iijima et al. (1979)
Fusarium moniliforme	Fusariocins A and C	Antitumor	T. Ito et al. (1981)
Fusarium nivale	Fusarenon X (trichothecene)	Antitumor	Ueno et al. (1969)
Fusarium nivale	Nivalenol (trichothecene)	Antitumor	Tatsuno (1968)
Fusarium niveum	Fusarenone	Antitumor	Ueno et al. (1975)

(continued)

TABLE IV (continued)

Organism	Compound	Type of activity	Reference
Fusarium niveum	Nivalenol	Antitumor	Tatsuno (1968)
Fusarium oxysporum	Bikaverin	Antitumor	Fuska et al. (1974d)
Fusarium rigidisculum	Anguidine	Antitumor	von Stahelin et al. (1968)
Fusarium roseum	Fusarenone	Antitumor	Ueno et al. (1969)
Fusarium roseum	Bleomycinic acid	Antitumor	Umezawa et al. (1974)
Fusarium roseum	Nivalenol	Antitumor	Tatsuno (1968)
Fusarium sambucinum	Anguidine	Antitumor	von Stahelin et al. (1968)
Fusarium scirpi	Anguidine	Antitumor	von Stahelin et al. (1968)
Fusarium solani	Anguidine	Antitumor	von Stahelin et al. (1968)
Fusarium sporotrichiella var. poae	Poine	Antitumor	Elpidina (1959)
Fusarium tricinctum	Anguidine	Antitumor	von Stahelin et al. (1968)
Fusarium sp.	Anguidine	Antitumor	Hartmann et al. (1978)
Gilmaniella humicola	Microline	Antitumor	Bollinger and Zardin-Tartaglia (1976)
Gliocladium fimibricatum	Gliotoxin	Antitumor, antiviral	Weindling and Emerson (1936)
Gliocladium virens	Gliotoxin	Antitumor, antiviral	Weindling and Emerson (1936)
Gliocladium virens	Gliocladic acid	Antitumor	Itoh et al. (1982)
Helminthosporium dematoideus	Cytochalasin B (macrolide)	Antitumor	Katagiri and Matsuura (1971)
Helminthosporium sativum	Helminthosporal	Antitumor	de Mayo et al. (1961)
Helminthosporium zonatum	Bleomycinic acid	Antitumor	Umezawa et al. (1974)
Helminthosporium sp.	3-amino-3'-deoxyadenosine	Antitumor	Suhadolnik et al. (1969)
Hormiscium sp.	Cytochalasins A–H, J	Antitumor	Katagiri and Matsuura (1971)
Hyalodendron sp.	Hyalodendrin A	Antitumor, antiviral	Strunz (1975)
Isaria atypicola	Polysaccharides	Antitumor	Nakajima et al. (1983)
Metarrhizium anisopliae	K582M	Antitumor, antiviral	Kondo (1980)
Metarrhizium anisopliae	Cytocholasins D and C (macrolide)	Antitumor	Katagiri and Matsuura (1971)
Myriococcum albomyces	ATCC 16425	Antitumor	Tendler (1969)

Myrothecium roridum	Verrucarins and roridins (trichothecenes)	Antitumor, antiviral	Bloem et al. (1983a)
Myrothecium roridum	Roridan A and Roridin L-2 PD113,325 and PD113,326	Antitumor	Bloem et al. (1983b)
Myrothecium roridum		Antitumor	Smitka (1984)
Myrothecium verrucria	Verrucarins and roridins (trichothecenes) Roridans D, E, H, J	Antitumor, antiviral	Jarvis et al. (1978, 1980)
Nigrosabulum sp.	Cytochalasins A–H, J	Antitumor	Katagiri and Matsuura (1971)
Nigrospora sphaeria	Aphidicolin	Antitumor, antiviral	Starratt and Loschiava (1974)
Oospora colorans	Oosporein	Antitumor	Glasby (1976)
Paecilomyces fulvus	Cell extract	Antitumor, antiviral	Tamura et al. (1968a)
Paecilomyces fulvus	ATCC 28799	Antiviral	Sokoloff and Toda (1967a)
Paecilomyces lilacinum	Leucinostatin	Antitumor	Ishiguro and Arai (1976)
Penicillium aurantiogriseum	Chaetoglobosin	Antitumor	Springer et al. (1976)
Penicillium aurantio-violaceum	Hadacidin (amino acid)	Antitumor	Gitterman et al. (1962)
Penicillium aurantio-virens	Chaetoglobosins A–F	Antitumor	Springer et al. (1976)
Penicillium brefeldianum	Brefeldin A	Antitumor, antiviral	Harri et al. (1963)
Penicillium brefeldianum	Frequentin	Antitumor	Fuska et al.(1972)
Penicillium brevi-compactum	Mycophenolic acid	Antitumor, antiviral	Williams et al. (1968)
Penicillium brevi-compactum	Statolon	Antiviral	Kleinchmidt and Ellis (1975)
Penicillium brevi-compactum	Penicillic acid	Antitumor, antiviral	Suzuki et al. (1971)
Penicillium brunneum	Rugulosin (anthroquinone)	Antitumor	Ueno et al. (1971)
Penicillium canescens	Rugulosin	Antitumor	Ueno et al. (1971)
Penicillium chrysogenum	Penicillic acid	Antitumor, antiviral	Suzuki et al. (1971)
Penicillium chrysogenum	Double-stranded nucleic acid	Antiviral	Sutherland et al. (1971)
Penicillium claviforme	Epoformin	Antitumor	Yamamoto et al. (1973)
Penicillium crustosum	Fatty acids	Antitumor	Ando et al. (1968)
Penicillium cryaneo-fulvum	Noxiversin (mycovirus, interferon)	Antiviral	Cooke and Stevenson (1965)
Penicillium cyaneum	Cyanein (=brefeldin A)	Antitumor	Betina et al. (1964)

(continued)

TABLE IV (continued)

Organism	Compound	Type of activity	Reference
Penicillium cyclopium	Cyclopin	Antiviral	Naficy (1965)
Penicillium cyclopium	Penicillic acid	Antitumor, antiviral	Suzuki et al. (1971)
Penicillium cyclopium	Rugolosin	Antitumor	Ueno et al. (1971)
Penicillium diversum	Trihydroxytetralones	Antitumor	Fujimoto and Satoh (1986)
Penicillium frequentans	Frequentin (carbocyclic compound)	Antitumor	Shibata et al. (1965)
Penicillium frequentans	Hadacidin	Antitumor	Gitterman et al. (1962)
Penicillium funiculosum	Helenine (mycovirus, interferon)	Antiviral	Shope (1966)
Penicillium glabrum	Hadicidin	Antitumor	Dulaney and Gray (1962)
Penicillium griseoroseum	Noxiversin	Antiviral	Murray et al. (1958)
Penicillium herquei	Duclauxin	Antitumor	Fuska et al. (1974d)
Penicillium islandicum	Cyclochorotine (peptide)	Antitumor	Uraguchi et al. (1961)
Penicillium islandicum	Luteoskyrin (anthraquinone)	Antitumor	Ueno et al. (1967)
Penicillium lilacinum	Leucinostatin (peptide)	Antitumor, antiviral	Arai et al. (1973)
Penicillium luteum	Sterigmatocystins	Antitumor	Bradner et al. (1975)
Penicillium martensii	Penicillic acid	Antitumor, antiviral	Suzuki et al. (1971)
Penicillium obscurum	Gliotoxin	Antitumor, antiviral	Weindling and Emerson (1936)
Penicillium olivino-viride	Penicillic acid	Antitumor, antiviral	Suzuki et al. (1971)
Penicillium palitans	Frequentin	Antitumor	Fuska et al. (1972)
Penicillium palitans	Penicillic acid	Antitumor, antiviral	Suzuki et al. (1971)
Penicillium piscarium	Penicillic acid	Antitumor, antiviral	Suzuki et al. (1971)
Penicillium puberulum	Penicillic acid	Antitumor, antiviral	Suzuki et al. (1971)
Penicillium purpurescens	Hadacidin (amino acid)	Antitumor	Gitterman et al. (1962)
Penicillium purpurogenum	Rubratoxin B	Antitumor	Umeda et al. (1970)
Penicillium roqueforti	Mycophenolic acid	Antitumor	Eagel et al. (1982)
Penicillium rugulosum	Rugulosin (anthroquinone)	Antitumor	Ueno et al. (1967)
Penicillium simplicissimum	Brefeldin A (macrolide)	Antitumor, antiviral	Harri et al. (1963)
Penicillium stipitatum	Duclauxin	Antitumor	Kuhr et al. (1973)
Penicillium stipitatum	Fatty acids	Antitumor	Fuska et al. (1974a)
Penicillium stoloniferum	Mycophenolic acid	Antitumor, antiviral	Williams et al. (1968)

Penicillium stoloniferum	Statolan (mycovirus, interferon)	Antiviral	Kleinschmidt and Ellis (1975)
Penicillium tardum	Fatty acids	Antitumor	Ando et al. (1968)
Penicillium terlikowskii	Fatty acids	Antitumor, antiviral	Richtsel et al. (1964)
Penicillium terlikowskii	Gliotoxin	Antitumor, antiviral	Weindling and Emerson (1936)
Penicillium thomii	Hadicidin	Antitumor	Dulaney and Gray (1962)
Penicillium thomii	Penicillic acid	Antitumor, antiviral	Suzuki et al. (1971)
Penicillium turbatum	Hyalodendrin A	Antitumor, antiviral	Strunz (1975)
Penicillium turbatum	A 26771A (epidithiopiperazinedione)	Antiviral	Michel et al. (1974)
Penicillium variabile	Rugulosin	Antitumor	Ueno et al. (1971)
Penicillium vermiculatum	Vermistatin, vermicillin	Antitumor	Fuska et al. (1979a,b)
Penicillium vermiculatum	Vermiculine	Antitumor	Jones et al. (1984)
Penicillium verrucosum	Penicillic acid	Antitumor, antiviral	Suzuki et al. (1971)
Penicillium viridicatum	Penicillic acid	Antitumor, antiviral	Suzuki et al. (1971)
Penicillium sp.	Decumbin (=brefeldin A)	Antitumor	Singleton and Bohonos (1958)
Penicillium sp. (ATCC 20411 and 20412)	Pterin deaminase	Antitumor	Kusakabe et al. (1976)
Phoma exigua	Cytochalasins A–H, J	Antitumor	Katagiri and Matsuura (1971)
Phoma lingam	Phomamide	Antitumor	Ferezou et al. (1979)
Phoma lingam	Sirodesmin	Antitumor	Ferezou et al. (1980)
Phoma sorghina	Tenuazonic acid	Antitumor, antiviral	Miller et al. (1963)
Phoma sp.	Cytochalasins A–H, J	Antitumor	Katagiri and Matsuura (1971)
Phoma sp.	Sphaeropsidin	Antitumor	Upjohn (1969)
Phomopsis paspali	Cytochalasins A–H, J	Antitumor	Katagiri and Matsuura (1971)
Piricularia oryzae	Tenuazonic acid	Antitumor, antiviral	Miller et al. (1963)
Pithomyces chartarum	Sporidesmins A, B, D, E, F, G, and H (dithioketopiperazine)	Antitumor	Ralman et al. (1978)
Rhodosporidium toruloides	Phenylalanine ammonia-lyase	Antitumor	McGuire (1987)
Rhodotorula glutinis	Phenylalanine ammonia-lyase	Antitumor	Abell et al. (1973)
Rhodotorula rubra	Phenylalanine ammonia-lyase	Antitumor	McGuire (1987)
Sepedonium ampullosporum	Fatty acids	Antitumor	Ando et al. (1968)
Sepedonium ampullosporum	Monoglycerides	Antitumor	Tamura et al. (1968c)

(continued)

TABLE IV (continued)

Organism	Compound	Type of activity	Reference
Sepedonium chrysospermum	SL-3364	Antitumor	Ger. Patent 2,011,582
Sirodesmium diversum	Sirodesmins A, B, C, and G	Antiviral	Curtis et al. (1977)
Stachybotrys chartarum	Roridins E and H	Antitumor	Harri et al. (1962)
Stachybotrys chartarum	Satratoxins F and G	Antitumor	
Stachybotrys chartarum	Satratoxin H	Antitumor	Eppley et al. (1980)
Stachybotrys chartarum	Verrucarin	Antitumor	ElKady et Moubash (1982)
Stemphylium botryosum	Stemlon	Antiviral	Cole and Planterose (1973)
Stemphylium radicinum	Pyrenophorin	Antitumor	Nozoe et al. (1965)
Trichoderma lignorum	Trichodermin	Antitumor	Godtfredsen and Vangedal (1965)
Trichoderma viride	Unnamed	Antiviral	Sokoloff and Toda (1967)
Trichoderma viride	Dermadine	Antitumor	Coats et al. (1971)
Trichoderma viride	Gliotoxin	Antitumor, antiviral	Richtsel et al. (1964)
Trichoderma viride	Gliocladic acid	Antitumor	Itoh et al. (1982)
Trichoderma viride	Trichodermin	Antitumor	Godtfredsen and Vangedal (1965)
Trichoderma viride	L-Lysine oxidase	Antitumor	Kusakabe et al. (1980b)
Trichothecium roseum	Crotocin (trichothecene)	Antitumor	Glaz et al. (1966)
Trichothecium roseum	Trichothecin (trichothecene)	Antitumor, antiviral	Arima et al. (1968)
Verticillium cinnabarinus	Chaetomin	Antitumor, antiviral	Trown (1968)
Verticillium cinnabarinus	Melinacidins I–IV	Antitumor	Argoudelis and Reusser (1971)
Verticillium lateritium	Melinacidins I–IV	Antitumor	Argoudellis (1972)
Verticillium psalliotae	Oosporein	Antitumor	Glasby (1976)
Verticillium teneum	Chaetocin	Antitumor, antiviral	Hauser et al. (1972)
Verticillium sp.	Verticillins A and B (dithiodioxopiperazine)	Antitumor	Katagiri et al. (1970), Minato et al. (1973)
Verticimonosporium diffractum	Vertisporin	Antitumor	Minato et al. (1975)
Zygosporium masonii	Cytochalasin D (macrolide)	Antitumor	Katagiri and Matsuura (1971)
Zygosporium masonii	Zygosporin A	Antitumor	Minato and Matsumoto (1970)

Basidomycetes

Agaricales	Culture filtrates	Antiviral	Villard et al. (1980)
Agaricales nebularis	Nebularine (purine nucleoside)	Antitumor	Lofgren and Luning (1953)
Amanita phalloides	Cell extracts	Antitumor	Villa and Agostoni (1972)
Armillariella mellea	Extracts	Antiviral	Kandefer-Szerszen et al. (1980)
Auricularia auricula	Heteroglycan polysaccharides	Antitumor	Ukai et al. (1983)
Auricularia auricula	D-Glucans	Antitumor	Misaki et al. (1981)
Boletus edulis	Polypeptide	Antitumor	Lucas et al. (1957)
Boletus edulis	Extracts	Antiviral	Lucas et al. (1957)
Calvatia bovista	Cell extracts	Antitumor	Lucas et al. (1958–1959)
Calvatia craniformis	Calvatic acids	Antitumor	Umezawa et al. (1975)
Calvatia cyathiformis	Cell extracts	Antitumor	Lucas et al. (1958–1959)
Calvatia gigantea	Calvacin	Antitumor	Lucas et al. (1958–1959)
Calvatia lilacina	Calvatic acid	Antitumor	Gasco et al. (1974)
Clitocybe illudens	Illudins M and S (sesquiterpene)	Antitumor	Anchel et al. (1950)
Coltricia perennis	Polysaccharides	Antitumor	Ohtsuka et al. (1977)
Coltricia pusilla	Polysaccharides	Antitumor	Ohtsuka et al. (1977)
Coprinus nycthemerus	Fermentation products	Antitumor	Espenshade and Griffith (1966)
Coriolus biformis	Protein-bound polysaccharides	Antitumor	Hotta et al. (1981a)
Coriolus conchifer	Protein-bound polysaccharides	Antitumor	Hotta et al. (1981b)
Coriolus consors	Coriolin (sesquiterpene)	Antitumor	Takeuchi et al. (1969)
Coriolus consors	Coriolin C	Antitumor	Glasby (1976)
Coriolus consors	Diketocoriolin B (sesquiterpene)	Antitumor	Takeuchi et al. (1971)

(continued)

TABLE IV (continued)

Organism	Compound	Type of activity	Reference
Coriolus consors	Protein-bound polysaccharides	Antitumor	Hotta et al. (1981a)
Coriolus hirsutus	Diketocoriolin B	Antitumor	Ueno et al. (1980)
Coriolus hirsutus	Protein-bound polysaccharides	Antitumor	Hotta et al. (1981a)
Coriolus pargamenus	Diketocoriolin B	Antitumor	Ueno et al. (1980)
Coriolus pargamenus	Protein-bound polysaccharides	Antitumor	Hotta et al. (1981b)
Coriolus pubescens	Protein-bound polysaccharides	Antitumor	Hotta et al. (1981a)
Coriolus unicolor	Monoglycerides	Antitumor	Tamura et al. (1968c)
Coriolus versicolor	Polysaccharide	Antitumor	Ueno et al. (1980),
Coriolus versicolor	ATSO (polysaccharide)	Antitumor	Naruse and Takeda (1974)
Coriolus versicolor	Coriolan (polysaccharide)	Antitumor	Naruse and Takeda (1974)
Coriolus versicolor	Coriolan (polysaccharide)	Antitumor	Ito et al. (1979)
Coriolus versicolor	PS-K (polysaccharide)	Antitumor	Hirase et al. (1970, 1976a,b)
Coriolus versicolor	Polysaccharide	Antitumor	G.B. Patent 2,007, 246
Coriolus versicolor	PSP (polysaccharopeptide)	Antitumor	Yang et al. (1987)
Coriolus versicolor	Protein-bound polysaccharides	Antitumor	Hotta et al. (1981a)
Corticium centrifugum	Polysaccharides	Antitumor	Komatsu et al. (1973)
Corticium rolfsii	Fermentation products	Antitumor	Espenshade and Griffith (1966)
Cortinellus shiitake	Mushroom extract	Antiviral	Tsunoda and Ishida (1970)
Crepidotus sp.	Polysaccharides CPS (glucan)	Antitumor	Nakayoshi (1967)
Crinipellis stipitaria	Crinipellin	Antitumor	Kupka et al. (1963)
Cryptoderma yamanoi	Polysaccharides	Antitumor	Ohtsuka et al. (1977)
Cyclomyces fuscus	Polysaccharides	Antitumor	Ohtsuka et al. (1977)
Daedaleopsis tricolor	Polysaccharides	Antitumor	Ohtsuka et al. (1977)
Dictyophora indusiata	Mannan	Antitumor	Ukai et al. (1983)

Favolus aveolarius	Polysaccharides	Antitumor	Ohtsuka et al. (1977)
Flammulina velutipes	Flammulin (polyene)	Antitumor	Komatsu et al. (1963)
Flammulina velutipes	Polysaccharides	Antitumor	Komasuka et al. (1968)
Flammulina velutipes	Polysaccharide	Antitumor	Komatsu et al. (1973)
Flammulina velutipes	EA3 and EA5	Antitumor	Ikekawa et al. (1982)
Fomes fomentarius	Polysaccharides	Antitumor	Ohtsuka et al. (1977)
Fomes juniperus	Fomecin A	Antiviral	Anchel et al. (1952)
Fomitopsis roseus	Polysaccharides	Antitumor	Ohtsuka et al. (1977)
Ganoderma applanatum	Polysaccharide (glucan G-2)	Antitumor	Sasaki et al. (1971)
Ganoderma lucidum	Heteroglycan, GL-1	Antitumor	Miyazaki and Nishijima, (1981)
Ganoderma tsugae	Polysaccharides	Antitumor	Ohtsuka et al. (1977)
Gloeophyllum trabeum	Unnamed	Antitumor	Espenshade and Griffith (1966)
Glocoporus amorphus	Polysaccharides	Antitumor	Ohtsuka et al. (1977)
Grifola frondosa	β-D-Glucan	Antitumor	Ohno et al. (1984, 1985b, 1986a,b)
Hohenbuehlia geogenius	Culture filtrate	Antitumor	Riondel et al. (1981)
Hydnum erinaceum	Fermentation products	Antitumor	Gregory et al. (1966)
Hymenochaete mougeotii	Polysaccharides	Antitumor	Ohtsuka et al. (1977)
Inonotus kanehirae	Polysaccharides	Antitumor	Ohtsuka et al. (1977)
Inonotus sciurinus	Polysaccharides	Antitumor	Ohtsuka et al. (1977)
Irpex consors	Fermentation products	Antitumor	Espenshade and Griffith (1966)
Irpex flavus	Fermentation products	Antitumor	Espenshade and Griffith (1966)
Lampteromyces japonicus	Illudin M and S (sesquiterpene)	Antitumor	Yoshida et al. (1962), Komatsu et al. (1961)
Lentinus edodes	KS-2-B	Antitumor	Kirin Brewery Co. Ltd. (1980)
Lentinus edodes	Lentinan (polysaccharide)	Antitumor	Togami et al. (1982)
Lentinus lepideus	Fermentation products	Antitumor	Espenshade and Griffith (1966)
Lentinus trabeum	Rugulovasines A and B	Antitumor	Abe (1972)

(continued)

TABLE IV (continued)

Organism	Compound	Type of activity	Reference
Lenzites saepiaria	Fermentation products	Antitumor	Espenshade and Griffith (1966)
Lenzites trabea	Fermentation products	Antitumor	Espenshade and Griffith (1966)
Leucoagaricus naucina	Basidalin	Antitumor	Iinuma et al. (1983)
Merulius niveus	Fermentation products	Antitumor	Espenshade and Griffith (1966)
Merulius tremellosus	Merulinic acids A–C	Antitumor	Giannetti et al. (1978)
Microporus flabelliformis	Polysaccharides	Antitumor	Ohtsuka et al. (1977)
Mucida (Collybia) radicata	Fermentation products	Antitumor	Espenshade and Griffith (1966)
Nematoloma fasicuclae	Nematolin	Antitumor	Ito et al. (1970)
Oudemansiella mucida	Fermentation products	Antitumor	Anke et al. (1979)
Oudemansiella radicata	Polysaccharides	Antitumor	Ohtsuka et al. (1977)
Panus conchataus	Fermentation products	Antitumor	Espenshade and Griffith (1966)
Panus conchatus	Panepoxydion	Antitumor	Kis et al. (1970)
Panus rudis	Panepoxydion	Antitumor	Kis et al. (1970)
Phellinus igniarius	Polysaccharides	Antitumor	Ohtsuka et al. (1977)
Phellinus linteus	Polysaccharide	Antitumor	Ikekawa et al. (1982)
Phellinus robustus	Polysaccharides	Antitumor	Ohtsuka et al. (1977)
Phlebia albida	Oosporein	Antitumor	Takeshita and Anchel (1965)
Phlebia mellea	Oosporein	Antitumor	Takeshita and Anchel (1965)
Phlebia radiata	Merulinic Acids A–C	Antitumor	Giannetti et al. (1978)
Pholiota formosa	Fermentation products	Antitumor	Espenshade and Griffith (1966)
Pholiota nameko	Polysaccharide	Antitumor	Komatsu et al. (1973)
Pholiota nameko	Polysaccharide	Antitumor	Komasuka et al. (1968)
Piptoporus betulinus	Polysaccharides	Antitumor	Ohtsuka et al. (1977)
Pleurotus mutilis	Pleuromutilin	Antiviral	Kavanagh et al. (1951)
Pleurotus ostreatus	Polysaccharide	Antitumor	Yoshioka et al. (1975)
Pleurotus passeckerianus	Fermentation products	Antitumor	Espenshade and Griffith (1966)
Pleurotus passeckerianus	Pleuromutilin	Antiviral	Kavanagh et al. (1951)

(Polyporus) biennis	Fermentation products	Antitumor	Espenshade and Griffith (1966)
Polyporus distortus	Fermentation products	Antitumor	Espenshade and Griffith (1966)
Polyporus obtusa	Fermentation products	Antitumor	Espenshade and Griffith (1966)
Polyporus pescaprae	Polysaccharides	Antitumor	Ohtsuka et al. (1977)
Polyporus sp.	Fermentation products	Antitumor	Gregory et al. (1966)
Poria cocos	Pachymic and tumulosic acids	Antitumor	Valisolalao et al. (1980)
Poria cocos	Pachymaran, pachyman (polysaccharide)	Antitumor	Chihara et al. (1970a)
Poria cocos	H11, D-glucan	Antitumor	Kanayama et al. (1983)
Poria corticola	Fermentation products	Antitumor	Espenshade and Griffith (1966)
Poria corticola	Poricin (acidic protein)	Antitumor	Schillings and Ruelius (1968)
Poria subacida	Fermentation products	Antitumor	Espenshade and Griffith (1966)
Poria xantha	Fermentation products	Antitumor	Espenshade and Griffith (1966)
Rhizoctonia repens	Ergosterol peroxide	Antitumor	Arditti et al. (1972)
Rigidoporus durus	Polysaccharides	Antitumor	Ohtsuka et al. (1977)
Rigidoporus ulmarius	Polysaccharides	Antitumor	Ohtsuka et al. (1977)
Schizophyllum commune	Schizophyllan (polysaccharide)	Antitumor	Komatsu et al. (1969)
Schizophyllum commune	Scleroglucan	Antitumor	Okuda et al. (1972)
Stereum subpileatum	Fermentation products	Antitumor	Espenshade and Griffith (1966)
Strobilurus tenacellus	Strobilurins	Antimicrobial Antineoplastic	Anke and Oberwinkler (1977)
Tremella fuciformis	Emitanin	Antitumor	Sakakida and Ikekawa (1980)
Tricholoma aggregatum	Polysaccharides	Antitumor	Kamatsu et al. (1973)
Tricholoma panaeolum	Fermentation products	Antitumor	Espenshade and Griffith (1966)
Tricholoma matsutake	Emitanin	Antitumor	Sakakida and Ikekawa (1980)
Ustilago zeae	Fermentation products	Antitumor	Gregory et al. (1966)

(continued)

TABLE IV (continued)

Organism	Compound	Type of activity	Reference
Volvariella volvacea	Lectin	Antitumor	Lin and Chou (1984)
Volvariella volvacea	Polysaccharides	Antitumor	Misaki et al. (1986)
Ascomycetes			
Amauroascus aureus	Acetylaranotin	Antiviral	Trown et al. (1968)
Amauroascus aureus	Apoaranotin	Antiviral	Neuss et al. (1968)
Arachniotus aureus	Aranotins	Antiviral	Trown et al. (1968)
Arachniotus aureus		Antiviral	Nagarajan et al. (1968)
Arachniotus flavolutens	Aranoflavins A and B (macrolide)	Antitumor	Glasby (1976)
Byssochlamys fulva	Unnamed	Antiviral	Sokoloff and Toda, (1967b)
Calonectria nivalis	Calonectrin	Antitumor	Gardner et al. (1972)
Calonectria nivalis	15-Deacetylcalonectrin	Antitumor	Gardner et al. (1972)
Chaetomium aureus	Oosporein	Antitumor	Divekar et al. (1959)
Chaetomium cochliodes	Chaetomin (dithiodioxopiperazine)	Antitumor, antiviral	McInnes et al. (1976)
Chaetomium cochliodes	Chaetoglobosins A–F	Antitumor	Umeda et al., (1975)
Chaetomium elatum	Chaetomin	Antitumor, antiviral	Trown (1968)
Chaetomium funiculum	Chaetomin	Antitumor, antiviral	Trown (1968)
Chetomium globosum	Chaetomin (dithiodiozopiperazine)	Antitumor, antiviral	Trown (1968)
Chaetomium globosum	Chaetoglobosins A–F	Antitumor	Umeda et al. (1975)
Chaetomium minutum	Chaetocin (dithiodiozopiperazine)	Antitumor, antiviral	Hauser et al (1970, 1972)
Chaetomium spirale	Chaetomin	Antitumor, antiviral	Trown (1968)

Organism	Product	Activity	Reference
Cochliobolus sativas	Polysaccharides	Antitumor	Komatsu et al. (1973)
Cordyceps militaris	Cordycepin	Antitumor	Cunningham et al. (1950)
Cordyceps militaris	3-Amino-3'-deoxyadenosine	Antitumor	Guarino and Kredien (1963)
Eupenicillium brefeldianum	Bredinin	Antitumor, antiviral	Mizuno et al. (1974)
Eupenicillium brefeldianum	Brefeldins A and C		Sunagawa et al. (1983)
Gibberella fujikuroi	Bikaverin (benzoxanthone)	Antitumor	Fuska et al. (1974d)
Morchella hortensis	Fermentation products	Antitumor	Espenshade and Griffith (1966)
Nectria coccinea	Ascochlorin (terpene)	Antitumor	Aldridge et al. (1972)
Peziza vesiculosa	Polysaccharides	Antitumor	Ohno et al. (1985a)
Peziza vesiculosa	Vesiculogen	Antitumor	Suzuki et al. (1982)
Pseudeurotium ovalis	Hyalodendrin A	Antitumor	Sigg and Stoll (1969)
Pseudeurotium ovalis	SL 1846	Antitumor	Sigg and Stoll (1969)
Pyrenophora avenae	Pyrenophorin (lactone)	Antitumor	Nozoe et al. (1965)
Pyrenomycetes	(+)-R-Avellaneol	Antitumor	Mair et al. (1982)
Rosellinia necatrix	Cytochalasins A–H, J	Antitumor	Katagiri and Matsuura (1971)
Saccharomyces cerevisiae	Immunity factor	Antitumor	Karson and Ballou (1978)
Saccharomyces cerevisiae	D-Mannan	Antitumor	Lewisohn (1940)
Saccharomyces sp.	Cell extract	Antitumor	Bradner et al. (1958)
Saccharomyces sp.	Zymosan (protein carbohydrate complexes)	Antitumor	
Sclerotium glucanicum	Scleroglucan polysaccharide	Antitumor	Singh et al. (1974)

(continued)

TABLE IV (continued)

Organism	Compound	Type of activity	Reference
Sclerotinia sclerotiorum	Polysaccharides (sclerotinan)	Antitumor	Komatsu et al. (1973)
Sordaria araneosa	SL 2266	Antitumor	Hauser and Sigg (1971)
Yeasts	L-Asparaginase	Antitumor	Broome (1965)
Phycomycetes			
Cunninghamella echinulata	9-Hydroxyacronycine	Antitumor	Betts et al. (1974)
Cunninghamella echinulata	Iomatiol, lapachol	Antitumor	Otten and Rosazza (1981)
Cunninghamella elegans	Acronycin	Antitumor	Betts et al. (1974)
Cunninghamella elegans	3'-Hydroxyrotenone	Antitumor	Sariasleni and Rosazza (1983)
Cunninghamella elegans	14-α-Hydroxywithaferin A	Antitumor	Rosazza et al. (1978)
Cunninghamella elegans	1',2'-Dihydrorotenone	Antitumor	Sariasleni and Rosazza (1983)
Mucor albo-ater	Pterin deaminase	Antitumor	Kusakabe et al. (1976)
Mucor lamprosporus	Pterin deaminase	Antitumor	Kusakabe et al. (1976)
Rhizopus arrhizus	Pterin deaminase	Antitumor	Kusakabe et al., (1976)
Rhizopus japonicus	Pterin deaminase	Antitumor	Kusakabe et al. (1976)
Rhizopus nigricans	Rhizopin	Antiviral	Das and Pal (1974)
Rhizopus oryzae	3'-Hydroxyverrucarin A	Antitumor	Pavanasasivam and Jarvis (1983)
Rhizopus oryzae	16-Hydroxyverrucarins A and B	Antitumor	Pavanasasivam and Jarvis (1983)
Rhizopus sp.	Rhizoxin	Antitumor	Kiyoto (1986)

The toxic trichothecenes are a chemically related group of fungal secondary metabolites produced in culture by various species of Fungi Imperfecti. They belong to the series of sesquiterpenoid mycotoxins classified as 12,13-expoxytrichothecenes (Eppley et al., 1980), which show marked selectivity and specificity of biological activity. Verrucarins and roridins extracted from fermentation broth and mycelia of *Myrothecium verrucaria* and *Myrothecium roridum* constitute a special subgroup characterized by a macrocyclic ester, or ester-ether, bridge structure in epoxytrichothecenes. They have strong antifungal and cytostatic properties *in vitro*, and inhibit the growth of sarcoma 37 and Ehrlich ascites carcinoma in mice and Walker carcinoma in rats (Bloem et al., 1983a). Verrucarin A is powerfully cytotoxic to primary chick embryo fibroblast monolayer and is also effective against Newcastle disease virus and herpes simplex *in vitro*, especially with the agar-diffusion plaque-inhibition method (Tamura et al., 1968c; ElKady and Moubash, 1982). The effective dosage of verrucarin A is low compared with tenuazonic acid (Griffin and Chu, 1983) and trichothecin (Arima et al., 1968). Roridin L-2, PD112,325, and PD113,326 exhibit good activity against P-388 lymphocytic leukemia in mice. PD113,325 and P113,326 are shown to be 12'-hydroxy-2'-(E)-verrucarin J and a stereoisomer of satroatox H (Smitka, 1984).

Anguidine, fusarenon X, nivalenol, and trichothecin are members of the 12,13-epoxytrichothecene group. Anguidine, also known as diacetoxysciropenol, is able to inhibit protein synthesis in HeLa cells and rabbit reticulocytes lysates (Liao et al., 1976). It has activity *in vivo* against the transplantable mouse tumor P-388 lymphatic leukemia, and phase I clinical trials of anguidine in the United States were completed. The fungi which have been shown to be capable of elaborating anguidine include *Fusarium anguioides*, *Fusarium concolor*, *Fusarium equisetti*, *Fusarium sambucinum*, *Fusarium scirpi*, and *Fusarium tricinctum* (Liao et al., 1976; Hartmann et al., 1978; von Stahelin et al., 1968). Fusarenon X and nivalenol produced by *Fusarium nivale* are closely related to anguidine. They inhibit DNA protein synthesis in HeLa cells, KB cells, human lung cells, and Ehrlich ascites tumor cells (Ueno, 1985). The trichothecin produced by *Trichothecium roseum* is an antibiotic with antifungal and antiviral activity (Arima et al., 1968).

3. Epidithiopiperazinedione Antibiotics

The epidithiopiperazinediones are an important group of fungal secondary metabolites of great pharmacological and chemical interest.

This group includes the gliotoxins, the sporidesmins, the aranotins, the chaetocins, the verticillins, the melinacidins, the hyalodendrins, the sirodesmins, oryzachlorin, and chaetomin (Michel et al., 1974). They all have similar biological activities, such as inhibiting the growth of bacteria, fungi, and viruses; they also have toxicity for animals, plants, and tissue cultures derived therefrom. Because of their high toxicity to mammalian cells, their practical use in therapy has been limited (Cole and Cox, 1981).

Gliotoxins, aranotins, and related compounds have the ability to inhibit the multiplication of RNA viruses both in tissue culture and in animals (Cole and Cox, 1981). They are active in tissue culture against coxsackie virus A21, parainfluenza virus types 1 and 3, poliovirus types 1–3, and strains of rhinovirus, and they protect mice against lethal infections caused by coxsackie virus A21 and influenza virus B/Md. Gliotoxin is produced by *Gliocladium fimbriatum*, *Aspergillus fumigatus*, and various species of *Trichoderma* and *Penicillium*. Aranotins are produced by *Arachniotus aureus* and *Aspergillus terreus* grown on synthetic medium in submerged culture (Miller et al., 1968).

Melinacidins are a mixture of at least four closely related compounds isolated from the culture broth of *Acrostalagmus cinnabarinus* var. *melinacidinus* (Reusser, 1968). They show inhibition to the growth of cultures of L-1210 and KB leukemia cells (Argoudelis, 1972).

Oryzachlorin has been isolated from the culture filtrates of *Aspergillus oryzae*. It has activity against Newcastle disease virus *in vitro* and cytotoxicity against chick embryo fibroblast cells. According to Kato et al. (1969b), the compound contains the epidithiopiperazine-dione moiety.

Antibiotics of A-26771-A and A-26771-C are produced by *Penicillium tubatum* (ATCC 28797). The chemical structure of both antibiotics has been determined as 3-benzyl-6-(hydroxymethyl)-1:4-diomethyl-3:6-epidotho-2:5-piperazinedione. The major difference between A-26771-A and A-26771-C is that the latter compound contains two sulfur atoms more than the former. They possess both antibacterial and antiviral activity. The antiviral activity has been demonstrated in tissue culture against certain viruses including polioviruses and coxsackie virus (Michel et al., 1974).

4. *Glycoprotein*

An extracellular glycoprotein complex produced from *Candida albicans* is effective in inhibiting the growth of solid sarcoma 180 tumors (Saltarelli, 1980).

5. Fatty Acids

Fatty acid fractions extracted from mycelia of *Penicillium crustosum*, *Penicillium tardum*, *Penicillium stipitatum*, *Cephalosporium dispyri*, and *Sepedonium ampulosporum* inhibit *in vivo* the growth of Ehrlich ascitic cells in mice and extend the life span of experimental animals (Endo et al., 1968, 1969; Kato et al., 1969a).The inhibitory effects are due mainly to the fractions containing palmitic, oleic, lineoleic, stearic, linolenic, and lauric acids (Fuska et al., 1974a).

6. Monoglycerides

Monoglyceride fractions obtained from mycelia of *Sepedonium ampullosporum*, *Cercospora oryzae*, and *Coriollus unicolor* (a Basidiomycete) have been found to suppress growth of Ehrlich ascite cells in mice (Tamura et al., 1968c). The main effective components are monolaurine and monooleine (Kato et al., 1969a).

7. Fumagillin

Fumagillin is an antiphage agent (antibiotic H-3) isolated by Eble and Hanson (1951) from *Aspergillus fumigatus*. It is also able to inhibit the growth of sarcoma 180 tumors implanted in mice (DiPaolo et al., 1958–1959).

8. Trihydroxytetralones

Trihydroxytetralones synthesized from juglone by *Penicillium diversum* var. *aureum* are cytotoxic to Yoshida sarcoma cells in tissue culture (Fujimoto and Satoh, 1986).

9. Leucinostatin

The peptide antibiotic leucinostatin has been isolated from the culture filtrate of *Penicillium lilacinum* and found to have broad-spectrum activity not only toward microorganisms but also for tumor cells. It is cytotoxic to HeLa cells and L-1210 leukemia cells, and also shows some inhibitory effect on Ehrlich subcutaneous solid tumor (Ishiguro and Ari, 1976).

10. Vermiculine

Fuska et al., (1974c) have studied the possible cancerostatic effect of vermiculine produced by *Penicillium vermiculatum* by using ascites tumors, such as Ehrlich carcinoma, lymphadenoma L-5178Y, and sarcoma 37. Vermiculine reduces both protein and nucleic acid synthesis, and has a cytotoxic effect on HeLa cells.

11. Aphidicolin

Bucknall et al. (1973) have described an antiviral antibiotic, aphidicolin, isolated from filtrates of *Cephalosporium aphidicola* (ATCC 28300). Starratt and Loschiavo (1974) described a strain of *Nigrospora sphaeria* (ATCC 26952) which also produced aphidicolin. This antibiotic is a tetracyclic diterpenoid active against iodedeoxyuridine (IUdR)-resistant herpes virus, which acts by inhibiting the synthesis of viral RNA.

12. Triornicin

One of the siderophores isolated from *Epicoccum purpurascens* is triornicin, which exhibits slight antitumor activity in mice injected with Ehrlich ascites tumor cells (Frederick et al., 1981b).

13. Cytochalasins

Several imperfect fungi have been found to produce a group of chemically related metabolites which have unusual effects on mammalian cells in tissue culture (Carter, 1967; Aldridge et al., 1973). The cytochalasins B and D are reported to possess antitumor and antiinflammatory properties (Katagiri and Matsuura, 1971). Cytochalasin B is obtained from culture filtrates of *Helminthosporium dematioideum* and *Phoma* sp. and has been shown to be identical to phomin (Aldridge et al., 1967), while cytochalasin D elaborated by *Metarrhizium anisopliae* and *Zygosporium masonii* is identical to zygosporin A (Aldridge and Turner, 1969; Minato et al., 1972).

14. Pterin Deaminases

Several species of *Aspergillus* and *Penicillium* yield pterin deaminases when grown on wheat bran at 28°C for days. The enzymes catalyze the hydrolytic deamination of pteron, pteroic acid, and folic acid, yielding the corresponding 2,4-dihydroxy compounds called lumazines. They inhibit growth of L-5178Y leukemia cells *in vitro* and significantly increase the average life span of the mice injected intraperitoneally with L-1210 leukemia cells. The enzymes also inhibit growth of B16 melanoma cells injected subcutanously into mice (Kusakabe et al., 1979c).

15. L-Lysine-α-oxidase

A new enzyme catalyzing the α-oxidative deamination of L-lysine, L-lysine-α-oxidase, is produced by *Trichoderma harzianum* and *Trichoderma viride*. The enzyme inhibits the growth of L-5178Y mouse leukemia cells *in vitro* and L-1210 mouse leukemia cells *in vivo*, and also shows an inhibitory effect on the *in vivo* DNA and RNA synthesis in human carcinoma ovarian cells (Khaduev *et al.*, 1987).

16. Phenylalanine Ammonia-lyase

Phenylalanine ammonia-lyase inhibits growth of both murine and human leukemic cells *in vitro* (Abell *et al.*, 1973). It is produced by *Rhodotorula glutinis*, *Rhodotorula rubra*, and *Rhodosporidium toruloides* (McGuire, 1987).

17. Substances Having Both Antitumor and Antiviral Properties

Tenuazonic acid produced by *Alternaria alternata*, brefeldin A by *Penicillium brefeldianum*, and mycophenolic acid by *Penicillium brevi-compactum* and *Penicillium stoloniferum* have been reported to have both anticancer and antiviral activities. Brefeldin A appears to be identical to ascotoxine isolated from *Ascochyta imperfecta*, cyanein from *Penicillium cyaneum*, and decumbin from *Penicillium sp.* (Fuska and Prokosa, 1976). Tenuazonic acid is effective against human adenocarcinoma HAd#1 implanted in chicken embryos and various animal tumors such as Ridgway osteogenic sarcoma, Ehrlich ascites carcinoma, and Friend leukemia virus (Hata, 1977). Tenuazonic acid also inhibits the cytopathic effects of a wide spectrum of viruses, such as measles, poliovirus MEF-1, and coxsackie B, p-influenza 3, Salisbury, vaccinia, and herpes simplex HF viruses (Miller *et al.*, 1963). Mycophenolic acid is active against several murine solid tumors and Walker adenocarcinoma 256 in rats (Williams *et al.*, 1968; Sweeney *et al.*, 1972); it inhibits the cytopathic effects of two DNA viruses, vaccinia and herpes simplex, and a measles RNA virus (Cline *et al.*, 1969; Franklin and Cook, 1969). Brefeldin A is cytotoxic against HeLa cells and Ehrlich carcinoma cells (Cole and Cox, 1981) and is also effective against herpes simplex DNA virus and Newcastle disease RNA virus (Tamura *et al.*, 1968a).

Ascochlorin is obtained from the filter cake of the fermented broth of *Ascochyta viciae*. It is effective against both DNA and RNA viruses in the agar-diffusion plaque-inhibition method and also shows cytotoxicity to chick embryo fibroblast monolayer and HeLa cells at a concentration of 0.3 µg/ml in tube culture (Tamura et al., 1968c).

Bredinin is an imidazole nucleoside isolated from the culture filtrate of *Penicillium brefeldianum*. It shows selective cytotoxicity against L-5178Y cells derived from malignant lymphoma of the mouse and also inhibits the growth of vaccinia virus *in vitro* (Mizuno et al., 1974).

In 1967, United States patent 3,303,094 was issued to B. Sokoloff and Y.Toda (1967a) for an antiviral antibiotic substance isolated from *Paecilomyces todicus* (ATCC 28700), later reidentified as *Paecilomyces fulvus* (Stolk and Samson, 1971). This antibiotic is active against influenza virus PR8 and influenza virus MF1 in embryonated eggs. It also shows a strong inhibiting activity against the transplanted tumors of mouse sarcoma 180 and Ehrlich mouse carcinoma.

18. Other Antitumor Substances

Duclauxin and PSX-1 have been isolated from the culture filtrates of *Penicillium stipitatum* (Fuska et al., 1974b). Czapek–Dox medium with glucose has been proved to be the most suitable medium for their production. Both metabolites show an inhibitory effect on Ehrlich ascites carcinoma, lymphadenoma L-5178Y, and sarcoma 37 (Kuhr et al., 1973; Fuska et al., 1974a). Duclauxin is also produced by *Penicillium duclauxi*.

Chlamydocin with pronounced cytostatic activity is a macrocyclic polypeptide produced by *Diheterospora chlamydosporia*. It is found in the filtrates of the culture growing in a common nutrient medium at 28°C (Close and Huguenin, 1974; Glasby, 1976).

A polysaccharide consisting mainly of a glucan with β-1,3-linked glycerol residues is extracted from the cell walls of *Aspergillus oryzae* (ATCC 48022) and is shown to possess cytostatic activity (Sasaki and Uchida, 1987).

Regulin is a fermentation product of *Aspergillus restrictus* in submerged cultures and has been found to have antitumor activity in certain induced tumors in animals (Olson and Goerner, 1966).

Botryodiplodin, with a gross chemical structure of 2-hydroxy-3-methyl-4-acetyltetrahydrofuran, has been isolated from *Botryodiplodia theobromae* and exhibits antileukemia activity (McCurry and Abe, 1973).

Dactylarin is an antiprotozoal antibiotic of the geodin group produced by *Dactylaria lutea* which also exhibits cytotoxic activity on HeLa cells and Ehrlich ascites cells (Horakova et al., 1974; Miko et al., 1979).

Quadrone isolated from *Aspergillus terreus* inhibits the growth of cancerous human nasopharynx cells and is relatively nontoxic in mice (Calton et al., 1978).

Bikaverin, which is also known as lycopersin, anthraquinon, passiflorin, mycogonin, and "fungal vacuolation factor," is reported to inhibit the growth of Ehrlich ascites carcinoma, lymphoma L-5178Y, sarcoma 37, lymphadenoma NK/LY, and HeLa cells in culture (Fuska et al., 1979a,b). It has been isolated from Fusarium oxysporum, Fusarium lycopersici, Gibberella fujikuroi, and Mycogone jaapii (Iijima et al., 1979). A new cytoxic substance, fusariocin C, is isolated from the culture filtrate of Fusarium moniliforme. It is cytotoxic to HeLa cell culture and leukemia L-1210 cells and shows some antitumor activity in vivo for the ascitic form of Ehrlich carcinoma (Ito, 1979; T. Ito et al., 1981).

Hadacidin in an L-aspartic acid analog obtained from Penicillium frequentans and Penicillium purpurescens. It inhibits the growth of human adenocarcinoma HAd#1 and transplantable animal tumors such as Ehrlich carcinoma and Walker carcinosarcoma 256 (Gitterman et al., 1962).

Asterriquinone has been isolated from Aspergillus terreus as one of the intracellular metabolic products. It appears to inhibit the growth of Ehrlich ascites carcinoma, ascites hepatoma AH-13, Yoshida sarcoma, and L-1210 mouse leukemia in vivo. The compound is synthesized from tryptophan in the culture medium; the isopentenyl unit is derived from mevalonic acid (Yamamoto et al., 1976).

Gliocladic acid is isolated from the fermentation broths of Gliocladium virens and Trichoderma viride and inhibits sarcoma 37 tumor cells implanted in ICR/JCR mice (Itoh et al., 1982).

19. *Other Antiviral Substances*

Funiculosin has been isolated in a crystalline form from the filter cake of the fermented broth of Penicillium funiculosum grown in a medium composed of glucose, lactose, and corn steep liquor as the chief nutrients. It inhibits both DNA and RNA viruses as tested in the infected primary chick embryo fibroblast cell monolayer (Ando et al., 1969).

In 1974, United States patent 3,819,832 was issued to N. N. Joshi for antiviral agent FWH-775 and the method of its production. The agent was obtained from a strain of Aspergillus niger (ATCC 16508) isolated from swine embryo kidney tissue culture infected with Batts V-13, a swine enterovirus. FWH-775 is believed to be an amino acid derivative of low molecular weight effective against both vaccinia and influenza viruses in laboratory animals (Joshi, 1974).

Cylindrochlorin is elaborated by *Cylindrocladium* sp. It is a chlorine-containing antibiotic shown to have antiviral activity against Newcastle disease virus (Kato et al., 1969a).

Stemphylium botryosum (ATCC 20183) produces a substance, designated "stemlon," which inhibits encephalomyocarditis virus, Semliki forest virus, and coxsackie virus in mice (Cole and Planterose, 1973).

B. BASIDIOMYCETES

1. Fermentation Products

Gregory et al (1966) screened more than 7000 cultures of Basidiomycetes for antitumor activity against three rodent tumor systems. Strains representing 20 genera produced materials in fermentation cultures that had inhibitory effects upon sarcoma, mammary adenocarcinoma 755, or leukemia L-1210 *in vivo*. These results were later confirmed and the cultures were identified as *Coprinus nycthemerus, Corticium rolfsii, Irpex consors, Irpex flavus, Lentinus lepideus, Lenzites saepiaria, Lenzites trabea, Merulius niveus, Mucidula radicata, Pholiota formosa, Pleurotus passeckerianus, Polyporus distortus, Polyporus obtusus, Poria corticola, Poria subacida, Poria xantha, Stereum subpileatum,* and *Tricholoma paneolum* (Espenshade and Griffith, 1966).

The culture filtrate of *Hohenbuehelia geogenius* inhibits the growth of Ehrlich ascites carcinoma L-1210 and a slow-growing spontaneous mammary tumor in ps strain mice (Riondel et al., 1981).

2. Antitumor Polysaccharides

Antitumor polysaccharides have been obtained from a number of Basidiomycetes, including *Coriolus consors, Coriolus versiocolor, Corticium centrifugum, Crepidotus* sp., *Flammulina velutipes, Ganoderma applantatum, Phellinus linteus, Pholiota nemeko, Pleurotus ostreatus, Poria cocos, Schizophyllum commune,* and *Tricholoma aggregatum* (Ikekawa et al., 1982). United States patent 3,759,896 has been issued to Komatsu et al. (1973) in connection with a process for the production of antitumor polysaccharides consisting mainly of -(1,3)-linked D-glucoses, their derivatives and partially hydrolyzed products, prepared from culture filtrates or by extracting the fruiting bodies, sclerotia, and mycelia of the fungi with hot water or an alkaline aqueous solution.

United States patent 4,051,314 issued to Ohtsuka et al. (1977) relates to antitumor mucopolysaccharides produced from cultured broths of

Coriolus versicolor, Coriolus conchifer, Coriolus pargamenus, Coriolus hirsutus, Coriolus biformis, Coriolus consor, Coriolus pubescens, and other Basidiomycetes, including *Coltricia pusilla, Fomes fomentarius, Polyporus pescaprae, Fomitopsis rosea, Rigidoporus culmarius, Caltricia perennis, Favolus alveolarius, Ganoderma tsugae, Ridigoporus durus, Piptoporus betulinus, Microporus flavelliformis, Gloeoporus amorphus, Daedaleopsis tricolor, Hymenochaete mougeotii, Inonotus kanehirae, Phellinus igniarius, Inonotus sciurinus, Cryptoderma yamanei, Phellinus robustus,* and *Oudemansiella radicata.* The polysaccharide–protein complex has been found to be effective in the treatment of the solid types of Ehrlich carcinoma or sarcoma 180 cancerous tumors in mice. The effective substance exhibits neither cytotoxicity nor side effects commonly seen in connection with the use of conventional anticancer agents, such as a decrease in the number of leucocytes, anemia of the liver and other organs, atrophy of the spleen, loss of body weight, and loss of appetite.

Lentinan, extracted from an edible mushroom, *Lentinus edodes,* shows significant inhibition of the growth of subcutaneously implanted sarcoma 180 in laboratory animals (Kirin Brewery Co. Ltd., 1980). Its antitumor activity is alicited by the stimulation of host-mediated responses (Hata, 1977). The results of periodate of oxidation, Smith degradation, and methylation analysis by Sasaki and Takasuka (1976) indicate that the chemical structure of lentinan is a branched molecule with a backbone of (1,3)-linked-β-D-glucan and side chains of both β-D-(1,3)- and β-D-(1,6)-linked D-glucose residues, together with a few internal β-D-(1,6)- linkages.

Schizophyllan, obtained from *Schizophyllum commune,* is effective against a range of tumors, such as M-32, Ehrlich carcinoma, and Yoshida sarcoma, but is inactive against ascitic tumors (Komatsu et al., 1969). This polysaccharide is composed of three β-(1,3)-linked glucose residues, and a single glucose residue is attached through a β-(1,6)- linkage to one of the three β-(1,3)-linked residues. Neoschizophyllan, produced by an ultrasonication of schizophyllan, has low viscosity in the form of an aqueous solution, a pharmacological activity comparable to or even higher than that of schizophyllan, and remarkably low toxicity. Neoschizophyllan is absorbed into blood and tissues more rapidly than is schizophyllan (Kikumoto et al., 1978).

An acidic polysaccharide fraction with antitumor activity has been isolated from *Pleurotus ostreatus* (Yoshioka et al., 1975). This antitumor-active component consists of a skeleton of β-(1,3)-linked glucose residues, having branches of galactose and mannose residues, and also contains acidic sugars (Yoshioka et al., 1975).

Ikekawa et al. (1969) have reported that water-soluble extracts of *Ganoderma applanatum, Phellinus linteus,* and *Coriolus versicolor* inhibit the growth of sarcoma 180 implanted subcutaneously in mice. Sasaki et al. (1971) have obtained antitumor polysaccharide preparations G-Z and P-Z from water-soluble extracts of *G. applanatum* and *P. linteus,* respectively. G-Z consists of a mixture of β-(1,3)- and -(1,4)-linked D-glucose residues and P-Z has a β-(1,3)-linked D-glucose residue. These polysaccharide preparations are active against transplanted tumors in mice.

The water-soluble polysaccharide GL-1 is isolated from the fruit bodies of *Ganoderma lucidum*. GL-1 strongly inhibits the growth of the sarcoma 180 solid-type tumor. The essential structure for the antitumor activity is a branched glucan involving (1,3)-β-(1,4)-β and (1,6)-β linkages (Miyazaki and Nishijima, 1981).

The polysaccharide PS-K has been extracted by Hirase et al. (1970) from *Coriolus versicolor*. It was found to be effective against sarcoma 180 when administered orally. PS-K is a β-(1,4)-glucan containing 10% protein. Its water solubility and effectiveness with oral administration seems to be due to its protein-containing structure (Hata, 1977). In addition, Naruse and Takeda (1974) have also isolated two antitumor polysaccharides, ATSO and coriolan (Ito et al., 1979) from mycelia and culture filtrate of *C. versicolor* (Hayashida and Watanabe, 1983). Coriolan contains no nitrogen and is a branched structure possessing β-(1,3) and β-(1,6) linkages. It shows antitumor activity against sarcoma 180, Shionogi carcinoma 42, Ehrlich solid-tumor carcinoma, and pulmonary tumor 7423 in mice. The water-soluble polysaccharide D-II is isolated from the cultured mycelium of *C. versicolor*. D-II strongly inhibits the growth of sarcoma 180 transplanted into mice (Sugiura et al., 1980). Polysaccharopeptide produced by *C. versicolor* has a wide range of antitumor activities, inhibiting Ehrlich ascites tumors, leukemia P388, and sarcoma 180, and the growth of the cells of gastric tumor, lymph, and monocytic leukemia (Yang et al., 1987). Both United States patents 4,271,151 and 4,289,688 issued to Hotta et al. (1981a,b) relate to protein-bound polysaccharides which have an excellent antitumor activity not only by intraperitoneal administration but also by oral administration. The polysaccharides have been contemplated for use against a wide variety of mammalian cancers, including mammary cancer; gastrointestinal cancers, such as those of the esophagus, stomach, and large colon; lung cancer and brain tumors. The protein-bound polysaccharides can be obtained by extraction of the mycelia and/or fruit bodies of *Coriolus versicolor, Coriolus consors, Coriolus pubescens, Coriolus biformis, Coriolus hirsutus, Coriolus conchifer,* and *Coriolus pargamenus.*

Pachyman, a β-(1,3) linear glucan obtained from *Poria cocos*, has no antitumor effect, but it has been shown that the β-(1,3) chain has some β-(1,6) linked branches. Its derivative, pachymaran, secured by reducing pachyman with separation of the β-(1,6) linkages and conversion to a β-(1,3) linear glucan similar to lentinan, appears to have pronounced antitumor activity with no toxicity (Hamuro et al., 1978). The polysaccharide H_{11} isolated from the mycelia of *P. cocos* is a new (1,3)-(1,6)-β glucan and shows a remarkable antitumor effect against sarcoma 180 (Kanayama et al., 1983).

A polysaccharide extracted from the fruit bodies of *Grifola frondosa* is a glucan made up of β-(1,6)-linked glucose residues with branches of β-(1,3) linked glucose. This glucan inhibits the growth of sarcoma 180 tumor in ICR mice (Nanba and Kurodas, 1987b; Adachi et al., 1987).

A water-soluble, branched (1,3)-β-D-glucan of *Auricularia auricula-judae* has exhibited potent, inhibitory activity against implanted sarcoma 180 solid tumor in mice (Misaki et al., 1981).

The polysaccharides EA3, EA5, and EA501 isolated from a Japanese edible mushroom, *Flammulina velutipes*, show markedly high inhibition ratios against transplanted solid tumor (sarcoma 180) in ICR mice (Ikekawa et al., 1982).

3. Coriolins

Coriolin, coriolin B, and coriolin C are tricyclic antibiotics produced by cultivation of *Coriolus consors* under submerged conditions. Coriolin is active against Yoshida sarcoma cells and coriolin C inhibits the growth of Ehrlich ascites sarcoma and leukemia L-1210 cells (Glasby, 1976). Diketocoriolin B is an oxidized product of coriolin B and has a sesquiterpene skeleton; it is effective against Ehrlich carcinoma and leukemia L-1210 in mice (Nishimura et al., 1980b).

4. Illudins

Anchel et al. (1950) have described two terpene compounds, illudins M and S, with antibacterial and antitumor activities from *Clitocybe illudens*. Komatsu et al. (1961) have found an anticancer substance identical to illudin from the fruiting body of *Lampteromyces japonica*. Yoshida et al. (1962) have isolated from the dried fruiting bodies of L. *japonica* a tumor inhibitor active against Ehrlich mouse tumors and human tumor cells growing in tissue culture.

5. Flammulin

Komatsu et al. (1963) have isolated a compound from cultures of the edible mushroom *Flammulina velutipes* with basic protein characteristics, which they named "flammulin." It acts directly on tumor cells *in vitro*.

6. Pleuromutilin

Pleuromutilin is an antiviral antibiotic elaborated by *Pleurotus mutilis* and *Pleurotus passackerianus* (Kavanagh et al., 1951; Knauseder and Brandl, 1976). It is active against influenza PR8 viruses in a concentration of 2 mg/ml.

7. Poricin

An antitumor agent, poricin, isolated from aqueous mycelial extracts of a Basidiomycete, *Poria corticola*, appears to be an acidic protein which is active in inhibiting a number of tumors (Schillings and Ruelius, 1968).

8. Nebularine

Lofgren and Luning (1953) isolated the antibiotic nebularine from the filtrate of the minced fresh mushroom *Agaricus nebularis*. They found purine and ribose to be present among the products following the acid hydrolysis of nebularine. Biesele et al. (1955) found that the necleosidic nebularine is toxic toward sarcoma 180 cells *in vitro*.

9. Fomecin

Fomecin is a simple aromatic antibiotic isolated from *Fomes juniperinus* grown on a medium containing corn steep liquor (Anchel et al., 1952). It possesses antiviral activity, inhibiting influenza PR8 viruses at concentrations of 0.5 mg/ml.

10. Monoglycerides

Tamura et al. (1968c) have isolated from the mycelia of *Coriolus unicolor* antiviral monoglycerides which exhibit remarkable inhibitory activity against Ehrlich ascites tumor cells in ddY mice.

11. Mucoprotein

Lucas et al. (1957) have obtained crude water extracts of sporophores of *Boletus edulis*, which contained a mucoprotein having inhibitory effects on the growth of Crocker's mouse sarcoma 180 *in vivo*.

12. Tumor Inhibitors

Lucas et al. (1958–1959) have found the sporophores of *Calvatia gigantia*, *Calvatia craniformis*, *Calvatia bovista*, and *Calvatia cyathiformis* to be capable of producing tumor inhibitors effective both *in vivo* and *in vitro*. Orally administered edible mushrooms, including

Lentinus edodes, Grifola frondosa, Agaricus bisporus, Pleurotus ostreatus, Flammulina velutipes, Pholiota glutinosa, Tremella fuciformis, Auricularia minor, and *Volvariella volvacea* have suppressive activities against allergenic tumors (sarcoma 180 in ICR mice) and syngenic tumors (B16 melanoma or Lewis lung carcinoma in C57BL/6 mice; Meth-A tibrosarcoma in BALB/c mice (Mori et al., 1987).

13. Calvatic Acid

Umezawa et al. (1975) and Okuda and Fugiwara (1982) obtained a new antibiotic calvatic acid from fermentation of a mushroom, *Calvatia craniformis*, grown in a medium of 2% glucose, 0.5% peptone, 0.3% yeast extract, 0.3% KH_2PO_4, and 0.3% $M_gSO_4 \cdot 7H_2O$ in a reciprocal shaker at 28°C for 10 days. The calvatic acid inhibits the growth of Yoshida sarcoma in cell culture by 50% at 1.56 $\mu g/ml$. It also shows growth inhibition of mouse leukemia L-1210. Gasco et al. (1974) have also isolated the same compound from *Calvatia lilacina*.

14. Calvacin

Lucas et al. (1958–1959) developed a submerged fermentation process to produce the antitumor substance now known as "calvacin" from *Calvatia gigantia*. Roland et al. (1960) scaled up the fermentation process to the pilot-plant stage and recovered calvacin from beer on a cellulose ion-exchange column. Calvacin possesses a broad antitumor spectrum and is active against 13 of 24 various mouse, rat, and hamster tumors.

15. Triterpene Acids

Lanosteral-derived triterpene acids are isolated from *Poria coccos*. Their methyl esters are found to be cytotoxic *in vitro* on cultured hepatoma cells (Valisolalao et al., 1980).

16. Merulinic Acids

Merulinic acids A, B, and C isolated from the fruiting bodies of *Merulius tremellosus* and *Phlebia radiata* have been shown to inhibit Ehrlich carcinoma ascites cells (Giannetti et al., 1978).

C. Ascomycetes

1. Antitumor Polysaccharides

Scleroglucan, elaborated by *Sclerotium glucanicum*, has a main chain of (1,3)-β-D-glucopyranosyl units with every third or fourth unit carrying a (1,6)-β-D-glucopyranosyl group. It is effective against sarcoma 180 at a level of 0.5 mg/kg and is one of the most effective antitumor polysaccharides that have been reported (Singh et al., 1974). Removal of the (1,6)-β-D-glucopyranosyl side groups from scleroglucan by a single Smith degradation gives an inactive glucan (Singh et al., 1974). However, when lentinan, a polysaccharide elaborated by *Lentinus edodes* (a Basidiomycete), is debranched and then hydrolyzed, no loss of antitumor activity occurs. According to Sasaki and Takasuka (1976), this is due to the different branched structures between scleroglucan and lentinan. The former has a single D-glucosyl group as a branch, and therefore it is converted to an inactive linear polysaccharide by Smith degradation. In contrast, lentinan has β-D-(1,3)-linked side chains, and it exhibits antitumor activity after being treated by Smith degradation.

Polysaccharides extracted with cold aqueous sodium hydroxide from the fruiting bodies of *Peziza vesiculosa* give a water-soluble heteroglycan fraction and a water-insoluble glucan fraction. Both fractions have shown potent antitumor activity against the solid form of sarcoma 180 tumor in ICR mice (Ohno et al., 1985a).

The polysaccharide AS-I containing 99.1% sugar and 0.9% protein is extracted from mycelia of *Cochliobolus miyabeanus*. AS-I is a β-(1,3)-glucan with branches of -(1,6)-linked glucose units, and inhibits the growth of sarcoma 180 solid tumor in ICR mice, solid tumors of IMC carcinoma in CDF mice, and MM-46 carcinoma in C3H mice (Nanba and Kuroda, 1987a).

Zymosan, consisting chiefly of protein–carbohydrate complexes, is isolated from the cell walls of *Saccharomyces cerevisiae*. The antitumor action of zymosan is elicited by the enhancement of specific defense mechanisms of the host and not by a cytotoxic effect on tumor cells (Bradner et al., 1958; Hata, 1977). D-Mannan of *S. cerevisiae* x2180-1A-5 mutant strains shows growth-inhibitory activity against mouse-implanted sarcoma 180 and Ehrlich carcinoma solid tumors (Matsumoto et al., 1980).

Other Ascomycetes, such as *Cochliobolus sativas* and *Sclerotinia sclerotiorum*, also have been reported to produce polysaccharides with antitumor action (Komatsu et al., 1973).

2. Aranoflavins A and B

Aranoflavins A and B have been isolated from cultures of *Arachniotus flavoluteus* grown in a medium containing potato starch for 3 days at 26°C. The antibiotics are active against Yoshida sarcoma in rats (Mizuno et al., 1970).

3. Cordycepin

A nucleosidic compound has been isolated from cultures of *Cordyceps militaris* grown in a medium containing glucose (1%) with casein hydrolysate (0.5%) as the source of amino acids. It is produced after 20 days of incubation at 24°C. The compound has been named cordycepin (Cunningham et al., 1950) and is shown to antagonize the growth of the Ehrlich mouse ascites tumor cells *in vivo*.

4. Bredinin

Bredinin (4-carbamonyl-1-β-D-ribofuranosyl-imidazolium-5-olate) is an imidazole nucleoside isolated from the culture supernatant of *Eupenicillium brefeldianum*. It has a cytotoxic action on L-5178Y cells (mouse leukemia cell line) (Mizuno et al., 1974; Yoshioka and Nakatsu, 1975).

5. Antibiotic SL-1846

The antibiotic is extracted from culture filtrates of a strain of *Pseudeurotium ovalis* (ATCC 34920) cultivated in a nutrient solution containing glucose, peptone, yeast, and malt extracts and inorganic salts. It inhibits the proliferation of mouse mastocytoma P-815 *in vivo* and also alleviates amebiasis (Sigg and Stooll, 1969).

6. Aranotins

The antiviral substances, aranotins, have been isolated from cultures of *Arachniotus aureus*. They are a new class of antiviral dithiodiketopiperazines which are also present in the well-known metabolites gliotoxins and sporidesmins from Fungi Imperfecti (Nagarajan et al., 1968). By chromatography of the ethyl acetate extract of the broth culture of *A. aureus*, Neuss et al. (1968) have isolated five new substances, including aranotin and bisdithio-di-(methylthio)-acetylapoaranotin. These metabolites of *A. aureus* are able to inhibit virus multiplication in tissue culture.

7. Vesiculogen

Vesiculogen is extracted from the fruiting bodies of *Peziza vesiculosa*, showing antitumor effects against both the solid and ascites forms of sarcoma 180 in ICR mice (Suzuki et al., 1982).

8. Gliocladic Acid

Gliocladic acid isolated from *Chaetomium globosum* shows a moderate inhibition effect on sarcoma 37 (Itoh et al., 1982).

D. Phycomycetes

1. Pterine Deaminase

A water-soluble substance that inhibits the growth of leukemia L-5178Y cells in vitro has been isolated from cultures of *Mucor albo-ater*, *Mucor lamprosporus*, *Rhizopus arrhizus*, and *Rhizopus japonicus* grown on wheat bran at 28°C for 4 days. The substance has been shown to be a pterin deaminase which is distributed rather widely among molds, including species of the Fungi Imperfecti genera *Aspergillus* and *Penicillium* (Kusakabe et al., 1976).

2. Rhizopin

Das and Pal (1974) have obtained an antibiotic in powder form by freezing and thawing the culture filtrate of *Rhizopus nigricans* and have designated it "rhizopin." It is antifungal, antibacterial, and antiviral in nature.

3. 14-α-Hydroxywithaferin A

Cunninghamella elegans (ATCC 10028b) converts the steroid lactone, withaferin A, into 14-α-hydroxywithaferin A, which has the same level of antitumor activity as withaferin A against the sarcoma 180 tumor test system in mice (Rosazza, 1979).

4. Rhizoxin

A complex of antitumor antibiotics, WF-1360 and WF-1360 A, B, C, D, E, and F is produced by *Rhizopus* sp. (ATCC 20577). These compounds are cytotoxic when tested on P-388 leukemia cells *in vitro*. WF-1360 is highly active against leukemia L-1210 and melanoma B16. WF-1360 is found to be identical to rhizoxin (Kiyoto, 1986).

References

Abe, M. (1972). Rugulovasine. U.S. Patent 3,651,220.

Abell, C. W., Hodgins, D. S., and Stith, W. J. (1973). An in vitro evaluation of the chemotherapeutic potency of phenylanine ammonia-lyase. *Cancer Res.* **33**, 2529–2532.

Achimi, R., Mueller, B., and Tamm, C. (1974). Biosynthesis of the verrucarins and roridins. Part 1. The transformation of mevalonic acid into verrucarinic acid. Evidence for a hydrogen 1,2-shift. *Helv. Chim. Acta* **57**, 1442–1459.

Adachi, K., Nanba, H., and Kuroda, H. (1987). Potentiation of host-mediated antitumor activity in mice by β-glucan obtained from *Grifola frondosa* (Maitake). *Chem. Pharm. Bull.* **35**, 262–270.

Albertsson, J., Fex, T., and Svensson, C. (1981). X-ray study of cytochalasin M, a secondary metabolite from the fungus *Chalara microspora*. *Acta Chem. Scand., Ser. B* **B35**, 707–714.

Aldridge, D. C., and Turner, W. B. (1969). Structure of cytochalasin C and D. *J. Chem. Soc. C* pp. 923–928.

Aldridge, D. C., Armstrong, J. J., Speake, R. N., and Turner, W. B. (1967). Structure of cytochalasin A and B. *J. Chem. Soc.* p. 1667.

Aldridge, D. C., Borrow, A., Foster, R. G., Large, M. S., Spencer, H., and Turner, W. B. (1972). Metabolites of *Nectria coccinea*. *J. Chem. Soc. Perkin I:* pp. 2136–2141.

Aldridge, D. C., Greatbanks, D., and Turner, W. B. (1973). Revised structure of cytochalasin E and F. *J. Chem. Soc., Chem. Commun.* pp. 551–552.

Anchel, M., Hervey, M. A., and Robbins, W. J. (1950). Antibiotic substance from Basidiomycetes. VII. *Clitocybe illudens*. *Proc. Natl. Acad. Sci. U.S.A.* **36**, 300–305.

Anchel, M., Hervey, M. A., and Robbins,W. J. (1952). Antibiotic substances from Basidiomycetes. X. *Fomes juniperus* Schrenk. *Proc. Natl. Acad. Sci. U.S.A.* **38**, 655.

Ando, K., Suzuki, S., Suzuki, K., Kodama, K., Kato, A., Tamura, G., and Arima, K. (1968). Antitumor activity of fatty acids produced by fungi. *J. Antibiot.* **21**, 690–691.

Ando, K., Suzuki, S., Saeki, T., Tamura, G., and Arima, K. (1969). Funiculosin, a new antibiotic. I. Isolation, biological and chemical properties. *J. Antibiot.* **22**, 189–194.

Anke, H., Schwab, H., and Achenbach, H. (1980a). Tetronic acid derivatives from *Aspergillus panamensis*, production, isolation, characterization and biological activity. *J. Antibiot.* **33**, 931–939.

Anke, H., Kolthoum, I., and Loathch, H. (1980b). Metabolic products of microorganisms. 192. The anthraquinone of the *Aspergillus glaucus* group. II. Biological activity. *Arch. Microbiol.* **126**, 231–236.

Anke, H., Kemmer, T., and Hofle, G. (1981). Deflectins, new antimicrobial azaphilones from *Aspergillus deflectus*. *J. Antibiot.* **34**, 923–928.

Anke, T., and Oberwinkler, F. (1977). The strobilurins-new antifungal antibiotics from the Basidiomycetes *Strobilurus tenacellus*. *J. Antibiot.* **30**, 806–810.

Anke, T., Hecht, H., Schramm, G., and Stieglich, W.(1979). Antibiotics from Basidiomycetes. IX. Oudemansin, an antifungal antibiotic from *Oudemansiella mucida* (Schrader et Fr.) Hoehnel. *J. Antibiot.* **32**, 1112–1117.

Anke, T. Kupka, J., Schramm, G., and Stieglich., W. (1980). Antibiotics from Basidiomycetes. X. Scorodonin, a new antibacterial and antifungal metabolite from *Marasmius scorodonius* (Fr.) Fr. *J. Antibiot.* **33**, 463–467.

Arai, T., Mikami, Y., Fukoshima, K., Utsumi, T., and Yazawa, K. (1973). A new antibiotic, leucinostatin, derived from *Penicillium lilacinum*. *J. Antibiot.* **26**, 157–161.

Arditti, J., Ernst, R., Fisch, M. H., and Flick, B. H. (1972). Ergosterol peroxide from *Rhizoctonia repens:* Composition, confirmation, and origin. *J. Chem. Soc., Chem. Commun.* pp. 1217–1218.

Argoudelis, A. D., (1972). Melinacidins II, III and IV. New "3,6-epidithiodiketopiperazine" antibiotics. *J. Antibiot.* **25**, 171–178.

Argoudelis, A. D., and Reusser, F. (1971). Melinacidins, a new family of antibiotics. *J. Antibiot.* **24**, 383–389.

Arima, K., Takatsuki, A., Suzuki, S., Ando, K., and Tamura, G. (1968). Antiviral activity of trichothecin. *J.Antibiot.* **21**, 158–159.

Backens, S. Steffan, B., Steglich, W., Zechlin, L., and Anke, T. (1984). Naematolon. *Liebigs Ann.Chem.* pp. 1332–1342.

Beale, J. M., Jr., Chapman, R. L., and Rosazza, J. P. N. (1984). Studies on the biosynthesis of terrecyclic acid A, an antitumor antibiotic from *Aspergillus terreus*. *J. Antibiot.* **37**, 1376–1381.

Betina, V., Drobnica, L., Nemec, P., and Zemanova, M. (1964). Study of the antifungal activity of the antibiotic, cyanein. *J. Antibiot., Ser. A* **17**, 93–95.

Betts, R. E., Waters, D. E., and Rosazza, J. P. (1974). Microbial transformations of antitumor compounds. I. Conversion of acronycine to 9-hydroyxacronycine by *Cunninghamella echinulata*. *J. Med. Chem.* **17**, 599–602.

Biesele, J. J., Slauterback, M. C., and Margolis, M. (1955). Unsubstituted purine and its riboside as toxic antimetabolites in mouse-tissue cultures. *Cancer (Philadelphia)* **8**, 87–96.

Bloem, R. J., Smitka, T. A., Bunge, R. H., French, J. C., and Mazzola, E. P. (1983a). Roridin L-2, a new trichothecene. *Tetrahedron Lett.* **24**, 249–252.

Bloem, R. J., Bunge, R. H. and French, J. C. (1983b). Antibiotic roridin L-2 and its use. U.S. Patent 4,382,952.

Bollinger, P. and Zardin-Tartaglia, T. (1976). Isolierung und strukturaufklarung von mikrolin. *Helv. Chim. Acta* **59**, 1809–1820.

Borrow, A., Jeffereys, E. G., Kessel, R. H. J., Lloyd, E. C., Lloyd, P. B., and Nixon, J. S. (1961). The metabolism of *Gibberella fujikuroi* in stirred culture. *Can.J. Microbiol.* **7**, 227–276.

Boyd, M. R. (1986). National Cancer Institute drug discovery and development. In "Accomplishments in Oncology" (E. J. Frei and E. J. Freireich, eds.), Vol. 1, No. 1, pp. 68–76. Lippincott, Philadelphia, Pennsylvania.

Boyd, M. R. (1988). Strategies for the identification of new agents for the treatment of AIDS: A national program to facilitate the discovery and preclinical development of new drug candidates for clinical evaluation. In "AIDS: Etiology, Diagnosis, Treatment, and Prevention" (V. T. DeVita, Jr., S. Hellman, and S. A. Rosenberg, eds.), Lippincott, Philadelphia, Pennsylvania (in press).

Boyd, M. R., Shoemaker, R. H., Cragg, G. M., and Suffness, M. (1988a). New avenues of investigation of marine biologicals in the anticancer drug discovery program of the national cancer institute. In "Pharmaceuticals and the Sea" (C. W. Jefford, K. L. Reinhart, and L. S. Shield, eds.), pp. 27–44. Technomic Publ., Lancaster.

Boyd, M. R., Shoemaker, R. H., McLemore, T. L., Johnston, M. R., Alley, M. C., Scudierto, D. A., Monks, A., Fine, D. L., Mayo, J. G., and Chabner, B. A. (1988b). New Drug Development. In "Thoracic Oncology" (J. A. Roth, J. C. Ruckdeschel, and T. H. Weisenburger, eds.), Chapter 51. Saunders, New York (in press).

Bradner, W. T., Clarke, D. A. and Stock. C. C. (1958). Stimulation of host defense against experimental cancer. I. Zymosan and sarcoma 180 in mice. *Cancer Res.* **18**, 347–351.

Bradner, W. T., Bush, J. A., Myllymaki, R. W., Nettleton, D. E., Jr., and O'Herron, F. A. (1975). Fermentation, isolation, and antitumor activity of sterigmatocystins. *Antimicrob. Agents Chemother.* **8**, 159–163.

Brannon, D. R., Mabe, J. A., Molloy, B. B., and Day, W. A. (1971). Biosynthesis of dithiodiketopiperazine antibiotics: Comparison of possible aromatic amino acid precursors. *Biochem. Biophys. Res. Commun.* **43**, 588–594.

Breitenstein, W., and Tamm, C. (1977). Verrucarin K, the first natural trichothecene derivative lacking the 12,13-epoxy group. Verrucarins and roridins. *Helv. Chem. Acta* **60**, 1522–1517.

Brian, P. W. (1951). Antibiotics produced by fungi. *Bot. Rev.* **17**, 357–430.

Broadbent, D. (1966). Antibiotics produced by fungi. *Bot. Rev.* **32**, 219–242.

Broome, J. D. (1965). Antilymphoma activity of L-asparaginase *in vivo*: Clearance rates of enzyme preparations from guinea pig serum and yeast in relation to their effect on tumor growth. *J. Natl. Cancer Inst.* **35,** 967–974.

Brown, L. C., and Jacobs, J. J.(1975). Isolation of ergosterol peroxide from *Alternaria dianthicola. Aust. J. Chem.* **28,** 2317–2318.

Buchi, G., Kitaura, Y., Yuan, S. S., Wright, H. E., Clardy, J., Demain, A. L., Glinsukon, T., Hunt, N., and Wogan, G. N. (1973). Structure of cytochalasin E, toxic metabolite of *Aspergillus clavatus. J. Am. Chem. Soc.* **95,** 5423–5425.

Bucknall, R. A., Moores, H., Simms, R., and Hesp. B. (1973). Antiviral effects of aphidicolin, a new antibiotic produced by *Cephalosporium aphidicola. Antimicrob. Agents Chemother.* **4,** 294–298.

Bu'Lock, J. D. (1965). "The Biosynthesis of Natural Products: An Introduction to Secondary Metabolism." McGraw-Hill, New York.

Bu'Lock, J. D., and Leich, C. (1975). Biosynthesis of gliotoxin. *J. Chem. Soc., Chem. Commun.* pp. 628–629.

Calton, G. J., Ranieri, R. L., and Espenshade, M. A. (1978). Quadrone, a new antitumor substance produced by *Aspergillus terreus.* Production, isolation, and properties, *J. Antibiot.* **31,** 38–42.

Calton, G. J., Espenshade, M. A., and Ranieri, R. L. (1979). Antineoplastic agent. U.S. Patent 4,147,798.

Carey, S. T. and Nair, M. S. R. (1977). Metabolites from Pyrenomycetes VIII. Identification of three metabolites from *Nectria lucida* as antibiotic triprenyl phenols. *Lloydia* **40,** 602–603.

Carter, S. B. (1967). Effects of cytochalasins on mammalien cells. *Nature (London)* **213,** 261–264.

Chien, M. M. and Rosazza, J. P. (1979). Microbial transformations of natural antitumor agents. VIII. Formation of 8- and 9-hydroxy-ellipticines. *Drug Metab. Dispos.* **7,** 211–214.

Chihara, G., Hamuro, J., Maeda, Y., Arai, Y., and Fukuoka, F. (1970a). Antitumor polysaccharide derived chemically from natural glucan (Pachyman). *Nature (London)* **225,** 943–944.

Cline, J. C., Nelson, J. D., Gerzon, K., Williams, R. H., and Delong, D. C. (1969). In vitro antiviral activity of mycophenolic acid and its reversal by guanine-type compounds. *Appl. Microbiol.* **18,** 14–20.

Close, A., and Huguenin, R. (1974). Isolierung und Strukturaufklarung von chlamydocin. *Helv. Chim. Acta* **57,** 533–535.

Coats, J. H., Meyer, C. E., and Pyke, T. R. (1971). Antibiotic dermadin and a process for producing the same. U.S. Patent 3,627,882.

Cole, M., and Planterose, D. N. (1973). Stemlon and its production. U.S. Patent 3,712,944.

Cole, R. J., and Cox, R. H., eds. (1981). "Handbook of Toxic Fungal Metabolites." Academic Press, New York.

Cooke, P. M., and Stevenson, J. W. (1965). An antiviral substance from *Penicillium cyaneo-fulvum* Biourge. *Can. J. Microbiol.* **11,** 913–919.

Cunningham, D., Flashner, M., and Tanenbaum, S. W. (1979). Evidence for selective sulfhydryl reactivity in cytochalasin A-mediated bacterial inhibition. *Biochem. Biophys. Res. Commun.* **86,** 173–179.

Cunningham, K. G., Manson, W., and Spring, F. S. (1950). Corycepin, a metabolic product isolated from cultures of *Cordyceps militaris* (Linn) Link. *Nature (London)* **166,** 949.

Cuppoletti, J., Mayhew, E., and Jung, C. Y. (1981). Cytochalasin binding to Ehrich ascites tumor cells and its relationship to glucose carrier. *Biochim. Biophys. Acta* **642,** 392–404.

Curtis, P. J., Greatbanks, D., Hesp, B., Forbes-Cameron, A., and Freer, A. A. (1977). Sirodesmins A, B, C and G, antiviral epipolythiopiperazine-2,5-diones of fungal origin: x-ray analysis of sirodesmin A diacetate. *J. Chem. Soc., Perkin Trans. 1* pp. 180–189.

Das, C. R., and Pal, A. (1974). Rhizopin: An antibiotic produced by *Rhizopus nigricans*. *Indian Phytopathol.* **27**, 33–36.

de Mayo, P., Spencer, E. Y., and White, R. W. (1961). Helminthosporal, the toxin from *Helminthosporium sativum*. I. Isolation and characterization. *Can. J. Chem.* **39**, 1608–1612.

de Mayo, P., Spencer, E. Y., and White, R. W. (1962). The constitution of helminthosporal. *J. Am. Chem. Soc.* **84**, 494–495.

DePaolo, J. A., Torbell, D. S., and Moore, G. E. (1958–1959). Studies on the carcinolytic activity of fumagillin and some of its derivatives. *Antibiot. Annu.* pp. 541–546.

Divekar, P. V., Haskins, R. H., and Vining, L. C. (1959). Oosporein from an *Acemonium* sp. *Can. J. Chem.* **37**, 2097–2099.

Dulaney, E. L., and Gray, R. A. (1962). Penicillia that make N-formyl hydroxyaminoacetic acid, a new fungal product. *Mycologia* **54**, 476–480.

Eagel, G., von Milczewski, K. E., Prokopek, D., and Teuber, M. (1982). Strain-specific synthesis of mycophenolic acid by *Penicillium roqueforti* in blue-veined cheese. *Appl. Environ. Microbiol.* **43**, 1034–1040.

Eble, T. E., and Hanson, F. R. (1951). Fumagillin, an antibiotic from *Aspergillus fumgiatus* H-3. *Antibiot. Chemother.* **1**, 54–58.

ElKady, I. A., and Moubash, M. H. (1982). Cultural conditions control production of verrucarin, cytotoxic metabolite of *Stachybotrys chartarum*. *Zentralbl. Bakteriol., Parasitenkol. Infectimskr. Hyg., Abt. 2, Naturwiss.: Mikrobiol. Landwirtsch., Technol. Umweltschutzes* **137**, 241–245.

Elpidina, O. K. (1959). Antibiotic and antiblastic properties of poine. *Antibiotiki* **4**, 46–50.

Endo, K., Suzuki, S., Kodama, K., Kato, A., Tamura, G., and Arima, K. (1968). Antitumor activity of fatty acids produced by fungi. *J. Antibiot.* **21**, 690–691.

Endo, K., Kato, A., Tamura, G., and Arima, K. (1969). Studies on antiviral and antitumor antibiotics. VIII. Chemical study of fatty acids with antitumor activity isolated from fungal mycelia. Studies on antiviral and antitumor anbibiotics. VIII. *J. Antibiot.* **22**, 23–26.

Endo, Y., and Wool, I. G. (1982) Site of action of α-sarcin on eukaryotic ribosomes. *J. Biol. Chem.* **257**, 9054–9060.

Engstrom, G. W., DeLance, J. V., Richard, J. L., and Baetz, A. L. (1975). Purification and characterization of roseotoxin B, a toxic cyclodepsipeptide from *Trichothecium roseum*. *J. Agric. Food Chem.* **23**, 244–253.

Eppley, R. M., Mazzola, E. P., Stack, M. E., and Dreifuss, P. A. (1980). Structure of satratoxin F and satratoxin G, metabolites of *Stachybotrys atra*: Application of proton and carbon-13 nuclear magnetic resonance spectroscopy. *J. Org. Chem.* 2522–2523.

Espenshade, M. A., and Griffith, E. W. (1966). Tumor-inhibiting Basidiomycetes and cultivation in the laboratory. *Mycologia* **58**, 511–517.

Farber, J. M., and Sanders, G. W. (1986). Fusarin C production by North American isolates of *Fusarium moniliforme*. *Appl. Environ. Microbiol.* **51**, 381–384.

Ferezou, J. P., Quesneau-Thierry, A., Barbier, M., Kollman, A., and Bousquet, J. (1979). Structure and synthesis of phomamide, a new piperazine-2,5-dione related to the sirodesmines, isolated from culture medium of *Phoma lingam* Tode. *J. Chem. Soc., Perkin Trans. 1* pp. 113–115.

Ferezou, J. P., Quesneau-Thierry, A., Servy, C., Zissman, E., and Barbier, M. (1980). Sirodesmin PL biosynthesis in *Phoma lingam* Todes. *J. Chem.Soc., Perkin Trans. 1* pp. 1739–1746.

Fex, T. (1981). Structure of cytochalasin K, L, and M isolated from *Chalara microspora*. *Tetrahedron Lett.* **22**, 2703–2706.

Franklin, T. J., and Cook, J. M. (1969). The inhibition of nucleic acid synthesis by mycophenolic acid. *Biochem. J.* **113**, 515–524.

Freckman, W. G., Jacubowski, Z. L., Bunge, R. H., French, J. C., and Balta, L. E. (1984). Antibiotic of roridin E-2, U.S. Patent, 4,463,182.

Frederick, C. B., Stanislow, P. J., Vickery, P. E., Bentley, M. C., and Shive, W. (1981a). Production of isolation of siderophores from the soil fungus *Epicoccum purpurescens*. *Biochemistry* **20**, 2432–2436.

Frederick, C. B., Bentley, M. D., and Shive, W. (1981b). Structure of triornicin, a new siderophore. *Biochemistry* **20**, 2436–2438.

Fujimoto, Y., and Satoh, M. (1986). Studies on the metabolites of *Penicillium diversum* var. *aureum*. II. Synthesis and cytotoxic activity of trihydroxytetralones. *Chem. Pharm. Bull.* **34**, 4540–4544.

Fukamura, T. and Kato, K. (1980). L-lysine produced by means of two enzymatic reactions. *J. Agric. Chem. Soc. Jpn.* **54**, 647–653.

Fukushima, T. (1983). Studies on peptide antibiotics leucinostatins. II. The structure of leucinostatins A and B. *J. Antibiot.* **36**, 1613–1630.

Fuska, J. and Prokosa, B. (1976). Cytotoxic and antitumor antibiotics produced by microorganisms. *Adv. Appl. Microbiol.* **20**, 259–370.

Fuska, J., Nemec, P. and Kuhr, I. (1972). Vermiculine, a new antiprotozoal antibiotic from *Penicillium vermiculatum*. *J. Antibiot.* **25**, 208–211.

Fuska, J., Kuhr, I., and Koman, V. (1974a). Isolation of fatty acids from the mycelium of *Penicillium stipitatum* Thom and their effect on Ehrlich ascites carcinoma cells. *Folia Microbiol (Prague)* **19**, 301–306.

Fuska, J., Kuhr, I., Nemec, P., and Fuskova, A. (1974b). antitumor antibiotics produced by *Penicillium stipitatum* Thom. *J.Antibiot.* **27**, 123–127.

Fuska, J., Ivanitskaya, I., HoraKova, K., and Kuhr, I. (1974c). The cytotoxic effects of a new antibiotic vermiculine. *J. Antibiot.* **27**, 141–142.

Fuska, J., Ivanitskaya, L. P., Makukho, L. V., and Volkova, L. Y. (1974d). Effect of vermicullin, PSX-1, bicaverin and daclauxin isolated from fungi on synthesis of nucleic acids in cells of some tumors. *Antibiotiki* **19**, 890–893.

Fuska, T., Nemec, P., and Fuskova, A. (1979a). Vermicillin, a new metabolite from *Penicillium vermiculatum* inhibiting tumor cells *in vitro*. *J. Antibiot.* **32**, 667–669.

Fuska T., Fuskov, A., and Nemec, P. (1979b). Vermistatin, an antibiotic with cytotoxic effects, produced from *Penicillium vermiculatum*. *Biologia (Bratislava)* **34**, 735–739.

Galmarini, O. L., Mastronardi, I. O., and Priestap, H. A. (1974).Two novel metabolites of *Aspergillus fonsecaeus*. *Experientia* **30**, 586.

Gardner, D., Glen, A. T., and Turner, B. (1972). Calonectrin and 15-deacetylcalonectrin, new trichothecenes from *Calonectria nivalis*. *J. Chem. Soc. Perkin I* pp. 2576–2578.

Gasco, A., Serafino, A., Mortarini, V., and Menziani, E. (1974).An antibacterial and antifungal compound from *Calvatia lilacina*. *Tetrahedron Lett.* pp. 3431–3432.

Gatenbeck, S., and Sierakewicz, J. (1973). Microbial production of tenuazonic acid analogues. *Antimicrob. Agents Chemother.* pp. 308–309.

German Patent 2,011,582.

Giannetti, B. M., Steglich, W., Quack, W., Anke, T., and Oberwinkler, F. (1978). Antibiotics from Hasidiomycetes. VI. Merulinic acids A,B, and C, new antibiotics from *Merulius tremellosus* and *Phlebia radiata*. *Z. Naturforsch., C. Biosci.* **33**, 807–816.

Gitterman, C. O., Dulangy, E. L., Kaczka, E. A., Hendlin, D., and Woodruff, H. B. (1962). The human tumor-egg host system. II. Discovery and properties of a new antitumor agent, Hadacidin. *Proc. Soc. Exp. Biol. Med.* **109**, 852–855.

Glasby, J. S. (1976). "Encyclopedia of Antibiotics." Wiley, New York.

Glaz, E. T., Csanyi, E., and Gyimesi, J. (1966). Supplementary data on crotocin—an antifungal antibiotic. *Nature (London)* **212**, 617–618.

Glinsukon, T., and Suvannapur, A. (1982). Changes in plasma volume and protein concentration in rats induced by cytochalasin E. *Toxicol. Lett.* **14**, 157–162.

Godtfredsen, W. O. and Vangedal, S. (1965). Trichodermin, a new sesquiterpene antibiotic. *Acta Chem. Scand.* **19**, 1088–1102.

Goetz, M. A., Lopez, M., Monaghan, R. L., Chang, R. S. L., Lotti, V. J., and Chen, T. B. (1985). Asperlicin, a novel non-peptidal cholecystokinin antagonist from *Aspergillus aliaceus*. Fermentation, isolation and biological properties. *J. Antibiot.* **38**, 1633–1637.

Graf, W., Robert, J. L., Vederas, J. C., Tamm, C., Solomon, P. H., Miura, I., and Nakanishi, K. (1974). Biosysnthesis of The Cytochalasins. Part III. ^{13}C-NMR. of Cytochalasin B (Phomin) and Cytochalasin D. Incorporation of [1-^{3}C]-Sodium Acetate. *Helv. Chim. Acta* **57**, 1801–1815.

Green, S. (1968). Antineoplastic activity of aflatoxin B_1. *Nature (London)* **220**, 931–932.

Gregory, F. J., Healy, E. M., Agersbory, H. P. K., Jr., and Warren, G. H. (1966). Studies on antitumor substances produced by Basidiomycetes. *Mycologia* **58**, 80–90.

Griffin, G., and Chu, F. S. (1983). Toxicity of alternaria metabolites, alternariol, alternariol methyl, altenuene, tenuazonic acid. *Appl. Environ. Microbiol.* **46**, 1420–1442.

Guarino, A. J., and Kredien, N. M. (1963). Isolation and identification of 3'amino-3'-deoxyadenosine from *Cordyceps militaris*. *Biochim. Biophys. Acta* **68**, 317–319.

Gupta, M., Chatterjee, T., Sengupta, S., and Majumdar, S. K. (1984). Structure of new mycotoxin (MT81). *Indian J. Chem., Sect. B* **23B**, 393–394.

Hamuro, J., Rollinghoff, M., and Wagner, H. (1978). β(1→3) glucan-mediated augmentation of alloreactive murine cytotoxic T-lymphocytes in vivo. *Cancer Res.* **38**, 3080–3085.

Harri, E., Loeffler, W., Sigg, H. P., Stahelin, H., Stoll, C., Tamm, C., and Wiesinger, D. (1962). Uber die Verrucarine und Roridine, eine Gruppe von cytostatisch hockwirksamen Antibiotica aus *Myrothecium*—Arten. *Helv. Chim. Acta* **45**, 839–853.

Harri, E., Loeffler, W., Sigg, H. P., Stazhelin, H., and Tamm, C. (1963). Uber die Isolierung neuer Stoffwechselprodukte aus *Penicillium brefeldianum* Dodge. *Helv. Chim. Acta* **46**, 1235–1243.

Hartman, G. H., Richter, H., Weiner, E. M., and Zimmerman, W. (1978). On the mechanism of action of cytostatic drug anguidine and of the immunosuppressive agent ovallcin, two sesquiterpense from fungi. *Planta Med.* **34**, 231–252.

Hata, T. (1977). Fungal carcinogenesis and carcinostasis. In "Recent Advances in Medical and Veterinary Mycology" (K. Iwata, ed.), pp. 299–316. University Park Press, Baltimore, Maryland.

Hauser, D., and Sigg, H. P. (1971). Isolierung und Abbau von Sordarin. *Helv. Chim. Acta* **54**, 119–120.

Hauser, D., Weber, H. P., and Sigg, H. P. (1970). Isolierung and Strukturaufklanung von chaetocin. *Helv. Chim. Acta* **53**, 1061–1073.

Hauser, D., Loosli, H. R., and Niklaus, P. (1972). Isolierung von 11, aa'-Dihydroxychaetocin aus *Verticillium tenerum*. *Helv. Chim. Acta* **55**, 2182–2187.

Hayashi, T., Takatsuki, A., Tamura, G. (1974). The action mechanism of brefeldin A. I. Growth recovery of *Candida albicans* by lipids from the action of brefeldin A. *J. Antibiot.* **27**, 65–72.

Hayashida, S., and Watanabe, Y. (1983). Production of intracellular melanoidin-decolor enzyme and extracellular antitumor polysaccharides by *Coriolus versicolor*. *J. Fac. Agric., Kyushu Univ.* **28**, 1–17.

Hino, M. Nakayama, O., Tsurumi, Y., Adachi, K., Shibata, T., Terano, H., Kohsaka, M., Aoki, H., and Imanaka, H. (1985). Studies of an immunodulator, swainsonine. *J.Antibiot.* **38**, 926–935.

Hirase, S., Nakae, Y., Otsuka, S., Ueno, S., Yoshi Kumi, C., Ohara, M., Hirose, F., Fujii, T., and Ohmura, Y. (1970). Studies on antitumor activity of polysaccharides. I. *Proc. Jpn. Cancer Assoc., 29th Annu. Meet.* 288.

Hirase, S., Nakai, S., Akatsu, T., Kobayashi, A., Oohara, M., Matsunaga, K., Fujii, M., Kidaira, S., Fujii, T., Furusho, T., Ohmura, Y., Wada, T., Yoshikumi, C., Ueno, S., and Ohtsuka, S. (1976a). Structural studies on the antitumor active polysaccharides from *Coriolus versicolor* (Basidiomycetes). I. Fractionation with barium hydroxide. *J. Pharm. Soc. Jpn.* **96**, 413–418.

Hirase, S., Nakai, S., Akatsu, T., Kobayashi, A., Oohara, M., Matsunega, K., Fujii, M., Kodaira, S., Fujii, T., Furusho, T., Ohmura, Y., Wada, T., Yoshjumi, C., Veno, S., and Ohtsuka, E. (1976b). Structural studies on the antitumor active polysaccharides from *Coriolus versicolor* (Basidiomycetes). II. Structures of β-D-glucan moieties of fractionated polysaccharides. *J. Pharm. Soc. Jpn.* **96**, 419–424.

Horakova, K., Navarova, J., Nemec, P., and Kettner, M. (1974). Effect of dactylarin on HeLa cells. *J. Antibiot.* **27**, 408–412.

Hotta, T., Enomoto, S., Yoshikumi, C., Ohara, M., and Uneo, S. (1981a). Protein-bound polysaccharides. U.S. Patent 4,271,151.

Hotta, T., Enomoto, S., Yoshikumi, C., Ohara, M., and Ueno, S. (1981b). Protein-bound polysaccharides. U.S. Patent 4,289,688.

Hutchinson, C. (1983). Comparative biochemistry of fatty acid and macrolide antibiotic (brefeldin A). *Tetrahedron Lett.* **39**, 3507–3513.

Iijima, I., Taga, N., Miyazaki, M., and Tanaka, T. (1979). Synthesis utilizing the β-carbonyl system. Part 5. A synthesis directed towards the fungal xanthone bikaverin. *J. Chem. Soc., Perkin Trans.* **1**, pp. 3190–3195.

Iinuma, H., Nakamura, H., Naganawa, H., Masuda, T., Takano, S., Takeuchi, T., Umezawa, H., Iitaka, Y., and Obayashi, A. (1983). Basidalin, a new antibiotic from Basidiomycetes. *J. Antibiot.* **36**, 448–450.

Ikekawa, T., Nakanishi, M., Uehara, N., Chirara, G., and Fukuoka, F. (1968). Antitumor action of some Basidiomycetes, especially *Phellinus linteus*. *Gann* **59**, 155–157.

Ikekawa, T., Uehara, N., Maeda, Y., Nskanishi, M., and Fukuoka, F. (1969). Antitumor activity of aqueous extracts of edible mushrooms. *Cancer Res.* **29**, 734–735.

Ikekawa, T., Ikeda, Y., Yoshioka, Y., Nakanishi, K., Yokoyama, E., and Yamazaki, E. (1982). Studies on antitumor polysaccharides of *Flammulina velutipes* (Curt. ex Fr.) Sing. II. The structure of EA3 and further purification of EA5. *J. Pharm. Dyn.* **5**, 576–581.

Ishiguro, K., and Arai, T. (1976). Action of the peptide antibiotic leucinostatin. *Antimicrob. Agents Chemother.* **9**, 893–898.

Ito, S., Kurita, H., and Sato, Y. (1970). Cytotoxic antibiotic nematolin. Japanese Patent 45-16795.

Ito, H., Hidaka, H., and Sigiura, M. (1979). Effects of coriolan, an antitumor polysaccharide, produced by *Coriolus versicolor* Iwade. *Jpn. J. Pharmacol.* **29**, 953–957.

Ito, T. (1979). Fusariocin C, a new cytotoxic substance produced by *Fusarium moniliforme*. *Agric. Biol. Chem.* **43**, 1237–1242.

Ito, T., Arai, T., Ohashi, Y., and Sasada, Y. (1981). Structure of fusariocin C, a cytotoxic metabolite from *Fusarium moniliforme*. *Agric. Biol. Chem.* **45**, 1689–1692.

Ito, Y., Shibata, T., Arita, M., Sawai, H., and Ohno, M. (1981). Cordycepin. *J. Am. Chem. Soc.* **103**, 6739.

Itoh, Y., Takahashi, S., and Arai, M. (1982). Structure of gliocladic acid. *J. Antibiot.* **35**, 541–542.

Jarvis, Stahly, B. B., et Curtis, C. R. (1978). Antitumor activity of fungal metabolites: Verrucarin B-9, 10-epoxides. *Cancer Treat. Rep.* **62**, 1585–1586.

Jarvis, B. B., Stahly, G. P., Parahasasivam, G., and Mazzola, E. P. (1980). Structure of roridin J, a new macrocyclic trichothecene from *Myrothecium verrucaria*. *J. Antibiot.* **33**, 256–258.

Jarvis, B. B., Midiwo, J. O., and Tuthill, D. (1981). Interaction between antibiotic trichothecene and *Baccharis mesopotamica*. *Science* **214**, 460–462.

Jarvis, B. B., Vrudhula, V. M., and Pavanasasivam, G. (1983). Trichoverritone and 16-hydroxyroridin L-2, new trichothecenes from *Myrothecium roridum*. *Tetrahedron Lett.* **24**, 3539–3542.

Jarvis, B. B., Yatawara, C. S., Greene, S. L., and Vrudhula, V. M. (1984). Production of verrucarol. *Appl. Environ. Microbiol.* **48**, 673–674.

Jones, D., Anderson, H. A., Russell, J. D., and Fraser, A. R. (1984). Vermiculine, a metabolic product from *Talaromyces wortmanii*. *Trans. Br. Mycol. Soc.* **83**, 718–721.

Jong, S. C., and Gantt, M. J. (1987). "ATCC Catalogue of Fungi/Yeasts," 17th ed. (with an index to industrial applications). American Type Culture Collection, Rockville, Maryland.

Joshi, N. N. (1974). Antiviral agent FWH-775 and method of production. U.S. Patent 3,819,832.

Kamasuka, T., Momoki, Y., and Sakai, S. (1968). Antitumor activity of polysaccharide fractions prepared from some strains of Basidiomycetes. *Gann* **59**, 443–445.

Kanayama, H., Adachi, N., and Togami, M. (1983). A new antitumor polysaccharide from mycelia of *Poria cocos*. *Chem. Pharm. Bull.* **31**, 1115–1118.

Kandefer-Szerszen, M., Kawecki, Z., Salata, B., and Witek, M. (1980). Mushrooms as a source of substances with antiviral activity. *Acta Mycol.* **16**, 215–220.

Karson, E. M. and Ballou, C. E. (1978). Biosynthesis of yeast mannan. Properties of a mannosylphosphate transferase in *Saccharomyces cerevisiae*. *J. Biol. Chem.* **253**, 6484–6492.

Katagiri, K., and Matsuura, S. (1971). Antitumor activity of sytochalasin D. *J. Antibiot.* **24**, 722–723.

Katagiri, K., Sato, K., Kayakawa, S., Matsushima, T., and Minato, H. (1970). Verticillin A, a new antibiotic from *Verticillium* sp. *J. Antibiot.* **23**, 420–422.

Kato, A., Saeki, T., Sukuki, S., Ando, K., Tamura, G., and Arima, K. (1969b). Oryzachlorin, a new antifungal antibiotic. *J. Antibiot.* **22**, 322–326.

Kato, A., Ando, K., Kodama, K., Tamura, G., and Arima, K. (1969a) Studies on antiviral and antitumor antibiotics. X. Identification and chemical properties of anti-tumor active monoglycerides from fungal mycelia. Studies on antiviral and antitumor antibiotics. X. *J. Antibiot.* **22**, 77–82.

Kato, A., Ando, K., Tamura, G., and Arima, K. (1970). Cylindrochlorin, a new antibiotic produced by *Cylindrocladium*. *J. Antibiot.* **23**, 168–169.

Kato, K., Inagaki, T., Shibagaki, H., Yamauchi, R., Okuda, K., Sano, T., and Ueno, Y. (1983). Structure analysis of β-D-glucan extracted with aqueous zinc chlorid from fruit body of *Grifola frondosa*. *Carbohydr. Res.* **123**, 259–269.

Kavanagh, F., Hervey, D., and Robbins, W. J. (1951). Antibiotic substances from Basidiomycetes. VIII. *Pleurotus mutilis* (Fr.) Sacc. and *Pleurotus passeckerianus* Pilat. *Proc. Natl. Acad. Sci. U.S.A.* **37**, 570.

Kawasaki, I. (1972). Penicillic acid. *Jpn. J. Exp. Med.* **42**, 327.

Khaduev, S. Kh., Zhukova, O. S., Dobrynin, Ya. V., Soda, K., and Berezov, T. T. (1987). The effect of L-lysine-α-oxidase from *Trichoderma harzianum* and *Trichoderma viride* on the *in vitro* DNA and RNA snythesis in tumor cells. *Byull. Eksp. Biol. Med.* **4**, 458–460.

Kikumoto, S., Yamamoto, O., Komatsu, N., Kobayashi, H., and Kamasuka, T. (1978). Method of producing neoschizophyllan having novel pharmacological activity. U.S. Patent 4,098,661.

Kirin Brewery Co. Ltd. (1980). Antitumor agent KS-2-B from *Daedalea dickensii* G.B. Patent 2,023,131.

Kis, Z., Close, A., Sigg, H. P., Hruban, L., and Snatzke, G., (1970). Die struktur von Panepoxydion und verwosdten Pilzmetaboliten. *Helv. Chim. Acta* **53**, 1577–1597.

Kiyoto, S. (1986). An new antitumor complex, WF-1360, WF-1360, A, B, C, E and F. *J. Antibiot.* **39**, 762–772.

Kleinschmidt, W. J., Cline, J. C., and Murphy, E. B. (1964). Interferon production induced by statolon. *Proc. Natl. Acad. Sci. U.S.A.* **52**, 752–774.

Kleinschmidt, W. J., and Ellis, L. (1975). Fungal viruses and interferon induction. *Dev. Ind. Microbiol.* **16**, 128–133.

Knauseder, F., and Brandl, E. (1976). Pleuromutilins: Fermentation, structure and biosynthesis. *J. Antibiot.* **29**, 125–131.

Kodama, K., Kusakabe, H., Machida, H., Midorikawa, Y., Shibuya, S., Kuninaka, A., and Yoshino, H. (1979). Isolation of 2'-doxyceformycin and cordycepin from wheat bran culture of *Aspergillus nidulans* Y176-2. *Agric. Biol. Chem.* **43**, 2375–2377.

Komasuka, T., Momaki, Y., and Sakai, S. (1968). Antitumor activity of polysaccharide fractions prepared from some strains of Basidiomcyetes. *Gann* **59**, 443–445.

Komatsu, N., Ogata, Y., Nakasawa, S., Yamamoto, I., Hamada, M., Terakawa, H., and Yamamoto, T. (1961). Anticancer substance from the fruit body of *Lampteromyces japonicus*. *Jpn. J. Bacteriol.* **16**, 746–747 (in Japanese).

Komatsu, N., Terakawa, Y., Nakanishi, K., and Watanabe, Y. (1963). Flammulin a basic protein of *Flammulina Velutipes* with antitumor activities. *J. Antibiot., Sec. A.* **16**, 139–143.

Komatsu, N., Okubo, S., Kikumoto, S., Kimura, K., Saito, G., and Sakai, S. (1969). Host-mediated antitumor action of *Schizophyllum commune*. *Gann* **60**, 137–144.

Komatsu, N., Kikumoto, S., Kimura, K., Sakai, S., Kamasuka, T., Momaki, Y., Takada, S., Yamamoto, T., and Sugayama, J. (1973). Process for manufacture of polysaccharides with antitumor action. U.S. Patent 3,759,896.

Kondo, S. (1980). Novel peptide type antibiotic and the process for the production thereof; antibiotic K582 M-A and M-B. U.S. Patent 4,221,705.

Kovalenko, A. G., Vasilev, V. N., and Votselko, S. K. (1978). Study of antiviral substances produced by *Candida tropicalis* K-41 yeast. II. Physiochemical characteristics of polysaccharide preparations. *Mikrobiol. Zh.* **40**, 767–772.

Kuhr, I., Fuska, J., Sedmera, P., Podozil, M., Vokoun, J., and Vanek, S. (1973). An antitumor antibiotic produced by *Penicillium stipitatum* Thom; its identity with duclauxin. *J. Antibiot.* **26**, 535–536.

Kumano, N., Kurita, K., and Oka, S. (1973). Inhibition of the development of 3-methylcholan-threne-induced mouse tumor with yeast mannan preparation. *Gann* **64**, 529.

Kumano, N., Kurita, K., Y., Ishikawa, T., Nakai, Y., and Konno, K. (1976). Antitumor yeast polysaccharide in the liver and the spleen of mice (radioautographic study). Sci. Rep. Res. Inst. Tohoku Univ. C. **23**, 1–4.

Kupka, J., Anke, T., and Oberwinkler, F. (1979). Antibiotics from Basidiomycetes. VII. Crinipellin, a new antibiotic from the Basidiomycetous fungus *Crinipellis stipitaria* (Fr.) Pat. *J. Antibiot.* **32**, 130–135.

Kureha Kagaku Kogyo KK. (1973). Anticarcinogenic polysaccharides from Basidiomycetes. G.B. Patent 1,331,513.

Kusakabe, H., Kodama, K., Midorikawa, Y., Machida, H., Kuninaka, A., and Yoshino, H. (1976). Process for producing pterin deaminase having antitumor activity. U.S. Patent 3,930,955.

Kusakabe, H., Kodama, K., Machida, H., and Kuninaka, A. (1979). Antitumor activity of a pterin deaminase. *Agric. biol. Chem.* **43**, 1983–1984.

Kusakabe, H., Kodama, K., Kuninaka, A., Yoshino, H., and Soda, K. (1980a). Effect of L-lysine-α-oxidase growth of mouse leukemia cells. *Agric. Biol. Chem.* **44**, 387–392.

Kusakabe, H., Kodama, K., Kuninaka, A., Yoshino, H., Misono, H., and Soda, K. (1980b). A new antitumor enzyme, L-lysine-α-oxidase from *Trichoderma viride*. *J. Biol. Chem.* **255**, 976–981.

Lemarinier, S., and Jaquet, J. (1975). Anticancerous action *in vitro* of aflatoxin B_1 on the epithelioma T8 of Wistar rats. *Bull. Acad. Vet. Fr.* **48**, 423–430.

Lesage, S., and Perlin, A. S. (1978a). Synthesis and stereochemistry of 2-pyrone derivatives related to some fungal metabolites. *Can. J. Chem.* **56**, 2889–2896.

Lesage, S., and Perlin, A. S. (1978b). ^{13}C-nuclear magnetic resonnance spectra of oxirenes. Configuration of 1,2-epoxypyropylside chains in fungal metabolites. *Can. J. Chem.* **56**, 3117–3120.

Lewisohn, R. (1940). Effect of intravenous injections of yeast extract on spontaneous breast adenocarcinomas in mice. *Proc. Soc. Exp. Biol. Med.* **43**(3), 558–561.

Liao, L. L., Grollman, A. P., and Horwitz, S. B. (1976). Mechanism of action of the 12, 13-epoxy-trichothecene, anguidine, an inhibitor of protein synthesis. *Biochim. Biophys. Acta* **454**, 273–284.

Lin, J. Y., and Chou, T. B. (1984). Isolation and characterization of a lectin from an edible mushroom, *Volvariella volvacea*. *J. Biochem.* (Tokyo) **96**, 35–40.

Loeffler, W., Mauli, R., Rusch, M. E., and Staheln, H. (1964). Anguidine and derivatives, new antibiotic and antitumor products. French Patent 1,372,122.

Lofgren, N. M., and Luning, B. (1953). The structure of nebularine. *Acta Chem. Scand.* **7**, 225.

Low, I., Jahn, W., Wieland, T., Sekita, S., Yoshihira, K., and Natori, S. (1979). Interaction between rabbit muscle actin and several chaetoglobosins or cytochalasins. *Anal. Biochem.* **95**, 14–18.

Lucas, E. H., Ringler, R. L., Byerrum, R. V., Stevens, J. A., Clarke, D. A., and Stock, C. C. (1957). Tumor inhibitors in *Boletus edulis* and other Holobasidiomycetes. *Antibiot. Chemother.* (Washington, D.C.) **7**, 1–4.

Lucas, E. H., Byerrum, R. V., Clarke, D. A., Reilly, H. C., Stevens, J. A., and Stock, C. C. (1958–1959). Production of oncostatic principles *in vivo* and *in vitro* by species of the genus *Calvatia*. *Antibiot. Annu.* pp. 493–496.

Magan, N., Cayley, G. R., and Lacey, J. (1984). Production of altenuene, alternariol and alternariol monomethyl ether. *Appl. Environ. Microbiol.* **47**, 1113–1117.

Maheshwari, R. K., and Gupta, B. M. (1973). A new antiviral agent designated 6-MFA from *Aspergillus flavus*. *J. Antibiot.* **26**, 320–327, 328–334, 335–338.

Maheshwari, R. K., Gupta, B. M., Ghosh, S. N., and Gupta, N. P. (1978). Antiviral agent (6-MFA) from *Aspergillus ochraceus*-sensitivity of arboviruses in experimentally infected mice. *Indian J. Med. Res.* **67**, 183–189.

Mair, M. S. R., Carey, S. T., and Anathasubramanian, L. (1982). Biogenesis of the antitumor antibiotic, (+)R-avellaneol. *J. Nat. Prod.* **45,** 644–645.

Mankowski, Z. T., Diller, I. C., and Fisher, M. E. (1957). The effect of various fungi on mouse tumor with special reference to sarcoma 37. *Cancer Res.* **17,** 382–386.

Matsumoto, T., Takanohoshi, M., Okubo, Y. Suzuki, M., and Suzuki, S. (1980). Growth-inhibitory activity of the D-mannan of *Saccharomyces cerevisiae* X2180-1A-5 mutant strain against mouse-implanted Sarcoma 180 and Ehrich-carcinoma solid tumor. *Carbohydr. Res.* **83,** 363–370.

Matsushima, T. (1970). Antibiotic cylindrocladin. Japanese Patent 45-18278.

McCapra, F., Scott, A. I., Delmotte, P., and Delmotte-Plaquee, J. (1964). The constitution of monorden, an antibiotic. *Tetrahendron Lett.* **15,** 869–875.

McCurry, P. M., Jr., and Abe, K. (1973). Stereochemistry and synthesis of the antileukemic agent botryodiplodin. *J. Am. Chem. Soc.* **95,** 5824–5825.

McGuire, J. C. (1987). Phenylalanine ammonia lyase-producing microbial cells. U.S. Patent 4,636,466.

McInnes, A.G., Taylor, A., and Walter, J.A. (1976). Chaetonin. *J. Am. Chem. Soc.* **98,** 6741.

Michel, K. H., Chaney, M. O., Jones, N. D., Hoehn, M. M., and Nagarajan, R. (1974). Epipolythioperazinedione antibiotics from *Penicillium turbatum*. *J. Antibiot.* **27,** 57–64.

Miko, M., Drobnica, L., and Chance, B. (1979). Inhibition of energy metabolism in Ehrlich ascites cells treated with dactylarin *in vitro*. *Cancer Res.* **39,** 4242–4251.

Miller, F. A., Rightsel, W. A., Ehrlich, J., French, J. C., Bartz, Q. R., and Dixon, G. J. (1963), Antiviral activity of tenuazonic acid. *Nature (London)*, **200,** 1338–1339.

Miller, P. A., Trown, P. N., Fulmor, W., Morton, G. O., and Karlines, J. (1968). An epidithioperazinedione antiviral agent from *Aspergillus terreus*. *Biochem. Biophys. Res. Commun.* **33,** 219–221.

Minato, H., and Matshushima, T. (1972), Production of zygosporins D, F, and G. Japanese Patent 47-22394.

Minato, H., and Matsumoto, M. (1970). Studies on the matabolites of *Zygosporium masonii*. I. Structure of Zygosporin A. *J. Chem. Soc.* **117,** 38–45.

Minato, H., Katayama, T., Hayakawa, S., and Katagiri, K. (1972). Identification of ilicicolins with oscochlorin and LL-Z1271. *J. Antibiot.* **25,** 315–316.

Minato, H., Nakato, M., and Katayama, T. (1973). Studies on the metabolites of *Verticillium* sp. structures of verticillins A, B, & C. *J. Chem. Soc., Perkin Trans. 1* pp. 1819–1825.

Minato, H., Katayama, T., and Tori, K. (1975). Vertisporin, a new antibiotic from *Verticimonosporium diffractum*. *Tetrahedron Lett.* **30,** 2579–2582.

Misaki, A., Kakuta, M., Sasaki, T., Tanaka, M., and Miyaji, H. (1981). Studies on interrelation of structure and antitumor effects of polysaccharides: Antitumor action of periodate-modified, branched (1,3)-β-D-glucan of *Auricularia auricula-judae* and other polysaccharides containing (1 3)-glycoside. *Carbohydr. Res.* **92,** 115–129.

Misaki, A., Nasu, M., Sone, S., Kishide, E., and Kinoshita, C. (1986). Comparison of structure and antitumor activity of polysaccharides isolated from Fukurotake, the fruiting body of *Volvariella volvacea*. *Agric. Biol. Chem.* **50,** 2171–2183.

Mitra, S., and Roy, D. K. (1963). Effect of jawaharene upon mouse tumors. *Naturwissenschaften* **50,** 308.

Miyazaki, T., and Nishijima, M. (1981). Studies on fungal polysaccharides. XXVII. structural examination of a water-soluble, antitumor polysaccharides of *Ganoderma lucidum*. *Chem. Pharm. Bull.* **29,** 3611–3616.

Mizuno, K., Ando, T., and Abe, J. (1970). Aranoflavin, a new antibiotic. *J. Antibiot.* **23,** 493–496.

Mizuno, K., Tsujino, M., Takeda, M., Hayashi, M., Atsumi, K., Asano, K., and Matsuda, T. (1974). Studies on bredinin. I. Isolation, characterization and biological properties. *J. Antibiot.* **27,** 775–782.

Mizuno, T., Ohsawa, K., Hagiwara, H., and Kuboyama, R. (1986). Fraction and charcterization of antitumor polysaccharides from Maitake, Grifola frondosa. *Agric. Biol. Chem.* **50,** 1679–1688.

Mizuno, T., and Hazama, T. (1986). Fraction, formolysis and antitumor activity of fibrous polysaccharides from fruit bodies of Ganoderma lucidum. *Bull. Fac. Agric., Shizuoka Univ.* **36,** 77–83.

Mohr, P., Tamm, C., Zuercher, W., and Zehnder, M. (1984). Trichothecenes. *Helv. Chim Acta* **67,** 406–412.

Mori, K., Toyomosu, T., Nanba, H., and Kurada, H. (1987). Antitumor action of fruit bodies of edible mushrooms orally administered to mice. *Mushroom J. Trop.* **7,** 121–126.

Murray, E. G. D., Denton, G., Stevenson, J. W., and Diena, B. B. (1958). A toxin-inactivating substance (noxiversin) from Penicillium cyaneo-fulvum Biourge. *Can. J. Microbiol.* **4,** 593–609.

Naficy, K. (1965). Cyclopin. *Ann. N.Y. Acad. Sci.* **130,** 449–459.

Nagarajan, R., Huckstep, L. L., Lively, K. H., DeLong, D. C., Marsh, M. M., and Neuss, N. (1968). Aranotin and related metabolites from Arachniotus aureus. I. Determination of structure. *J. Am. Chem. Soc.* **90,** 2980–2982.

Nakajima, K., Hirata, Y., Uchida, H., Kimizuka, Y., Taniguchi, T., Obayasi, A., and Tanabe, O. (1983). Polysaccharides having anticarcinogenic activity and method for producing same. U.S. Patent 4,409,385.

Nakamura, S., Nii, F., Inoue, S., Nakanishi, I., and Shimizu, M. (1974). Anti-influenzal effect of rugulosin. *Jpn. J. Microbiol.* **18,** 1–7.

Nakayoshi, H. (1967). Studies on antitumor activity of a polysaccharide produced by a Basidiomycete (chahiratake) I. Chemical properties. *Jpn. J. Bacteriol.* **22,** 641–648.

Nanba, H., and Kuroda, H. (1987a). The chemical structure of an antitumor polysaccharide in mycelia of Cochliobolus miyabeanus. *Chem. Pharm. Bull.* **35,** 1285–1288.

Nanba, H., and Kuroda, H. (1987b). Potentiating effect of β-glucan from Cochliobolus miyabeanus on host-mediated antitumor activity in mice. *Chem. Pharm. Bull.* **35,** 1289–1293.

Nanba, H., and Kuroda, H. (1987c). Potentiation of host-mediated antitumor activity by a β-glucan derived from mycelia of Cochliobolus miyabeanus. *Chem. Pharm. Bull.* **35,** 1523–1530.

Naruse, S., and Takeda, S. (1974). Studies on antitumor activity of basidiomycete polysaccharide. II. Antitumor effects of polysaccharides prepared from cultures of Basidiomycetes. *Mie Med. J.* **23,** 207–231.

Neuss, N., Boeck, L. D., Brannon, D. R., Cline, J. C., DeLong, D. C., Gorman, M., Huckstep, L. L., Lively, D. H., Mobe, J., Marsh, M. M., Molloy, B. B., Nagaragan, R., and Stark, W. M. (1968). Aranotin and related metabolites from Arachniotus aureus (Eidam) Schroeter. IV. Fermentation, isolation, structure elucidation, biosynthesis and antiviral properties. *Antimicrob. Agents Chemother.* pp. 213–219.

Nishimura, Y., Koyama, Y., Umezawa, S., Takeuchi, T., Ishizuka, M., and Umezawa, H. (1980a). Chemical modification of the ester group of diketocoriolin B. *J. Antibiot.* **33,** 393–403.

Nishimura, Y., Koyama, Y., Umezawa, S., Takeuchi, T., Ishizuka, M., and Umezawa, H. (1980b). Syntheses of coriolin, 1-deoxy-1-ketocoriolin and 1,8-dideoxy 1,8-diketocoriolin from coriolin B. *J. Antibiot.* **33,** 404–407.

Notori, S. (1977). Toxic cytochalasins. In "Mycotoxins in Human and Animal Health" (J. V. Rodricks, C. W. Hesseltine, and M. A. Mehlman, eds.), pp. 559–581. Pathotox, Park Forest South, Illinois.

Nozoe, S., Hirae, K., Tsuda, K., Ishibashi, K., Shirasaka, M., and Grove, J. F. (1965). The structure of pyrenophorin. Tetrahedron Lett. **51**, 4675–4627.

Ohno, H., Suzuki, I., Oikawa, S., Sato, K., Miyazaki, T., and Yadomae, T. (1984). Antitumor activity and structural characterization of glucans extracted from cultured fruit bodies of Grifola frondosa. Chem. Pharm. Bull. **32**, 1142–1151.

Ohno, H., Mimura, H., Suzuki, I., and Yadomae, T. (1985a). Antitumor activity and structural characterization of polysaccharide fractions extracted with cold alkali from a fungus Peziza versiculosa. Chem. Pharm. Bull. **33**, 2564–2568.

Ohno, N., Suzuki, I., Sato, K., Oikawa, S., Miyazaki, T., and Yadomae, T. (1985b). Purification and structural characterization of an antitumor β-1,3-glucan isolated from hot water extract of the fruit body of cultured Grifola frondosa. Chem. Pharm. Bull. **33**, 4522–4527.

Ohno, N., Adachi, Y., Suzuki, I., Sato, K., Oikawa, S., and Yadomae, T. (1986a). Characterization of the antitumor glucan obtained from liquid-cultured Grifola frondosa. Chem. Pharm. Bull. **34**, 1709–1715.

Ohno, N., Iino, K., Oikawa, S., Sato, K., Ohsawa, M., and Yadomae, T. (1986b). Fractionation of acid antitumor β-glucan of Grifola frondosa by anion-exchange chromatography using urea solutions of low and high ionic strengths. Chem. Pharm. Bull. **34**, 3328–3332.

Ohtsuka, S., Uneo, S., Yoshikumi, C., Hiroshi, F., Ohmura, Y., Wada, T., Fujii, T., and Takahashi, E. (1977). Polysaccharides and method for producing same. U.S. Patent 4,051,314.

Oka, S., Kimano, N., Sato, K., Tamari, K., Matsuda, K., Hirai, H., Oguma, T., Ogawa, K., Koyooka, S., and Miyao, K. (1969). Antitumor activity of some plant polysaccharides. Gann **60**, 287.

Okamoto, K., Kobayashi, Y., Yoshida, K., Nozaki, Y., Kawai, Y., Kwanao, H., Mayumi, T., and Takao, H. (1978). Teratogenic effects of bredinin, a new immunosuppressive agent, in rats. Congenital Anomalies **18**, 227–233.

Okuda, T., and Fugiwara, A. (1982). Calvatic acid production by the Lycoperdaceae. Trans. Mycol. Soc. Jpn. **23**, 235–238.

Okuda, T., Yoshioka, Y., Ikekawa, T., Chihara, G., and Nishioka, K. (1972). Anticomplementary activity of antitumor polysaccharides. Nature (London), New Biol. **238**, 59–60.

Olson, B. H. (1963). Alpha sarcin. U.S. Patent 3,104,204.

Olson, B. H., and Goerner, G. L. (1965). Alpha sarcin, a new antitumor agent. I. Isolation, purification, chemical composition, and the identity of a new amino acid. Appl. Microbiol. **13**, 314–321.

Olson, B. H., and Goerner, G. L. (1966). Process for the production of regulin by Aspergillus restrictus and resulting product. U.S. Patent 3,230,153.

Olson, B. H., Harvey, C. L., Junek, A. J., and Hennings, J. C. (1963). Restrictocin. U.S. Patent 3,104,208.

Olson, B. H., Jennings, J. C., Roga, V., Junek, A. J., and Schuurmans, D. M. (1965). Alpha sarcin, a new antitumor agent. II. Fermentation and antitumor spectrum. Appl. Microbiol. **13**, 322–326.

Otten, S. L., and Rosazza, J. P. (1981). Conversion of lapachal by Cunninghamella echinulata. J. Nat. Prod. **44**, 562–568.

Owen, F., and Bhuyna, B. K. (1965). Biological properties of new antibiotic, U-13,933. Antimicrob. Agents Chemother. pp. 480–483.

Pavanasasivam, G., and Jarvis, B. B. (1983). Microbial transformation of macrocyclic trichothecenes. *Appl. Environ. Microbiol.* **46**, 480–483.

Perlman, D. (1979). Microbial production of antibiotics. In "Microbial Technology" (H. J. Peppler and D. Perlman, eds.), Vol. 1, pp. 241–280. Academic Press, New York.

Pero, R. W., Hatran, D., and Blois, M. C. (1973a). Isolation of the toxin, altenuisol, from the fungus, *Alternaria tenuis* Auct. *Tetrahedron Lett.* **12**, 945–948.

Pero, R. W., Posner, H., Blois, M., Harvan, D., and Spaulding, J. W. (1973b). Altertoxin I and II, altenuene and alternariol toxicity. *Environ. Health Perspect.* **4**, 87.

Porter, J. N. (1976). Cultural conditions for antibiotic-producing microorganisms. In "Methods in Enzymology" (J. H. Hash, ed.), Vol. 43, pp. 3–23. Academic Press, New York.

Ralman, R., Safe, S., and Taylor, A. (1978). Sporidesmins. Part 17. Isolation of sporidesmin H and sporidesmin J. *J. Chem. Soc. Perkin I* pp. 1476–1478.

Reilly, H. C., Stock, C. C., Buckley, S. M., and Clark, D. A. (1953). The effect of antibiotics upon the growth of sarcoma 180 *in vitro*. *Cancer Res.* **13**, 684–687.

Reusser, F. (1968). Mode of action of melinacidin, an inhibitor of nicotinic acid biosynthesis. *J. Bacteriol.* **96**, 1285–1290.

Richtsel, W. A., Schneider, R. G., Sloan, B. J., Graf, P. R., Miller, F. A., Bartz, Q. P., and Ehrlich, J. (1964). Antiviral activity of gliotoxin and gliotoxin acetate. *Nature (London)* **204**, 1333–1334.

Riondel, J., Beriel, H., Dardas, A., Carraz, G., and Oddoux, L. (1981). Studies of antitumor activity of culture filtrate of *Hohenbuehelia geogenius* (D.C. ex. Fr.) Sing. *Arzneim-Forsch.* **31**, 293–299.

Roland, J. F., Chmielewicz, Z. F., Weiner, B. A., Gross, A. M., Boening, O. P., Luck, J. V., Bardos, T. J., Reilly, H. C., Sugiura, K., Stock, C. C., Lucas, E. H., Byerrum, R. V., and Stevens, J. A. (1960). Calvacin, a new antitumor agent. *Science* **132**, 1897.

Rosazza, J. P. (1979). Conversion of lapachal to dehydro-α-lapachone by *Curvularia lunata*. *Appl. Environ. Microbiol.* **38**, 311–313.

Rosazza, J. P., Nicholas, A. W., and Gustafson, M. E. (1978). Microbial transformations of natural antitumor agents. 7. 14-alpha-hydroxylation of withaferin-A by *Cunninghamella elegans* (NRRL 1393). *Steroids* **31**, 671–679.

Rose, A. H., ed. (1979). "Secondary Products of Metabolism." Academic Press, London.

Roth, R. F., Arcus, Y. M., and Knight, C. A. (1980). Proteolytic action of *Aspergillus niger* extract on influenza virus. *Intervirology* **14**, 167–172.

Rottman, F., and Guarino, A. J. (1964). Inhibition of purine biosynthesis *de novo* in *Bacillus subtilis* by cordycepin. *Biochim. Biophys. Acta* **80**, 640.

Sakai, S., Takada, S., Kamasuka, T., Momoki, Y., and Sugayama, J. (1968). Antitumor action of some glucans, especially on its correlation of their chemical structure. *Gann* **59**, 507–512.

Sakakida, K., and Ikekawa, T. (1980). Emitanin, with antitumor activity produced by *Tricholoma matsutake*, *Volvariella volvacea* or *Tremella fusiformis*. U.S. Patent 4,177,108.

Saltarelli, C. G. (1980). Sarcoma 180 tumor inhibitor produced by cultures of *Candida albicans*. U.S. Patent 4,182,753.

Sariasleni, F. S., and Rosazza, J. P. (1983). Microbial transformation of natural antitumor agents: Products of rotenone and dihydrotenone transformation by *Cunninghamella blakesleeana*. *Appl. Environ. Microbiol.* **45**, 616–621.

Sasaki, S., and Uchida, K. (1987). Isolation and characterization of cell wall polysaccharides from *Aspergillus oryzae*. *Agric. Biol. Chem.* **51**, 2595–2596.

Sasaki, S., Kodama, K., Uchida, K., and Yoshino, H. (1985). Antitumor activity of cell walls of microorganisms. *Agric. Biol. Chem.* **49**, 2807–2808.

Sasaki, T., and Takasuka, N. (1976). Further study of the structure of lentinan, an antitumor polysaccharide from Lentinus edodes. Carbohydr. Res. **47,** 99–104.

Sasaki, T., Arai, Y., Ikekawa, T., Chihara, G., and Fukuoka, F. (1971). Antitumor polysaccharides from some polyporaceae, Ganoderma applanatum (Pers.) Pat. and Phellinus linteus (Berk et Curt.) Aoshima. Chem. Pharm. Bull. **19,** 821–826.

Sawada, J., Omura, S., Nakayoshi, H., Okumura, K., and Kitahara, T. (1968). Antibiotic funginon and a process for producing, using Aspergillus clavatus. U.S. Patent, 3,361,629.

Schillings, R. T., and Ruelius, H. W. (1968). Poricin, an acidic protein with antitumor activity from a Basidiomycete. II. Crystallization, composition and properties. Arch. Biochem. Biophys. **127,** 672–679.

Sekita, S., Yoshihira, K., Natori, S., Udagawa, S., Sakabe, F., Kurata, H., and Umeda, M. (1982). Production, isolation and some cytological effects of chaetoglobosins A-J. Chem. Pharm. Bull. **30,** 1609–1617.

Shearer, W. T., and Moore, E. G. (1981). Cytochalasin B stimulates complement dependent calcium uptake in antibody treated cells. Cell. Immunol. **61,** 62–77.

Shibata, S., Ogihara, Y., Tokutake, N., and Tanoka, O. (1965). Duclauxin a metabolite of Penicillium duclauxi (Delacroix). Tetrahedron Lett. pp. 1287–1288.

Shope, R. E. (1966). An antiviral substance from Penicillium funiculosum. IV. Inquiry into the mechanism by which helenine, exerts its antiviral effect. J. Exp. Med. **123,** 213–227.

Sigg, H. P., and Stoll, C. (1969). Antibiotic SL 1846. U.S. Patent 3,465,079.

Singh, P. P., Whister, R. L., Tokuzen, R., and Wakahara, W. (1974). Scleroglucan, an antitumor polysaccharide from Sclerotium glucanicum. Carbohydr. Res. **37,** 245–247.

Singleton, V. L., and Bohonos, N. (1958). Pecumbin, a new compound from a species of Penicillium. Nature (London) **181,** 1072.

Slater, G. P., Haskins, R. H., and Hodge, L. R. (1971). Metabolites from a Chrysosporium species. Can. J. Microbiol. **17,** 1576–1579.

Smitka, T. A. (1984). Two new trichothecenes, PD 113,325 amd PD 113,326. J. Antibiot. **37,** 823–828.

Sokoloff, B., and Toda, Y. (1967a). Antivirus antibiotic from Paecilomyces todicus and method of producing same. U.S. Patent 3,303,094.

Sokoloff, B., and Toda, Y. (1967b). Antiviral antibiotic from Trichoderma todica and method of producing same. U.S. Patent 3,323,996.

Sorenson, W. G., Sneller, M. R., and Larsh, H. W. (1975). Qualitative and quantitative assay of trichothecin: A mycotoxin produced by Trichothecium roseum. Appl. Microbiol. **29,** 653–657.

Springer, J. P., Clardy, J., Well, J., Cole, R. J., Kirksey, J. W., McFarlane, R. D., and Torgerson, D. (1976). Isolation and structure determination of the mycotoxin chaetoglobosin C, a new [13] cytochalasin. Tetrahedron Lett. **17,** 1355–1358.

Springer, J. P., Cox, R. H., Cutler, H. G., and Crumely, F. G. (1980). The structure of chaetoglobosin K. Tetrahedron Lett. **21,** 1905–1908.

Starratt, A. N., and Loschiavo, S. R. (1974). The production of aphidicolin by Nigrospora sphaerica. Can. J. Microbiol. **20,** 416–417.

Stolk, A. C., and Samson, R. A. (1971). Studies on Talaromyces and related genera. I. Hamigera gen. nov. and Byssochlamys. Persoonia **6,** 341–357.

Strunz, G. M. (1975). An epitetrathiodioxopiperazine with 3S, 6S configuration from Hyalodendron sp. Can. J. Chem. **53,** 295–297.

Sugiura, M., Ohno, H., Kunihisa, M., Hirata, F., and Ito, H. (1980). Studies on antitumor polysaccharides, especially D-11, from mycelium of *Coriolus versicolor*. *Jpn. J. Pharmacol.* **30**, 503–513.

Suhadolnik, R. J., Chassy, B. M., and Walker, G. R. (1969). Nucleoside antibiotics. III. Isolation, structural elucidation and biological properties of 3'-acetamido-3'deoxyadenosine from *Helminthosporium* sp. 215. *Biochim. Biophys. Acta* **179**, 258–267.

Sunagawa, M., Ohta, T., and Nozoe, S. (1983). Biosynthesis of brefeldin A, Introduction of oxygen at the C-7 position. *J. Antibiot.* **36**, 25–29.

Susuki, Y., Tanaka, H., Aoki, H., and Tamura, T. (1970). Ascotoxin (Decumbin), a metabolite of *Ascochyta imperfecta* Peck. *Agric. Biol. Chem.* **34**, 395–413.

Sutherland, S. E., Heath, J., and Bessell, C. J. (1971). Production of double strained ribonucleic acid. U.S. Patent 3,597,318.

Suzuki, I., Yadomae, T., Yonekubo, H., Nishijima, M., and Miyazaki, T. (1982). Antitumor activity of immunodulating material extracted from fungus *Peziza vesiculosa*. *Chem. Pharm. Bull.* **30**, 1066–1068.

Suzuki, S., Kimura, T., Saito, F., and Ando, K. (1971). Antitumor and antiviral properties of penicillic acid. *Agric. Biol. Chem.* **35**, 287–290.

Suzuki, S., Takaku, S., and Mori, T. (1976). Antitumor activity of derivatives of mycophenolic acid. *J. Antibiot.* **29**, 275–285.

Sweeney, M. J., Garzon, K., Harris, P. N., Holmes, R. H., Poore, G. A., and Williams, R. H. (1972). Experimental anti-tumor activity and preclinical toxicology of mycophenolic acid. *Cancer Res.* **32**, 1795–1802.

Tabuchi, T., Nakahara, T., Kodama, K., Vchiyama, H., and Sakai, S. (1981). Cytochalasin D. *Agric. Biol. Chem.* **45**, 1641.

Takahashi, S., Nitta, K., Okami, Y., and Umezawa, H. (1961). Identity of an anti-hela cell substance produced by *Aspergillus terreus* with terreic acid. *J. Antibiot.* **14**, 107.

Takahashi, C., Yoshihira, K., Natori, S., Umeda, M., Ohtsubo, K., and Saito, M. (1974). Toxic metabolites of *Aspergillus candidus*. *Experientia* **30**, 529.

Takaro Shiyo Co. Ltd. (1981). Polysaccharides having anticarcigenic activity and method for producing same. G.B. Patent 2,055,873A.

Takehara, M., Toyomasu, T., Mori, K., and Nakata, M. (1984). Isolation and antiviral activity of double stranded DNA from *Lentinus edodes*. *Kobe J. Med. Sci.* **30**, 25–34.

Takeshita, H. and Anchel, M. (1965). Production of oosporein and its leuco form by Basidiomycete species. *Science* **147**, 152–153.

Takeuchi, T., Iinuma, H., Iwanaga, J., Takehashi, S., Takita, T., and Umezawa, H. (1969). Coriolin, a new Basidiomycetes antibiotic. *J. Antibiot.* **22**, 215–217.

Takeuchi, T., Takahashi, S., Iinuma, H., and Umezawa, H. (1971). Diketocoriolin B, an active derivative of coriolin B produced by *Coriolus consors*. *J. Antibiot.* **23**, 631–635.

Tamura, G., Ando, K., Suzuki, S., Takatsuki, A. and Arima, K. (1968a). Antiviral activity of brefeldin A and verrucarin A. *J. Antibiot.* **21**, 160–161.

Tamura, G., Suzuki, S., Takasuki, S., Ando, K., and Arima, K. (1968b). Ascochlorin, a new antibiotic found by paper-disc agar-diffusion method. I. Isolation biological and chemical properties of Ascochlorin. *J. Antibiot.* **21**, 539–544.

Tamura, G., Kato, A., Ando, K., Hodama, K., Suzuki, S., and Arima, K. (1968c). Antitumor active monoglycerides produced by fungi. *J. Antibiot.* **21**, 688–689.

Tatsuno, T. (1968). Toxicologic research on substance from *Fusarium nivale*. *Cancer Res.* **28**, 2393–2396.

Tendler, M. D. (1969). Production of cellulase, antibiotic and antitumor substances by growing Eumyces ATCC 16425. U.S. Patent 3,438,864.

Togami, M., Takeuchi, I., Imaizumi, F., and Kawakami, M. (1982). Studies on Basidiomycetes. I. Antitumor polysaccharides from bagasse medium on which mycelia of Lentinus edodes (Berk.) Sing. had been grown. Chem. Pharm. Bull. **30**, 1134–1140.

Toth, J. O., Luu, B., and Ourisson, G. (1983). Production of ganoderic acids. Tetrahedron Lett. **24**, 1081–1084.

Trown, P. W. (1968). Antiviral activity of N,N'-dimethyl-epidithiapiperazinedione, a synthetic compound related to the gliotoxins, LL-S88a and B, chetomin and the sporidesmins. Biochem. Biophys. Res. Commun. **33**, 402–407.

Trown, P. W., Lindth, H. F., Milstrey, K. P., Gallo, V. M., Mayberry, B. R., Lindsay, H. L., and Miller, P. A. (1968). LL-S88a, and antiviral substance produced by Aspergillus terreus. Antimicrob. Agents Chemother. pp. 225–228.

Tsuda, Y., Kaheda, A., Tada, A., Nitta, K., Yamamoto, Y., and Iitaka, Y. (1978). Aspterric acid, a new sesquiterpenoid of carotane group, a metabolite from Aspergillus terreus. IFO-6123. X-ray crystal and molecular structure of its p-bromobenzoate. J. Chem. Soc., Chem. Commun. pp. 160–161.

Tsunoda, A., and Ishida, N. (1970). A mushroom extract as an interferon inducer. Ann. N.Y. Acad. Sci. **173**, 719–726.

Ueno, S., Yoshikumi, C., Hirose, F., Omura, Y., Wada, T., Fujii, T., and Takahashi, E. (1980). Method of producing nitrogen-containing polysaccharides. U.S. Patent 4,202,969.

Ueno, Y. (1985). Toxicology of mycotoxins. CRC Crit. Rev. Toxicol. **14**, 99–132.

Ueno, Y., Platel, A., and Fromageot, P. (1967). Interaction entre pigments et acides nucléiques. II. Biochim. Biophys. Acts **134**, 27.

Ueno, Y., Ueno, I., Tatsuno, T., Ohokubo, K., and Tsunoda, H. (1969). Fusarenon-X, a toxic principle of Fusarium nivale-culture filtrate. Experientia **25**, 1062.

Ueno, Y., Ueno, I., Sato, N., Iitoi, Y., Saito, M., Enomoto, M., and Tsunoda, H. (1971). Toxicological approach to (+) rugulosin, an anthraquinoid mycotoxin of Penicillium rugulosum Thom. Jpn. J. Exp. Med. **41**, 177–183.

Ueno, Y., Sawano, M., and Ishii, K. (1975). Production of trichothecene mycotoxins by Fusarium species in shake culture. Appl. Microbiol. **30**, 4–9.

Ukai, S., Kiho, T., Hara, C., Morita, M., Goto, A., Imaizumi, N., and Hasegawa, Y. (1983). Polysaccharides in fungi. XIII. Antitumor activity of various polysaccharides isolated from Dictyophora indusiata, Ganoderma japonicum, Cordycep cicadae, Auricularia auricula-judae, and Auricularia species. Chem. Pharm. Bull. **31**, 741–744.

Umeda, M., Saito, A., and Saito, M. (1970). Cytotoxic effects of toxic culture filtrate of Penicillium purpurogenum and its toxic metabolite, rubratoxin B, on hela cells. Comparative study among the effects of rubratoxin B, colcemid and vinblastine. Jpn. J. Exp. Med. **40**, 409–423.

Umeda, M., Saito, M., Yoshihira, K., Udagawa, S., Ohtsubo, K., Sekita, S., Natori, S., Sakabe, F., and Kurata, H. (1975). Chaetoglobosins A,B,C and F. Experientia **31**, 435.

Umezawa, H., Takahashi, Y., Shirai, T., and Fujii, A. (1974). Bleomycinic acid and process for preparing thereof. U.S. Patent 3,843,448.

Umezawa, H., Takeuchi, T., Iinuma, H., Ito, M., Ishizuka, M., Kurakata, Y., Umeda, Y., Nakanishi, Y., Nakamara, T., Obayashi, A., and Tanabe, O. (1975). A new antibiotic, calvatic acid. J. Antibiot. **28**, 87–90.

Umezawa, H., Takeuchi, T., Iinuma, H., and Tanabe, O. (1976). Production of a new antibiotic, calvatic acid. U.S. Patent 3,980,522.

Upjohn Company. (1969). The antibiotic sphaeropsidin and derivatives thereof. British Patent 1,159,502.

Uraguchi, K., Tatsuno, T., Sakai, F., Tsukioka, M., Sakai, Y., Yonemitsu, O., and Ito, H. (1961). Isolation of two toxic agents, luteoskyrin and chlorine-containing peptide, from the metabolites of *Penicillium islandicum* Sopp, with some properties thereof. *Jpn. J. Exp. Med.* **31**, 19–46.

Urry, W. H., Wehrmeister, H. L., Hodge, E. B., and Hidy. (1966). The structure of zearalenone. *Tetrahedron Lett.* **27**, 3109–3114.

Usui, T., Iwasaki, Y., Mizuno, T., Tanakara, M., Shinkai, K., and Arakawa, M. (1983). Isolation and characterization of antitumor active β-D-glucans from fruit bodies of *Ganoderma applanatum*. *Carbohydr. Res.* **115**, 273–280.

Valisolalao, J., Bang, L., Beck, J.-P., and Ourisson, G. (1980). Chemical and biochemical studies on Chinese drugs. V. Cytotoxicity of tritopenes and analogs from *Poria cocos* (polyporaceae) and related substances. *Bull. Soc. Chim. Fr.* **9**(10), Part 2, 473–477.

Vanderheghe, H., van Dijck, P., and de Somer, P. (1965). Identity of ramycin with fusidic acid. *Nature (London)* **205**, 710–711.

Vesonder, R. F., Tjarks, L. W., Rohwedder, W. K., Burmeister, H. R. B., and Laugal, J. A. (1979). Equisetin, an antibiotic from *Fusarium equiseti* NRRL 5537, identified as a derivative of N-methyl-2,4-pyrollidone. *J. Antibiot.* **32**, 759–761.

Villa, L., and Agostoni, A. (1972). Antitumor activity of total extract of 'Amanita phalloides' Fr. on Ehrlich ascites tumor. *Tumori* **58**, 45–48.

Villard, J., Oddoux, L., and Porte, M. (1980). Anticoxsackie virus activity *in vitro* in culture filtrates of Agaricales. *Ann. Pharm. Fr.* **37**, 143–152.

Vining, L. C., Kelleher, W. J., and Schwarting, A. E. (1962). Oospoein production by a strain of *Beauveria bassiana* originally identified as *Amanita muscaria*. *Can J. Microbiol.* **8**, 931–933.

von Stahelin, H., Kalbere-Rusch, M. E., Signer, E., and Lazary, S. (1968). Uber einige biologische Wirkungen des Cytostaticum Diacetoxyscirpenol. *Arzemein.-Forsch.* **18**, 989–994.

Weindling, R., and Emerson, O. H. (1936). The isolation of a toxic substance from the culture filtrate of *Trichoderma*. *Phytopathology* **26**, 1068–1070.

Wells, J. M., Cole, R. J., Cutler, H. C., and Spaulding, D. H. (1981). *Curvularia lunata*, a new source of cytochalasin A and B. *Appl. Environ. Microbiol.* **41**, 967–971.

White, J. P. and Johnson, G. T. (1971). Zinc effects on growth and cynodontin production of *Helminthosporium cynodontis*. *Mycologia* **63**, 548–561.

Williams, W. H., Lively, D. H., DeLong, D. C., Cline, J. C., Sweeney, M. J., Poore, G. A., and Larsen, S. H. (1968). Mycophenolic acid: Antiviral and antitumor properties. *J. Antibiot.* **21**, 463–464.

W. R. Grace & Co. (1980). Production of monordene by cultures of *Diheterospora chlamydospora*. G. B. Patent 2,013,672.

Yamamoto, Y., Kirijama, N., and Shimizu, S. (1976). Antitumor activity of asterriquinone, a metabolic product from *Aspergillus terreus*. *Gann* **67**, 623–624.

Yang, Q. Y., Zhou, J. S., Li, S. Y., Zheng, R. T., Hsu, L. Z., and Zou, Y. H. (1987). Antitumorigenetic functions of polysaccharopeptide of *Coriolus versicolor*. Personal communication.

Yoshida, T. O., Rising, J. A., and Nungester, W. J. (1962). A tumor inhibitor in *Lampteromyces japonica*. *Proc. Soc. Exp. Biol. Med.* **111**, 676–679.

Yoshioka, H., and Nakatsu, K. (1975). Studies on bredinin. II. The molecular structure of bredinin. *Tetrahedron Lett.* **46**, 4031–4034.

Yoshioka, Y., Ikekawa, T., Noda, M., and Fukuoka, F. (1972). Studies on antitumor activity of some fractions from Basidiomycetes. I. An antitumor acidic polysaccharide fraction of *P. ostreatus* (Fr.) Quel. *Chem. Pharm. Bull.* **20**, 1175–1180.

Yoshioka, Y., Emori, M., Ikekawa, T., and Fukuoka, F. (1975). Isolation, purification, and structure of components from acidic polysaccharides of *Pleurotus ostreatus* (Fr.) Quel. *Carbohydr. Res.* **43**, 305–320.

Biotechnology—The Golden Age

V. S. MALIK

Philip Morris Research Center
Richmond, Virginia 23261

    I. Introduction
   II. The Industrial Organism
  III. The Technology
      A. Recombinant DNA
      B. Reverse Genetics
      C. The Renaissance of Protein Chemistry
  IV. Microbial Degradation of Toxic Pollutants
   V. Enzymes: The Catalysts of the Future
  VI. The Energy
 VII. Engineering Tomorrow's Antibiotics
VIII. Crop Improvement
      A. Tissue Culture
      B. Plants
  IX. Human Proteins of Therapeutic Value
      A. Human Insulin
      B. Growth Hormone
      C. Human Interferon
      D. Tissue Plasminogen Activator
      E. Human Renin
      F. Interleukin-2
   X. Vaccines for the Future
  XI. Hybridomas, Monoclonal Antibodies, and Diagnostic Kits
 XII. Inherited Diseases
XIII. Embryo Transfer and Animal Husbandry
XIV. Future Prospects
      References

I. Introduction

You see things that are and say "why,"
I dream of things that never were and say "why not."

                        George Bernard Shaw

Biotechnology is the application of microorganisms and biological systems to the production of goods and services that are beneficial to human welfare. It is an integration of several disciplines including microbiology, biochemistry, genetics, and biochemical engineering. Biotechnology provides economical and efficient solutions to the rising problems of increasing energy costs, pollution, and depletion of world renewable resources. Microorganisms are being used for transforming naturally occurring molecules into pharmaceuticals. Microbial production of antibiotics, hydroxylations of steroids, conversion of sitosterol into androstadienedione or 9-α-hydroxyandrostenedione, hydrolysis of penicillins G and V into 6-aminopenicillanic acid, and enzymatic isomerization of glucose into fructose are a few examples of the use of microorganisms in the synthesis of products useful to humans (Malik, 1982a; Scott, 1987).

The potential for production of specific fats and sterols by fermentation is enormous. The molecular structure of fats can be altered by enzyme technology. Waste from the petroleum and chemical industries may be converted to useful single-cell protein lipids and biosurfactants (Kretschmer et al., 1982; Cooper and Paddock, 1984; Kawashima et al., 1983; Ristau and Wagner, 1983). Microbially produced polymeric materials such as polyhydroxybutyrate, polyesters, polyglycolic acid, poly-L-lactic acid, and polypivalolactone may be biocompatible and biodegradable, with potential applications in medicine. Acrylates, terpenes, phenolics, and a whole plethora of complex chemical structures that are difficult to synthesize chemically can be produced by fermentation in good yield. Separation of products from the fermentation broth is important and determines the cost of the final product (Fig. 1).

With genetically engineered cells, it would be possible to increase the production of lipids and industrial and diagnostic enzymes, and to synthesize new vaccines, human insulins, lymphokines, interferons, calf renin, human and animal growth hormones, hyaluronic acid, and microbial pesticides. Use of genetically engineered cells may lead to new waste disposal processes and create inexpensive energy sources. The emergence of such new technologies provides a growing opportunity for developing countries (Hommel et al., 1987; Knowles, 1987).

When Watson and Crick discovered the structure of DNA in 1953, the implications for industry were not foreseen. With many applications now clearly in sight and with the continuous refinement of techniques of recombinant DNA and hybridoma technologies, a new era of discoveries has begun. The tools of this revolutionary era are more powerful than is generally realized.

II. The Industrial Organism

Within this century, one third of all Nobel prizes in physiology and medicine have been bestowed upon microbiologists. Microbiologists

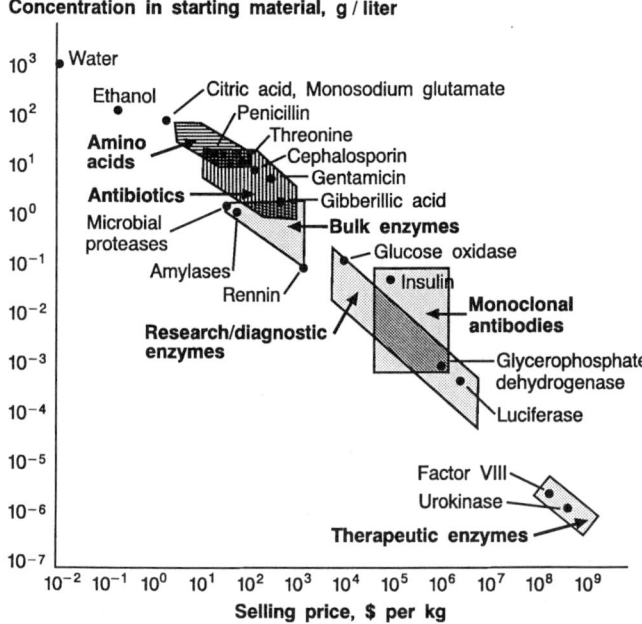

FIG. 1. The selling price of products is a strong function of product concentration and consequently the cost of separation and/or purification. The reactor cost is typically less than 25% of the total production cost. Three distinct categories are evident. From Haggin (1988), *Chemical and Engineering News*, July 11, p. 30. The figure originally appeared in *Bio/Technology* (copyright © 1984) and is reproduced with permission.

are solving many of the mysteries of life itself, because microbes are convenient and useful tools for the study of fundamental life processes common to all cells. Furthermore, microbes are being used for producing rare human and other proteins of commercial significance.

The microorganisms that are of industrial significance have thus far been isolated from nature. Nature is a good source of novel organisms with unexpected abilities to degrade toxic chemicals and to synthesize organic molecules of tremendous chemical complexity. Selection of the organism depends on the presumptive utility of the organism, the growth substrates and environment. Availability of a quick, reliable assay is the key to success in discovering a metabolite and the microbes associated with it. Imaginative methods for the isolation of organisms using rapid assays yield an array of organisms to investigate. Recombinant DNA technology allows addition of extra, desirable genetic material to the organisms isolated from nature, thus modifying their genetic make-up to meet industrial expectations.

The world's largest oil reserves in Saudi Arabia, as well as the flora and microflora of the seas that surround the land, represent a huge and accessible reservoir of feedstock for the production of compounds of considerable economic value. Such compounds can be made by (1) established fermentation processes and (2) harvesting the algal mass from the sea (and eventually growing the representative organism under controlled conditions by known techniques).

A variety of useful compounds ("specialty chemicals," i.e., those compounds costing over $1.00/lb) can be produced microbiologically from alkanes: (1) L-Amino acids, such as glutamate (produced in yields of up to 84 g/liter), lysine (34 g/liter), ornithine (9.3 g/liter, phenylalanine (10 g/liter), tyrosine (19 g/liter), and others in lower amounts; (2) organic acids, such as citrate (up to 183 g/liter), α-ketoglutarate (65.8 g/liter), or fumarate (up to 50 g/liter), and long-chain (C_{22}–C_{18}) dicarboxylic acids; (3) vitamins (riboflavin, pyridoxine hydrochloride, pyridoxal 5'-phosphate, and cyanocobalamin); (4) ergosterol, β-carotene, xanthophylls, antibiotics (β-lactams, corynecins, phenazine derivatives), and carbohydrates (polysaccharides, polyols); and (5) enzymes (protease, lipase) and single-cell protein (Fukui and Tanaka, 1980).

Algae are a largely untapped source of potentially useful material. They are photosynthetic, chlorophyll a-containing organisms and include seaweeds as well as unicellular and mutlicellular microscopic forms. Microalgae are components of the aquatic and marine phytoplankton and are able to grow in extreme environments, including hot deserts. Various marine organisms, particularly blue-green algae, could yield potentially marketable products such as β-carotene and other flavor-enhancing pigments, algal nitrogen-fixing fertilizers, phycobiliproteins, and cosmetic colorants. Carotenoids and other pigments are responsible for the pleasant appearance of ocean-dwelling salmon and shrimp as well as the yellow color of egg yolk. Yeast, algae, and other microbes are known to produce an array of interesting pigments and the super carotenoid-producing strains of these organisms can be used to feed poultry and fish to enhance their desirability as foodstuffs.

In addition, algae produce a wide array of biologically active compounds including toxins, plant growth regulators, sterols, sulfated polysaccharides, polyunsaturated lipids, and antibacterials. Algal products represent an attractive material to evaluate for other pharmacological properties as well. It is of interest to note that the largest potential markets for which natural products are being developed are as anti-cancer drugs with a market value (expressed in United States currency) of $400 million, antifungals ($192 million), cardiovascular drugs ($27 million), and antivirals ($24 million) (Metting and Pyne, 1986).

Algal mariculture is now technically feasible but the economic realities must be assessed. Lack of synthetic substitutes for the phycocolloids (agar, alginates, and carrageenin) and their increasing costs make mariculture an attractive means for their production. Genetic engineering of marine algae is a field wide open for exploitation and awaits exploration (Gellenbeck and Chapman, 1983).

> The most glaring lack of knowledge for the successful application of biotechnology to the production of specialty chemicals is in the identification and characterization of micro-organisms that perform particular chemical conversions. Often, when industrially useful reactions in micro-organisms have been identified, the micro-organism is so poorly understood that the application of new biotechnology is not possible. There are many opportunities for the specialty chemical industry to expand and improve its production capabilities using biotechnology, but before it can take advantage of these opportunities, useful micro-organisms, especially those that function at high temperature and pressure, will have to be screened and identified.
>
> For the specialty chemical industry to take full advantage of biotechnology, sharing of information between industrial chemists and biologists is needed. The sharing of information has to proceed beyond identification of specific steps in a chemical synthesis that are inherently expensive to discussion of the total process for the manufacture of a specialty chemical. Broad discussion could suggest a bioconversion that uses a less expensive starting material and that would replace several steps of the chemical process. Processes for the manufacture of many specialty chemicals could ultimately combine chemical and biological steps, thereby resulting in more economic and energy-efficient manufacturing.

This quotation from a report, *Commercial Biotechnology—An International Analysis* (p. 212) published by the United States Office of Technology Assessment in 1984, emphasizes the importance of selecting the organism that will produce the desired compounds most economically. Voluminous literature exists that describes the strategies in the search for bioactive microbial metabolites, selection of the organism with the desired properties, the techniques for the development of fermentation media and optimization of the process, and strain improvement by mutagenesis and random screening as well as by genetic recombination (Malik, 1979).

Genetic engineering of methane-producing microbes could be important, since methane production from the anaerobic fermentation of organic waste material could provide energy, with the residue used as fertilizer (Cram et al., 1987; Klass, 1984; Mullens and Dalton, 1987). The nitrogen-fixing, photosynthetic, thermophilic and osmophilic facultative autotrophs have not yet been exploited for producing valuable metabolites (Brock, 1986). Exploitation of the biochemical abilities of such organisms could be of economic significance. Thermophiles that utilize cellulose to produce acetic acid and organic solvents such as acetone could be valuable.

Genetic manipulation of industrially important microorganisms has been focused primarily on the improvement of the desirable metabolic by-products by mutation and directed selection. Although, in most cases, the reasons for this highly successful approach have been justified, the recent explosive advances in both cellular and molecular genetics and the rapid development of techniques for both in vitro and in vivo gene transfer for a growing number of prokaryotic and eukaryotic organisms strengthen the industrial commitment to apply recombinant DNA technology to fermentation process improvement.

Emerging strategies for producing proteins utilize recombinant DNA technology for producing proteins in microbes in large yields, from which they are easily purified and sold commercially. Human insulin, interferon, hepatitis B vaccine, human growth hormone, somatostatin, and tissue plasminogen activator already constitute a billion-dollar market. In choosing a host for producing a foreign protein, complexity of protein, number of disulfide bonds, number of subunits, and glycosylation pattern are factors.

Escherichia coli is being used to produce several commercial proteins, such as human insulin, growth hormone, and somatostatin. In *E. coli*, accumulation of protein occurs in cytoplasmic inclusion bodies. The protein is usually in an aggregated, denatured form. Inclusion bodies are easily processed to 90% purity in a single centrifugation step, followed by washing away of many contaminating proteins. Proteins are solubilized by using chaotropic agents such as urea or guanidine and detergents such as sodium dodecylsulfate (SDS). Using specific conditions and reducing agents, proteins are folded to correct disulfide bonding (Lambing et al., 1988). Each protein appears to be unique and requires specific conditions for its proper folding. Most possible secondary structures are often inactive. Different folding and disulfide bonding patterns of the same protein molecule produce different patterns on SDS–polyacrylamide gel electrophoresis (SDS–PAGE. Isoelectric focusing gels can show heterogeneity in products that appear homogeneous on SDS–PAGE.

All recombinant proteins have an N-terminal methionine. Methionine aminopeptidase has been cloned and it is no longer limiting in *E.*

coli that produces large amounts of protein. Inclusion of Mn^{2+} in the growth medium results in stimulation of methionine aminopeptidase activity and thereby removal of methionine from the N-terminus of proteins.

James Cregg (1987) of SIBIA, Inc., La Jolla, California, has developed *Pichia pastoris* as a host for the production of heterologous proteins. *Pichia pastoris* performs posttranslational modification of proteins. This yeast can be grown cheaply on methanol, up to 130 g dry cell weight per liter of medium. A single integrated copy of a gene can produce up to 2.5 g of protein (invertase) per liter.

Current state of the art in recombinant DNA methodology and genetic engineering allows development of vector and transformation systems for any organism. Even though *E. coli* is very versatile as a host for recombinant DNA work, some proteins may be better produced in other organisms. For example, *Aspergillus niger*, which produces glucoamylase in excess of 20 g/liter, could be a good host for producing valuable proteins secreted under the direction of controlling elements of its glucoamylase gene. There is no universal host for producing foreign proteins, and each protein may have its favorite host and specific conditions for its commercial production.

III. The Technology

A. Recombinant DNA

Gene isolation, mapping of genes on the chromosomes, and construction of genomic maps of various organisms are now routine. Genetics has become a tool that will ultimately lead to an understanding of molecular biology in the development of humans. In this effort, other organisms such as yeast, *Drosophila, Arabidopsis,* and the mouse are being studied as model systems (Koshland, 1988). In the last decade, developmental biology has undergone a dramatic change. This change is primarily due to two technical advances: (1) the ability to isolate and characterize wild-type and mutant alleles of genes that are expressed at specific developmental stages with major impetus from DNA sequencing (Prober *et al.*, 1987), and (2) the ability to generate transgenic organisms after introducing various mutant versions of the wild-type gene, with the subsequent study of their phenotypes.

Such studies allow for a definition of the effect of various genes. Recombinant DNA can now break out of its limits of defining only small DNA fragments. This change stems from additional technical advances, including:

(1) Recent developments in pulse field gel electrophoresis of large DNA molecules allow resolution of DNA containing up to 10 mega-base pairs. Thus, isolation of chromosomes in pure form is possible (Fangman, 1978; Michiels et al., 1987; Chu et al., 1987; Schwartz, 1983; Schwartz and Cantor, 1984; Carle and Olson, 1984; Gardiner et al., 1986; Van der Ploeg et al., 1984; McPeek et al., 1986; Gray et al., 1987). Pulse field gel electrophoresis uses two electrical fields that are alternatively applied at different angles (120°) for defined time periods. These pulse fields force large molecules to change direction while migrating through the gel pores. This improves separation of small DNA fragments and provides high-resolution electrophoresis of large DNAs, enabling examination of the entire genome in a single gel. A combination of various blotting and hybridization steps permits quick insight into chromosomal structure abnormalities, oncogenes, and other rearrangements. Gene location, chromosome mapping, and linkage analysis can now be performed with ever-increasing speed (Michiels et al., 1987; Moores, 1987; Gemmill et al., 1987; Umesono et al., 1983; Niwa et al., 1986).

(2) Another advancement of unparalleled importance is the availability of versatile vectors. There are now vectors that can be used to clone very large fragments (~250 kilobases) of DNA into yeast (Burke et al., 1987). These fragments are stable in yeast and replicate as linear molecules; that is, they are more stable than are large circular molecules. These large linear molecules have telomeres at their terminus, and by adding additional autonomously replicating sequences, their stability is further enhanced.

These technical advances will allow the purification of chromosomes from various organisms. Very large fragments of these chromosomes can be cloned and propagated in yeast and then transformed into homologous organisms. This will speed up mapping and sequencing of entire genomes. Knowledge so gained will not only allow analysis of genetic disease in humans but may permit manipulation of the events involved in the development of a fertilized egg.

Even though the methodology of recombinant DNA was discovered approximately two decades ago, the technology has become routine in many biological laboratories. Using these techniques, much biochemical information is being generated. The predictive power of recombinant DNA technology allows logical planning toward solution of complex biological problems. This technology, when combined with formal genetics, offers synergistic speed toward success in many fronts.

Recombinant DNA technology allows the scientist to attack long-standing complex biological problems. Determining gene structure through sequencing has provided clues to the function and evolution of many eukaryotic genes. DNA sequence analysis provides insights into the regulatory sites in the DNA or RNA molecule. The capability to modify precisely the genetic code through site-specific mutagenesis, using strategies based on synthetic oligonucleotide chemistry, may be used to generate precise mutations so that the effects of single amino acid substitutions can be evaluated (Malik, 1982b; Masui et al., 1984; Abdel-Meguid et al., 1987; Robinson and Austen, 1987; Kaback, 1987; Pantoliano et al., 1987; Carter and Wells, 1987; Gish and Eckstein, 1988).

Genomes of many phages, plasmids, viruses, human mitochondria, and tobacco chloroplasts have already been completely sequenced (Moores, 1987). By the end of this century, many more genomes (including those of humans) will be almost, if not completely, sequenced, thus allowing scientists access to the blueprint of life stored in their computers (Smith et al., 1987). The technology is so powerful that now no task appears intimidating and the major obstacle is the choice of the right problems to be attacked.

B. Reverse Genetics

1. Homologous Recombination

The isolation of a gene and the ability to make mutants of it at will at the nucleotide level is a triumph of modern biology. These in vitro, mutated genes can be introduced back into the chromosome to examine their effect. The practice of introducing mutant genes into organisms and then studying the organism's phenotype is the reverse of classical genetics. Using homologous recombination (Fig. 2), precise replacement of the resident gene is now possible in yeast and a few other prokaryotes (Lozanne and Spuditch, 1987). In higher organisms such as plants and animals, transformed DNA integrates randomly and is not specific to the homologous target site. This problem will be overcome in the near future, since direct selection for homologous recombination is possible (Smith and Berg, 1984). In a population of genetically transformed cells, targeted events can be differentiated from random insertions (Smithies, 1985; Jasin and Berg, 1988; Jasin et al., 1985).

Reverse genetics is a powerful tool for studying developmental biology and has practical applications to gene therapy. It can also be used for deleting undesirable DNA sequences, replacing defective sequences, and for protein engineering. (Mansour et al., 1988).

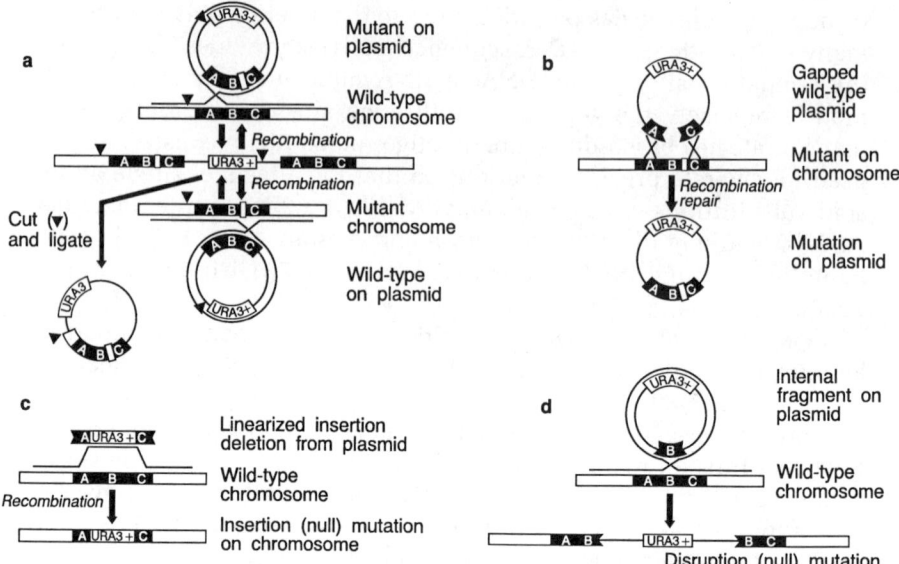

FIG. 2. The use of homologous recombination to transfer mutations into and out of the normal locus on a yeast chromosome. (a) Integration of a cloned gene (A–B–C) by homologous recombination into the mutant locus results in a heterogenetic duplication. The same duplication can be produced by homologous recombination of a mutant plasmid into the normal locus. Depending on the position of the crossover event, excision of the plasmid by homologous recombination from the duplication can result in either a mutant or a wild-type gene at the locus. If one digests the DNA containing the duplication with a suitable restriction endonuclease (shown to the left) and ligates the fragments, one can obtain the mutation by selection for vector markers in *Escherichia coli*. (b) A mutation on a yeast chromosome can be recovered by recombination repair after transformation with a suitably gapped plasmid carrying the wild-type gene. (c) Gene disruption can be accomplished by integration of a linear fragment containing an insertion or deletion containing a selectable marker.(d) Integrative gene disruption occurs when an internal fragment of a gene integrates by homologous recombination into the intact locus, splitting the gene into two partially duplicated but incomplete parts, one missing the amino-terminal coding region and the other missing the carboxyl terminal. Reproduced with permission from Botstein and Fink (1988).

2. Anti-Sense RNA

Since anti-sense RNA is complementary to the messenger RNA species, it inhibits gene expression by forming a double-stranded structure with the corresponding message. Anti-sense RNA regulates gene expression in phages and bacteria (Green, 1986). It can be used for preventing gene expression in plants (Van der Krol et al., 1988) and animals (Green, 1986; Knecht and Loomis, 1987). It is possible to construct an anti-sense C-DNA library (Fig. 3) and transform organisms to screen for desired phenotypes. Usually, anti-sense genes are constructed *in vitro* by reversing the orientation of the known gene, including the coding and regulatory regions, which are placed under the control of a strong promoter and other regulatory elements. The coding region can even have a deletion or insertion in the middle that might recombine with and replace the target gene with a reasonable stretch of a homologous DNA sequence. In this way, both anti-sense and homologous recombination can be exploited to make mutant cell lines of any organism. While homologous recombination allows gene replacement, anti-sense recombination results in gene turn-off. Anti-sense is most effective when selective reduction of tissue-specific expression at certain stages of development, regulated by a specific

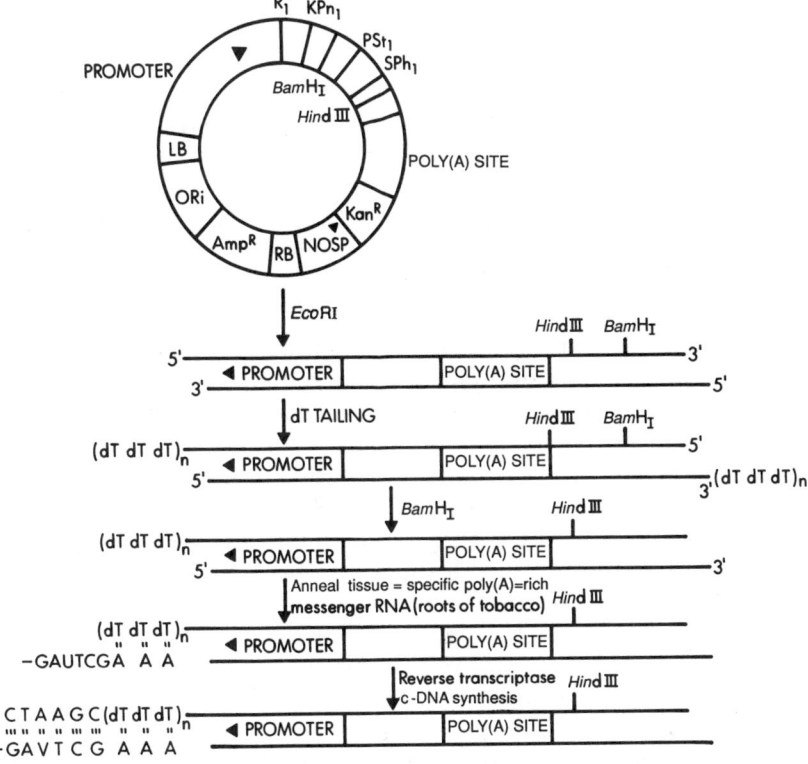

FIG. 3. Tissue-specific anti-sense c-DNA cloning enables researchers to construct libraries in predetermined anti-sense orientation. After colony-hybridization against c-DNA from leaf samples or any other differential hybridization, clones that do not hybridize can be used to transform plants followed by analysis of transformants for altered phenotype.

promoter, is required. Anti-sense recombination can be used to construct viral resistance (Beachy et al., 1988) and suppress activity of undesirable genes. Van der Krol et al. (1988) have used anti-sense RNA to suppress synthesis of petal-specific chalcone synthase, an enzyme required for pigment synthesis, thus generating transgenic plants with altered flower color. Stable and heritable suppression of the expression of nopaline synthase in tobacco plants expressing anti-sense RNA demonstrates that it is possible to use anti-sense technology to alter the phenotype of a mature plant (Rothstein et al., 1988). Reduction by anti-sense RNA of polygalacturonase may yield a tomato with prolonged shelf life and is being pursued by Calgene Inc.

C. The Renaissance of Protein Chemistry

With the advent of new gel matrices that can withstand pressure, dye ligand affinity, hydrophobic chromatography, fast protein liquid chromatography, and 2-dimensional gel electrophoresis, purification and characterization of any protein are possible (Scopes, 1987; Hewick et al., 1981). Furthermore, a few micrograms can yield a partial amino acid sequence that can lead to gene isolation. Repetitive degradation of proteins with phenylisothiocyanate and 4-N,N-dimethylaminoazobenzene-4'-isothiocyanate required more protein than could be easily isolated. However, the gas phase sequencer has created a true renaissance in protein characterization. The sensitivity of amino acid analyses has been enhanced for postcolumn derivatization with fluorescamine and for precolumn derivatization with O-phthaldialdehyde. With the increased sensitivity of newer techniques such as fast atom bombardment mass spectrometry, sequence determination of any protein is possible (Biemann and Scoble, 1987).

Modification of proteins to improve their characteristics is an expanding field these days. Genencor, Inc., South San Francisco, California, has scientists working on protein engineering of subtilisin. The gene for this serine protease of *Bacillus amyloliquefaciens* has been cloned, and several mutant variants have been cloned and sequenced (Carter and Wells, 1987). It is a commercial product and a good secretion model, and is made as a high-molecular-weight precursor, preprosubtilisin. Insertion of aspartic acid into the active site results in accumulation of preprosubtilisin in the cytoplasm. The catalytic activity is enhanced by inserting lysine at position 166 and glutamine at position 156. Methionine and cysteine contribute to the instability and undesirable sulfhydryl linkages in the protein. Methionine 50, 124, and 222 can be oxidized by peroxides to sulfoxides. These methionine residues can be changed by genetic engineering. Similarly, cysteine residues, which are involved in the formation of secondary sulfhydryl groups, can be replaced by genetic engineering, and the mutant proteins are easy to renature to obtain active enzyme (Seno et al., 1988).

IV. Microbial Degradation of Toxic Pollutants

Mixed cultures of microbes, with unexpected metabolic abilities to degrade recalcitrant environmental toxic waste, can be enriched from nature. These organisms can be further endowed with additional capabilities through genetic engineering and used to clean up toxic wastes such as dioxins, polychlorinated biphenyls, chlorinated phenols, and chlorinated benzenes. Mixed cultures are often efficient, but they may not use the toxic chemical as their primary energy source. In such instances, a second compound is used as the energy source and the toxic compound is slowly degraded in a fortuitous reaction called co-metabolism.

Novel microbial pathways slowly evolve to metabolize environmental chemicals if the selection pressure is low. The diverse catabolic reactions scattered among various microbes can be marshalled in one organism by the methods of recombinant DNA technology to metabolize chemicals resistant to biodegradation. By this technology, an existing pathway can be restructured or an entirely new one can be constructed.

Ramos et al. (1987) are trying to broaden the pathway in *Pseudomonas putida*, which degrades methylbenzoate and 3-ethylbenzoate but not 4-ethylbenzoate. The protein that stimulates synthesis of methylbenzoate metabolizing enzymes in *P. putida* does not recognize 4-ethylbenzoate. Furthermore, an intermediate of 4-ethylbenzoate metabolism is a suicide substrate and kills the enzyme catechol 2,3-dioxygenase, which is involved in the metabolism of 4-ethylbenzoate. A *P. putida* that has the mutant regulatory protein to recognize 4-ethylbenzoate and an engineered catabolic catechol 2,3-dioxygenase that is not killed by the suicide intermediate of the catabolic pathway have been constructed to degrade 4-ethylbenzoate (Fig. 4).

Even though two distinct pathways, an ortho and a meta, exist in microbes to degrade chloro- and methylaromatics, only one is used, depending upon the available substrate. Both pathways can be induced in the presence of both chloro- and methylaromatics, but metabolic intermediates are channeled down the wrong route, resulting in a termination of metabolism. By recruiting novel enzymes, strains of *P. putida* that simultaneously degrade a mixture of both chloro- and methylaromatics have been constructed (Eaton and Timmis, 1986a,b; Rojo et al., 1987; Mermod et al., 1986).

V. Enzymes: The Catalysts of the Future

Due to the availability of powerful expression systems in various microorganisms, the large-scale commercial production of useful enzymes could become increasingly attractive. Furthermore, catalytic properties of the enzymes can be improved by using methods of in vitro mutagenesis. Previously, the organisms for producing the enzymes were initially isolated from nature. Enzyme production was usually enhanced by media manipulation and mutagenesis of the organism (Table I).

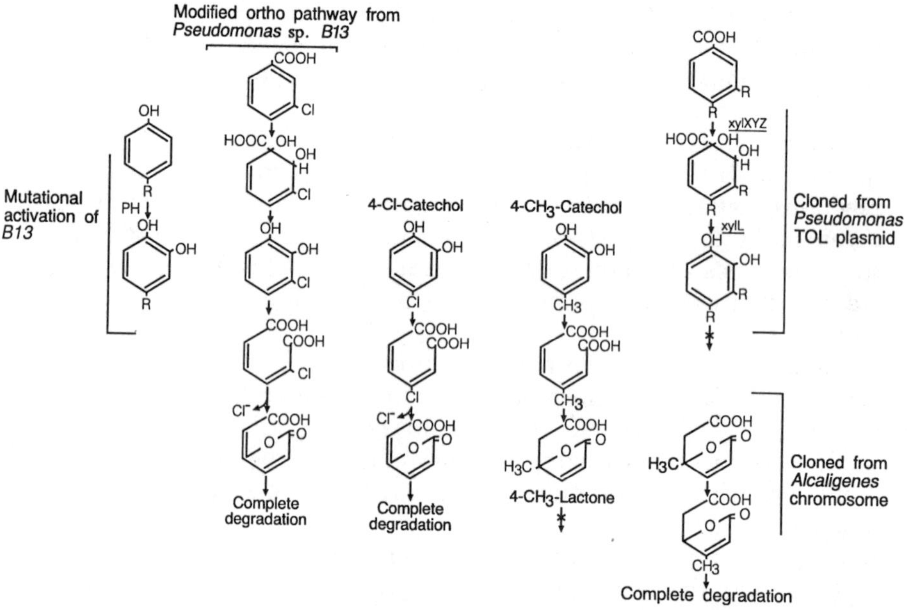

FIG. 4. Constructed hybrid pathway for the simultaneous degradation of chloro- and methylaromatics. The pathway is based on the modified ortho pathway for 3CB of *Pseudomonas* sp. *B13*. Introduction into *B13* of the TOL plasmid genes that code for toluate 1,2-dioxygenase (*xylXYZ*) and 1,6-dihydroxycyclohexa-2,4-diene-1-carboxylate dehydrogenase (*xylL*), together with that of the positive regulator of the *xylXYZL* operon (*xylS*), expands the degradation range to include 4CB and permits transformation of 4MB to 4-methyl-2-enelactone, which would accumulate as a dead-end metabolite. Recruitment of a 4-methyl-2-enelactone isomerase from *Alcaligenes eutrophus* allows transformation of 4-methyl-2-enelactone to 3-methyl-2-enelactone (bottom right), which is degraded by other enzymes of *B13*. Mutational activation of a phenol hydroxylase of *B13* further extends its degradation capacities to chloro- and methylphenols. R = CH_3, Cl, or H. Reproduced with permission from Rojo et al. (1987).

Now, genes for the desired enzymes can be cloned into suitable hosts to facilitate large yield improvements. Enzymes can be engineered to accumulate intracellularly or to be secreted in the medium (Kaiser et al., 1987). Phage λ PL promoter-controlled expression of the *Bacillus stearothermophilus* gene coding for a thermostable α-amylase in *E. coli* (Reinikainen et al., 1988) is being pursued and many enzymes are going to be produced by recombinant organisms in the near future. A highly thermostable acidophilic cellulase system could be useful for clarification of fruit juices at temperatures as high as 75°C, which would result in decreased microbial contamination. Such an enzyme has been discovered in a bacterium, *Acidothermus cellulolyticus*, isolated from Yellowstone's hot springs. The therapeutic use of enzymes will increase, and genes for some of the enzymes that are of plant or animal origin may be expressed in microbes (Vehmaanpera et al., 1987). Chymopapain and collagenase are presently used to dissolve cartilage from the slipped portion of a disc and to remove the dead skin, respectively. J. S. Emtage and P. S. Lowe, working at Celltech Ltd. in Slough, Berkshire, England, cloned and expressed the gene for prorennin, a natural precursor of the cheese-making enzyme rennin (chymosin). Prorennin was up to 5% of total cell protein in *E. coli* and was converted to active rennin by acid treatment. The product is similar to natural rennin, and partially breaks down the milk protein κ-casein. Rennin is currently obtained from the stomach of unweaned calves when they are slaughtered for veal.

VI. The Energy

In the cases of agricultural surpluses and abundance of organic waste material, large-scale biotechnology may be involved. Sugar cane in Brazil is used for producing more than 4.5 million tons of ethanol annually. Increasing petroleum prices and farm surpluses could make fermentation alcohol very attractive as a fuel. Genetically engineered ethanol-tolerant yeast that could produce high yields of ethanol and subsequently serve as cattle feed could be promising (D'Amore and Stewart, 1987).

The fact that the number of new plants for producing ethanol as fuel is growing emphasizes the importance of biomass. Fermentation of cheap, renewable resources is used for producing more than 6 million tons of ethanol a year worldwide. Fuel alcohol is used as an unleaded octane booster. Since substrate cost dominates most fermentation processes, cellulose may eventually be used as inexpensive substrate to produce ethanol, acetone, butanol, microbial polysaccharides, and

TABLE I
SOME INDUSTRIAL ENZYMES: THEIR SOURCES AND USES

Enzyme	Source	Use
Alkaline proteases	Microorganisms	Detergents
Rennin	Calf and lamb stomachs, microbes	Cheese production
Papain	Papaya	Meat; beer; leather; textiles; pharmaceuticals; tenderizer; digestive aid; dental hygiene; clarification of beer haze
Bromelin	Pineapple cannery residues	Meat; baking; pharmaceuticals
Ficin	Figs	Meat; beer; leather; pharmaceuticals
Pepsin	Hog stomachs	Cereals; pharmaceuticals; feeds
Trypsin	Hog and calf pancreases	Meat; pharmaceuticals
Amylases	Fungi, plants, recombinant organisms	Hydrolyze starch for ethanol production; detergents; baked goods; milk; cheese; beer; fruit juices; digestive aid; dental hygiene
Invertases	Yeast	Produce invert sugar; confectionery; distilled beverages
Glucose isomerases	Microorganisms	Convert glucose to fructose; in production of high-fructose corn syrup for soft drinks; also other beverages and foods

Enzyme	Source	Application
Pectinases	Fungi, tomatoes	Hydrolyze pectic substances; clearing of fruit juices; wine; coffee; cocoa
Glucose oxidases	Microorganisms	Oxidize glucose to gluconic acid; eggs; preservation of flavor and color in eggs; food and fruit juices; dental hygiene
β-Amylase	Microbes	Production of maltose and baked products; digestive aid
Pullulanase	Microbes	Production of beer, maltose, and glucose
Cellulase	Microbes	Ethanol production; digestive aid
Urokinase	Urine	Treatment of thrombosis
Asparaginase	*Escherichia coli*	Antitumor agent
Lipases	Hog kidneys and calf pancreases and glands; recombinant organisms	Hydrolyze fats and fatty acid esters; detergents; chocolate; cheese; feeds digestive aids
Catalases and lipoxidase	Livers	Decompose hydrogen peroxide; milk sterilization; production of cheeses; bleaching agent in baking
Penicillin acylase	*Escherichia coli*	Production of 6-aminopenicillanic acid; for synthesis of various β-lactam antibiotics

organic acids. Ethanol is already used as a feedstock in both India and Brazil for production of polyvinyl chloride, polyethylene, synthetic rubber, and other products.

Using fermentation processes and appropriate organisms, it would be possible to produce chemicals such as ethylene, propylene, butylene, and butadiene. Oxygenated hydrocarbons such as acetone, butanol, and ethanol were produced by fermentation early in this century and the economy of their production by fermentation is being reexamined. They may be competitive with oil- and gas-based processes if waste organic material can be used as the primary feedstock.

VII. Engineering Tomorrow's Antibiotics

Increased antibiotic production by amplification of the genes coding for rate-limiting enzymes in biosynthetic pathways will be possible in the near future. Furthermore, hybridization of microbial strains may produce novel antibiotics (Malik, 1981, 1986; Murakami, 1983; Martin and Gil, 1984; Cantoral et al., 1987). At present, penicillins and cephalosporins are hydrolyzed to obtain 6-aminopenicillanic acid (6APA) and 7-aminocephalosporanic acid (7ACA). Derivatives of both acids have been a source of numerous clinically useful analogs. Now enzymes that hydrolyze penicillins and cephalosporins can be introduced into the penicillin-producing strains so that they are secreted into the medium and hydrolyze the accumulating antibiotic. Enzymes that could make useful analogs from 6APA and 7ACA (Ruy and Ruy, 1987; McCullough, 1983) may also be introduced into the producing organisms to synthesize the clinically desirable product directly by fermentation. Amoxicillin was synthesized from D-2-p-hydroxyphenylglycine methyl ester and 6APA with enzymes from β-lactamase-deficient mutants of *Pseudonomas melanogenum* (Kawamori et al., 1983).

Genes that determine the synthesis of chloroperoxidase and chloramphenicol resistance may be expressed in corynecin-producing cultures to favor the accumulation of chloramphenicol (Nuell et al., 1988). Hybridization among various β-lactam, macrolide, and aminoglycoside producers may produce novel molecules with useful therapeutic properties. Isolation of genes involved in antibiotic synthesis has already begun (Malik, 1981, 1986; Samson, 1987). Purification of Isopenicillin-N-synthase and several other enzymes at Eli Lilly by routine chromatographic methods has been achieved. The N-terminus of the cyclase gene was blocked and a Herculean effort was used to sequence several internal peptides. Oligonucleotides were synthesized, based on amino acid sequence, and were used to isolate the corresponding gene from a cosmid library. In streptomyces, there is codon bias. The third codon is 90% GC. Expandase and hydroxylase activity were found to be located in a single polypeptide.

Our knowledge of the chemistry of β-lactam antibiotics has reached the level of sophistication whereby analogs can now be synthesized that are active against organisms resistant to lincomycin, aminoglycosides, and other antibiotics. Therefore, manipulation of the immunological system of the host, in combination with novel analogs of β-lactam antibiotics, may play a major role in the treatment of infectious diseases.

VIII. Crop Improvement

A. Tissue Culture

Secondary metabolites of plants may be produced in tissue culture (Hallewell, 1987; Marcotrigiano, 1986; Collinge, 1986; Eilert et al., 1987). A decade of repeated clonal selection in tissue culture by Kyoto University scientist Yamada has produced cell lines of Coptis japonica that produce 13.2% berberine on a dry-cell-weight basis (Fontanel and Tabata, 1988). The hybridization between Euphorbia millii and C. japonica could produce very high yields of both berberine and anthrocyanin. Isolated protoplasts of different species may be fused to select hybrids that produce commercial products and chemical structures representing precursors from both parents. Tomato golden mosaic virus carries its own DNA polymerase gene. This DNA polymerase gene can be inserted into plant chromosomes, and the origin of replication of the golden mosaic virus could be used to develop an autonomously replicating vector. An autonomously replicating, multicopy plasmid vector with a strong constitutive promoter for plant cells could help exploitation of plant tissue culture for making valuable metabolites. In vitro mutagenesis with somaclonal variation could be combined to obtain variability in the regenerated plant (Davies, 1986; Sabour, 1986; Goldberg, 1988).

Isolation of herbicide-resistant mutants from tobacco cell cultures has already been reported (Chaleff and Ray, 1984). Somaclonal variation may allow isolation of many useful mutants and clones that would be of economic significance (Rangaswami, 1987). Type and amount of variability may be more controllable in tissue culture than in the whole plant. Protoplast fusion could be used to generate hybrids among sexually incompatible plants (Pelletier, 1986; Haissig et al., 1987; Gupta and Durzan, 1987).

B. Plants

Plants may be made genetically resistant to herbicides, pesticides, frosts, and insecticides (Nevins, 1987). Plants or mutant organisms that express novel forms of natural resistance are the source of genes determining resistance. Many single structural genes occurring in nature have already been transferred and expressed in plants (Comai and Stalker, 1986). Such plants that have been engineered by introducing coding regions of single genes under promoters functional in plants are moving from laboratory to field trials. Within 5 years, we may have a genetically engineered plant in the marketplace (Umbeck et al., 1987).

Naturally occurring genetic variability among sexually compatible plants has already been exploited by plant breeders to obtain present-day, high-yielding varieties (Table II). However, the sexual incompatibility barrier may now be crossed by using protoplast fusion and modern methods of recombinant DNA technology.

Isolation of genes and their modification *in vitro* are routine these days. Therefore, genes of interest can be isolated, mutagenized, and then inserted into the plant genome, where they are stably inherited and expressed. Unlike in plant breeding, well-defined DNA sequences are added to the plant genetic material and their effect on the phenotype of the plant can thus be correlated. An important beginning in this direction has already been made by using the tumor-inducing (Ti) plasmid of *Agrobacterium tumefaciens* as a vector. The Ti plasmid inserts a fragment of its DNA called T-DNA into host plant nuclear DNA (David et al., 1984). The integration of T-DNA is directed by

TABLE II

Examples of Agriculturally Important Genes and Traits Transferred to Crop Plants by Interspecific or Intergenic Hybridization[a]

Crop species	Donor species	Trait
Avena sativa (oat)	A. sterilis	Increase yield 25–30%
Beta vulgaris (sugar beet)	B. procumbens	Sugar beet nematode resistance
Brassica napus (Swede turnip)	B. campestris	Club root resistance
Cucurbita pepo (pumpkin)	C. lundelliana	Mildew resistance
Gossypium hirsutum (cotton)	G. tomentosum	Nectariless (decreased incidence of boll rot)
Gossypium hirsutum	G. raimondii	Rust resistance
Lycopersicon esculentum (tomato)	L. hirsutum	Bacterial canker resistance
Lycopersicon esculentum	L. peruvianum	Nematode resistance
		(continued)

TABLE II (continued)

Crop species	Donor species	Trait
Lycopersicon esculentum	L. peruvianum	Jointless (facilitates clean fruit harvest without stems)
Lycopersicon esculentum	L. peruvianum	TMV resistance
Lycopersicon esculentum	L. pimpinellifolium	Fusarium wilt race 1 resistance
Nicotiana tabacum (tobacco)	N. glutinosa	TMV resistance
Nicotiana tabacum	N. longiflora	Blackfire resistance
Oryza sativa (rice)	O. nivora	Grassy stunt virus resistance
Ribes nigrum (black currant)	R. sanguineum	Mildew resistance
Ribes nigrum	R. grossularium	Gall mite resistance
Solanum tuberosum (potato)	S. acaule	Potato virus X resistance
Solanum tuberosum	S. demissum	Late blight resistance, leaf roll resistance, potato virus Y resistance
Solanum tuberosum	S. stoloniferum	Late blight field resistance, potato virus A resistance, potato virus Y resistance
Triticum aestivum (bread wheat)	Aegilops comosa	Stripe rust resistance
Triticum aestivum	Aegilops ovata	High kernel protein
Triticum aestivum	Aegilops speltoides	Stem rust resistance
Triticum aestivum	Aegilops squarrosa	Leaf rust resistance
Triticum aestivum	Aegilops umbellulata	Leaf rust resistance
Triticum aestivum	Agropyron elongatum	Leaf rust resistance, drought tolerance
Triticum aestivum	Secale cereale	Yellow rust resistance, powdery mildew resistance, winter hardiness, leaf rust resistance, stem rust resistance
Triticum aestivum	T. monococcum	Stem rust resistance
Triticum aestivum	T. timopheevi	Stem rust resistance
Triticum aestivum (durum wheat)	T. monococcum	Stem rust resistance
Zea mays (maize)	Tripsacum dactyloides	Northern corn leaf blight resistance

[a] Though selective, the examples given are representative of the plant families in which such transfers have been most successful. The two families dominating the list are the Gramineae (wheats, oat, rice, and maize) and the nightshade family, Solanaceae (tomato, potato, and tobacco). TMV, Tobacco mosaic virus.

specific sequences of about 25 base pairs located at both extremities of the T-region on the Ti plasmid. No functions located within the T-region are required for transfer or integration of DNA. DNA sequences of 50 kilobases or more inserted within the T-region are integrated into the plant genome. The plasmid-derived T-DNA codes for enzymes involved in opine biosynthesis and functions that inhibit plant differentiation (Kudrika, 1986). Since mobilization functions of a small, wide, host-range plasmid can substitute for the Ti plasmid's 25 base-pair, direct-repeat T-DNA borders, the *Agrobacterium* host range may be expandable (Buchanan-Wallaston et al., 1987; Gheysen et al., 1987).

Recently Ti plasmids have been engineered so as to allow insertion and expression of foreign genes into plant cells from which normal plants can be regenerated (Table III). Coding sequences of foreign genes are put under the control of promotors that are functional in plants. These methods are being used to transfer genes for desirable traits, such as resistance to diseases, insects, and herbicides, into plants, thus paving the way for genetic engineering of plants (Schell, 1987).

Transformation of protoplasts and their regeneration into plants of cereal crops is now possible (Cocking and Davey, 1987). Single-stranded DNA of gemini viruses can be used as a vector for introducing genes into monocots. Recombinant DNA can also be introduced by microinjection into ovules and pollen. Floral modifications and exten-

TABLE III

APPLICATIONS OF NEW PLANT VECTORS FROM PHARMACIA[a]

	pGA580	pGA482	pNOS	pNCN	pCaMVCN
Introduction via					
A. tumefaciens	+	+	−	−	−
Electroporation	+	+	+	+	+
Expression					
Promoter on vector	−	−	+	*	*
Promoter with insert	−	+	*	*	−
Promoter screening	+	−	−	*	−
Transient expression	+	+	+	+	+
Stable expression	+	+	−	−	−
Positive control	−	−	−	+	+
Monocots	−	−	−	−	+
Dicots	+	+	+	+	+
Cosmid rescue	−	+	−	−	−

[a], *, With modification of the vector using appropriate flanking restriction sites. Reproduced with permission of *Pharmacia Analects* (1988).

sion of the flowering periods in ornamental plants are also important. Chimeric, single-stranded RNA molecules packaged into viruslike particles could be vectors, simply sprayed on leaves. Genes determining synthesis and structure of storage proteins (e.g., gluten, zein) can be isolated, modified, and reinserted to improve nutritive value and to make better bread. Mutated genes will be used to introduce deletion mutations in plants that could facilitate plant genetics and breeding. Vectors that would integrate DNA sequences by homologous recombination are yet to be developed. Even though homologous recombination is prevalent during plant meiosis, it has yet to be carried out with genes modified *in vitro*.

Stable DNA-mediated transformation of soybean maize tissue culture (Christou et al., 1987, Fromm et al., 1987), Douglas fir (Dandekar et al., 1987), sunflower (Everett et al., 1987), *Asparagus officianalis* (Bytebier et al., 1987), flax (Basisa et al., 1987), *Brassica napus* (Fry et al., 1987), and many other plants has already been achieved (Sheerman and Bevan, 1988).

Transgenic plants of *B. napus*, one of the most important oil seed crops in the world, have already been regenerated. *Agrobacterium tumefaciens* binary vectors were used to insert mouse mutant dihydrofolate reductase coding sequences driven by the cauliflower mosaic virus 35S promoter into *B. napus*, which makes the plant resistant to methotrexate. This opens the way for modifying the biochemistry of edible oil production. Another biotechnology company, Agracetus in Middleton, Wisconsin, has transformed cotton plants and paved the way for introducing other traits, such as insect-killing toxin (Umbeck et al., 1987). Control of gene expression in specific plant tissues is also being addressed (Ecker, 1986; Koncz and Schell, 1986).

Nuclear and chloroplast transformation systems have been developed in the green, unicellular alga *Chlamydomonas reinhardii* (Boynton et al., 1988; Johnson et al., 1988). This will accelerate understanding of the function and regulation of algal genes (Rochaix, 1987; Rochaix and Erickson, 1988). A cell wall-deficient arginine auxotroph was complemented by transformation with a plasmid containing the DNA sequence of the yeast arg 4 locus. The yeast DNA was integrated into the nuclear DNA. Several vectors that use the yeast arg 4 locus as a selective marker and shuttle genes between *E. coli* and *C. reinhardii* have been constructed.

The nitrogen fixing ability of a plant is affected by the efficiency of carbon fixation during photosynthesis. Carbon assimilation in the C-3 photosynthetic group of plants is inhibited by atmospheric oxygen and by the simultaneous release of recently fixed CO_2 in photorespiration.

This is due to the fact that both CO_2 and O_2 are competitive substrates for the bifunctional enzyme, ribulose 1,5-bisphosphate carboxylase/oxygenase (Husic et al., 1987).

The oxygenation of ribulose bisphosphate does not only inhibit carboxylation but channels Calvin cycle intermediates in such a way that 25% of the fixed CO_2 is photorespired as CO_2. The gene for ribulose 1,5-bisphosphate carboxylase/oxygenase (Rubisco) has been isolated and sequenced from many different organisms (Pichersky et al., 1987a,b). In vitro modification of the cloned genes might yield an enzyme that will have little or no oxygenase activity. Introduction of a modified gene into plants may improve the efficiency of photosynthesis. Production of the mutant form of the Rubisco enzyme from higher plants by expressing cloned genes in E. coli has so far been unsuccessful. This could be due to failure of the large subunit to assemble with the mutant small subunit. The DNA sequences of all three proteins, large subunit and small subunit of Rubisco, and the protein that assembles the large and small subunits into active Rubisco enzyme may have to be expressed in the same E. coli cell in order to rescue the assembly of mutant Rubisco.

Zeins, the major storage proteins (50%) of maize seeds, are synthesized in large amounts. The isolation of C-DNA clones corresponding to different molecular-weight classes of zeins and DNA sequence analysis of these clones reveal gene structure typical of eukaryotes. The 5' flanking regions contain consensus sequences such as TATA and CCAT boxes; the 3' flanking regions have AATAAA-specifying polyadenylation sites. These zein genes have no introns and may be subjected to in vitro mutagenesis to obtain sequences that would increase the essential amino acids and nutritional value of zein proteins. Genes that, in nature, code for proteins with an abundance of essential amino acids can also be inserted into food crops (Cocking and Davey, 1987).

The phaseolin family of seed protein genes has been isolated and sequenced from *Phaseolus vulgaris* (Sengupta-Gopalan et al., 1985). The globulin glycoprotein phaseolin comprises up to 50% of the stored protein in the cotyledons of the French bean. These storage proteins are low in methionine and, due to glycosylation, they are not completely digested. However, mutation of certain sequences that affect sites of glycosylation in phaseolin might yield a gene that will code for a nutritive, easily digestible protein (Beachey et al., 1987). These genes, when transformed into tobacco plants, are properly regulated.

Soybeans are a source of sterols that are microbially converted to many useful steroid pharmaceuticals. The mycobacteria are used to transform β-sitosterols to androstadienedione and 9-α-hydroxyandrostenedione, which are important intermediates in the synthesis of steroid hormones. The genes for the biotransformation of sterols can be isolated from the mycobacterium or any other organism (Ide et al., 1987; Kuliopulos et al., 1987) and introduced into plants (Singh and Bhardwaj, 1986), or organisms that are a source of sterols under the control of native promoter (Malik, 1982c). This should allow a direct production of the desired steroid moiety. Agricultural biotechnology would thus include plants engineered to produce pharmaceuticals.

Thaumatin, a novel protein under development as a sweetener by Tate and Lyle, a United Kingdom sugar refiner, may soon be cleared for use in the United Kingdom. Thaumatin is already being commercialized in Japan under the trade name Talin and may soon have clearance as a flavor adjunct. Ripe fruit of Thaumatococcus danielli, a tropical plant found in West Africa, contains this sweet protein, which consists of 207 amino acid residues and has a molecular weight of about 22,000. The protein has two forms differing in amino acid sequence, but both possess eight disulfide bridges that confer stability in a variety of solvents over a wide pH and temperature range. Talin is 5,000 times sweeter than a comparable 4% sucrose aqueous solution. Scientists working at Unilever Research Laboratories in Holland have cloned and expressed in yeast the gene for thaumatin from T. danielli. Thaumatin may now be economically produced by fermentation processes utilizing tailored yeast as a producing organism. The thaumatin gene may also be introduced into sugar cane or beets.

The stage for producing herbicide-resistant crops has been set. DuPont scientists have introduced the gene for resistance to the herbicide sulfonylurea into tobacco plants. Sulfonylurea inhibits the activity of the first enzyme, acetolactate synthase, in the biosynthesis of the branched chain amino acids (Mazur et al., 1987). Genes that encode for acetolactate synthase resistant to sulfonylurea herbicides were isolated from bacteria and expressed in tobacco. Another herbicide, glyphosate, inhibits 5-enol pyruvoyl shikimate-3-phosphate (EPSP) synthase, an enzyme involved in the synthesis of aromatic amino acids. Monsanto scientists linked the EPSP synthase coding sequence to cauliflower mosaic virus promoter and introduced it into tomato plants. The modified plant produces about 20 times the normal level of EPSP synthase, making it resistant to glyphosate. Calgene used a mutant EPSP synthase gene isolated from bacteria to transform tobacco and tomato plants. Calgene has also developed tomato plants resistant to bromoxynil by introducing a herbicide-degrading enzyme. The gene was isolated from a soil bacterium that metabolizes bromoxynil (Cioppa et al., 1987; DeBlock et al., 1987).

Potato, tobacco, and tomato plants have been rendered resistant to another herbicide, phosphinothricin, sold by the German company Hoechst under the trade name Basta. The active ingredient in Basta is an analog of glutamine, which blocks glutamine synthetase. If the herbicide is acetylated, it cannot inhibit the plant enzyme.

A new petunia flower color generated by transformation of a mutant with a maize gene has already been achieved (Meyer et al., 1987).

Construction of plants resistant to viruses by use of anti-sense RNA technology appears promising. Tobacco mosaic virus coat protein genes have been expressed in tomato and tobacco plants from a strong constitutive cauliflower mosaic virus promoter (Tumer et al., 1987; Frier et al., 1987). These plants are cross-protected against infection by several strains of mosaic viruses. This approach may be used to generate plants resistant to their respective viruses (Harrison et al., 1987; Gerlach et al., 1987; Deom et al., 1987).

Many molecules that are produced by microbes have potential in controlling parasites and pests. *Bacillus thuringiensis var. israelensis* produces protein crystals during sporulation that are toxic to the larval stage of many mosquito and black fly species (Sgarella and Szulmajster, 1987; Schnell et al., 1984; Barton et al., 1987). Crystalline protein toxin of *B. thuringiensis* has been used to control crop insects for the last 20 years. Instead of producing the toxin by fermentations of *B. thuringiensis* and then spraying it onto crops, Monsanto scientists genetically engineered *Pseudomonas fluorescens* living in the roots of corn plants to secrete *B. thuringiensis* toxin to kill black cutworm (Obukowicz et al., 1987). Mycogen Corp. of San Diego, California, have isolated strains that produce proteins toxic to beetles and weevils. Chitinase, which is prevalent among streptomyces and bacteria, may be introduced into bacteria that are aggressive at colonizing a crop's root system. The chitinase may kill pathogenic root fungi by hydrolyzing their chitin (Fraley et al., 1988). Expression of the trypsin inhibitor gene of cowpea in tobacco plants enhances their resistance to tobacco budworm attack (Hilder et al., 1987).

Plant Genetic Systems, a Belgian company, and Monsanto scientists have engineered insect-tolerant transgenic tobacco and tomato plants. This has been achieved by inserting the insect control protein gene from *B. thuringiensis* into the plants. These engineered plants and their progeny are tolerant of lepidopteran larvae, since they express the insect control protein gene in association with a plant promoter.

Advanced Genetics Systems scientists have deleted a gene from certain bacteria that determines their ability to form ice crystals on the leaves of plants (Morreale et al., 1988). Such bacteria (called "ice-minus" bacteria), when sprayed on plant leaves, replace the original "ice-plus" bacteria, allowing the plants to survive colder temperatures. Ice-plus bacteria are relatively minor populations in the microbial world of plant leaves. As an example of this strategy, *Pseudomonas syringae,* from which the gene that promotes formation of ice crystals had been deleted, has been sprayed on the leaves and blossoms of frost-sensitive plants, such as strawberries. The mutant P. syringae, rather than the wild-type parent, colonized and protected the plant from frost damage.

Biotechnica International has constructed a strain of *Rhizobium* bacteria that is improved with respect to its ability to fix nitrogen. Manipulation of genes that control the synthesis of secretory proteins that solubilize phosphates and their constitutive expression in roots of various plants might revolutionize phosphate fertilizer technology. Similarly, other aspects of plant nutrition might be altered to afford greater economic benefits for agriculture. Lefebure *et al.* (1988) have shown that chinese hamster metallothionein functions in plants and is expressed from a cauliflower mosaic virus promoter. This may be used to concentrate cadmium in the roots or any other tissues of plants. Many bacteria could be the source of resistance genes to toxic elements (Summers, 1985; Trevors, 1987).

It is apparent that plant molecular biology finally has come of age. In this decade, many exciting developments are expected, and some of the fruits of this basic research may even be transported to the marketplace. Crops may be improved by introducing genes for increased resistance to herbicides, insects, pests, heavy metals, salty soils, and for production of proteins containing more lysine and other essential amino acids. Biotechnology will revolutionize agriculture. Plants with heritable resistance to insects and disease will be based on the incorporation of naturally occurring genes into crop plants. Food crops will be made more nutritious and easier to process. Most products will not require sophisticated farming practices and will be useful to farmers in the third-world, where the technology is most needed.

IX. Human Proteins of Therapeutic Value

Identification, analysis, and sequencing of proteins that exist in cells in minute amounts are now possible by prediction from the sequence of DNA. Such proteins can also be synthesized in large amounts at a high level of purity by expressing their genes in microbes (Masui *et al.,* 1984). Large fragments of DNA can be synthesized, purified rapidly, and cloned to produce biologically active gene products. By placing the synthetic human epidermal growth factor urogastrone gene under control of the promoter of the yeast glyceraldehyde-3-phosphate dehydrogenase gene, a biologically active urogastrone has been synthesized (Urdea et al., 1983). Genentech has also expressed in *E. coli cells* a plasminogen activator that could be used for dissolving blood clots. Many molecules of human origin such as lymphokines, erythropoietin, factor VIII, albumin, interferon, T-cell growth factor (interleukin-2), and colony-stimulating factor will be microbially produced and used to improve human health (Hallewell, 1987; Cullen et al., 1987).

Transgene, a French genetic engineering company, has successfully cloned coagulation factor IX, which is needed to treat some hemophiliacs. Conventional treatment against factor IX deficiency involves transfusion with bulk human blood plasma, which includes all coagulation factors and often hepatitis antigen. DNA technology would eliminate hepatitis transmission and the alleged causative agents of acquired immune deficiency syndrome (Cregg, 1987). A Transgene team made a 52-nucleotide-long probe deduced from the known amino acid sequence of bovine factor IX. This chemically synthesized probe was used to isolate a 1650-nucleotide-long human factor IX gene from a human C-DNA library.

It is essential that human proteins that are made with recombinant bacteria be of high purity and have posttranslational modifications (Lennarz, 1987). This can be achieved by altering the structure of protein by recombinant DNA and gene splicing methods in such a way that protein purification is optimized. The native proteins can then be recovered after purification. Sassenfeld and Brewer (1984) designed a polypeptide fusion for the purification of recombinant proteins. These authors produced human urogastrone with a C-terminal polyarginine fusion. Polyarginine-fused protein was substantially purified due to its basicity. Polyarginine was subsequently cleaved off with carboxypeptidase-B and further purified. Fusion proteins can be produced that bear a decapeptide extension at the N-terminus that can be bound by an antibody. After affinity purification the protein can be separated from the residue flag decapeptide by proteolysis. Rapid purification of a cloned gene product by genetic fusion and site-specific proteolysis has been described by Germino and Bastia (1984).

By the end of this century, biotechnology could be a $100 billion-a-year business. In medicine, several products are already on the market and hundreds more are in the pipeline, undergoing tests for safety, effectiveness, and profitability. The following sections describe some of the products already available.

A. Human Insulin

In 1982, genetically engineered human insulin (Humulin) was the first such product and was sold by Eli Lilly. The human insulin is produced by E. coli that contains genes that were chemically synthesized. This bacterially produced insulin can be obtained in unlimited quantities and is as effective as animal insulin for use in treatment of diabetes.

Denmark's Novo Industries produces human insulin by using a genetically modified baker's yeast. It secretes into the fermentation medium the proinsulin precursor, which is easily purified and converted into biologically active human insulin in a single enzymic step.

B. Growth Hormone

Human growth hormone is used for treatment of children with pituitary dwarfism. Bovine growth hormone may be used to increase production of milk in dairy cattle. Amgen, based in Thousand Oaks, California, is developing bovine growth hormone for Upjohn (Kalamazoo, Michigan), and porcine growth hormone for Smith, Kline & Beckman. In October 1985, Genentech marketed human growth hormone (protropin), which has an extra methionine on N-terminal end. *Pseudomonas aeruginosa* secretes and correctly processes human growth hormone (Gray et al., 1984). Scientists at Cetus Corp., Emeryville, California, have cloned and expressed methionine aminopeptidase, which cleaves off the amino-terminal methionine from genetically engineered proteins possessing an additional methionine at the N-terminal end. However, in March 1987, Eli Lilly marketed a growth hormone that is identical to the human growth hormone.

C. Human Interferon

Interferons are a family of proteins produced in minute amounts by human cells, which may be effective in the treatment of certain diseases. Today, interferons are produced by genetically engineered yeasts and bacteria. Interferons are being sold for treatment of a rare form of cancer called hairy cell leukemia.

D. Tissue Plasminogen Activator

Tissue plasminogen activator dissolves blood clots *in situ* and does not cause bleeding elsewhere in the body. Small blood clots in blood vessels can cause heart attacks and are treated by drugs that dissolve blood clots and prevent coagulation. Since these drugs diffuse throughout the body, they can cause bleeding elsewhere. Genentech scientists have engineered a yeast that has DNA sequences coding for human tissue plasminogen activator and produces this enzyme. Many other corporations are trying to commercialize this product. Analysts estimate that the world market would be $250–400 million a year.

E. HUMAN RENIN

Human renin is involved in controlling blood pressure and will be produced in microrganisms by using new technology.

F. INTERLEUKIN-2

The human protein interleukin-2 is produced by white blood cells. It stimulates growth of disease-fighting white blood cells.

The future of products produced via application of recombinant DNA is very bright and many more will be commercialized in the near future (Table IV).

X. Vaccines for the Future

Safe, effective, and inexpensive vaccines against diseases such as measles, herpes, polio, tuberculosis, leprosy, malaria, rheumatic fever, and pneumonia, along with polyvalent vaccines that immunize against several diseases, may be available. Most vaccines are produced from weakened or killed viruses and bacteria. Genetically engineered vaccines are now produced not from the whole organism but from the part of the organism that prompts the immune system to develop antibodies against that organism. Such a vaccine can be free of contaminants that cause undesirable side effects.

Subunit vaccines that contain one or more major antigens from an infectious agent, but none of its genetic material, could eliminate many problems associated with the use of whole organisms as vaccines. This is done by cloning the gene that codes for an antigenic protein and expressing it in an organism. This approach was used to produce antigens in *E. coli* that immunize swine against scours and ruminants against foot-and-mouth disease virus. Subunit vaccines are sometimes less effective than the inactivated whole organism vaccine. However, antigens which are highly immunogenic in animals may be produced in microorganisms such as *E. coli* or yeast. Cultured animal cells are being engineered in such a way that multicopy vectors such as vaccinia

TABLE IV
SOME PRODUCTS OF RECOMBINANT DNA

Product	Comments	Competitors
Human insulin	Commercially available for treatment of diabetes	Eli Lilly, Genentech, Novo
Human growth hormone	Commercially available to treat pituitary dwarfism	Eli Lilly, Genentech, KABI
Transplasminogen activator	Dissolves blood clots	Genentech, Genetics Institute
Factor VIII	Used in hemophilia for blood clot formation	Genentech, Genetics Institute Pharmacia, Chiron,
Superoxide dismutase	Heart attacks and other uses	Biotechnology General
Colony stimulating factor (CSF)	Simulate granulocyte (white cells) differentiation; G-CSF could be useful in treating leukemia to restore immune competence as adjunct to chemotherapy	Amgen, Immunex, Genetics Institute, Cetus
Epidermal growth factor	Fosters epidermal cell proliferation; first application may be for skin graft donor sites; healing of wounds and burns; for corneal repairs	Amgen, Chiron, Ethicon
Fibroblast growth factor	Stimulates growth of blood vessels; angiogenesis; may be used for treating wounds and burns	California Biotechnology
Nerve growth factor	Stimulates nerve growth and repair	
Platelet-derived growth factor	Stimulates division of fibroblastlike cells	Amgen
Skeletal growth factor	Stimulates bone cell growth	
Wound angiogenesis factor	Stimulates wound healing	
Tumor angiogenesis factor	Stimulates blood vessel proliferation in tumors	
Erythropoietin	To treat anemia in end-stage kidney disease; to treat chronic anemia and in elective surgery transfusions	Amgen, Genetics Institute, Biogen, Integrated Genetics
Interleukin-2	Useful for cancer therapy; in combination with LAK cells or tumor-infiltrating lymphocytes	Amgen, Cetus, Immunex
Gamma interferon	Antiviral and anti-cancer applications	Amgen, Biogen, Genentech
Porcine somatotropin	Improved feed efficiency in swine; leaner pork	Amgen, Biogen, Bio-Technology General, others
Bovine somatotropin	Improved milk yield in dairy cows	Amgen, Bio-Technology General, Genentech
AIDS[a] vaccine, hepatitis B vaccine, herpes vaccine	Prevention of AIDS,[a] hepatitis, and herpes	Repligen, Merck, Amgen, Biogen, Chiron, Genentech

[a] AIDS, Acquired immunodeficiency syndrome.

virus, with strong promoters, ribosome binding sites, and efficient terminators, are used to produce mammalian proteins and antigens. Expression of influenza HA in cells has already been achieved using both bovine papilloma virus and SV40 as vectors by Gething and Sambrook at Cold Spring Habor Laboratory. A herpes virus subunit vaccine has been produced by Larry Lasky (Genentech) by amplifying a herpes glycoprotein (gp) D gene integrated into Chinese hamster ovary cells. This was achieved by amplifying both methotrexate resistance and the gp D gene, which had been altered to delete its anchor peptide. Synthetic peptides that mimic the amino acid sequence of the reactive site of antigen may be either attached to immunogenic carrier proteins for use as vaccines or effective as a prelude to immunization with the living organism (Wilson, 1984).

Potential live vaccines using recombinant vaccinia viruses have been constructed for both hepatitis and herpes simplex. These recombinant vaccinia viruses express cloned genes of the hepatitis B virus surface antigen or the gp D gene from herpes simplex virus (Paoletti et al., 1984; Hogle, 1988). Both Lederle Laboratories and Molecular Genetics, Inc., have cloned the herpes simplex virus gene into E. coli to produce a protein designed to protect against both genital and oral herpes infections. Merck, Sharp & Dohme commercialized a new vaccine against hepatitis B, produced with the help of a genetically engineered yeast. A synthetic peptide malaria vaccine against *Plasmodium falsiparum* sporozoites is being tested in humans and an improved vaccine will be available in the future (Herrington, 1987). Aldolase, a key enzyme in glycolysis, synthesized during the maturation of the blood stages of the malarial parasite, could be a good antigen. The blood stages of the malaria parasite *P. falciparum* lack a citric acid cycle. The inhibition of aldolase activity, which is a limiting step in glycolysis, could inhibit the maturation of the parasite (Certa, 1988). Large, complex protein antigens with many sulfhydryl groups may be better produced in yeast or *Aspergillus*, while simple, small proteins are conveniently synthesized in E. coli (Cullen et al., 1987; Kitano et al., 1987; Husson and Young, 1987).

XI. Hybridomas, Monoclonal Antibodies, and Diagnostic Kits

Monoclonal antibodies are being increasingly used for diagnostic purposes. They are pure and are produced in unlimited quantities through hybridomas constructed by fusing antibody-producing white cells with cancer cells that are immortal. They are being used to detect infections such as meningitis, chlamydia, and acquired immunodeficiency syndrome, as well as mycotoxins, plant pathogens, and pollutants.

Monoclonal antibodies are tools to detect, treat, and eventually prevent parasitic diseases such as malaria, leischmaniasis, trichinosis, leprosy, schistosomiasis, and coccidiosis. Trypanosomes escape immune surveillance by altering some of their surface antigens. However, nonvariable antigenic proteins or the conserved immunosilent regions of variable antigens may be made antigenic by using synthetic peptide analogs. Malignancy may be diagnosed by monoclonal antibodies, since tumor-associated antigens have been recognized. Monoclonal antibodies have a future in diagnosing many diseases and spermatozoa sexing (Klausner, 1987).

Small pieces of DNA of a gene can be used to scan for part or all of that gene (Chirikjan, 1987). DNA sequences linked to Huntington's chorea have been identified using a probe. Carriers of this genetic defect, which does not appear until midlife and after childbearing years, can be identified before carriers have children (Caskey, 1987; White and Caskey, 1988). Several hereditary diseases, such as Duchenne's muscular dystrophy, hemophilia, and Huntington's disease, can be predicted by chromosomal analysis and analysis of the DNA restriction patterns of susceptible families.

Gen-Probe, a San Diego, California-based company, diagnoses disease by using probes that detect the specific RNA sequences of pathogens. A single strand of iodinated DNA binds to the pathogen's ribosomal RNA. In this way, tests for infections of *Legionella, Mycoplasma pneumoniae, Mycobacterium tuberculosis*, and *Mycobacterium aviumintracellular* have been developed. Such tests for *M. tuberculosis* and *Mycobacterium laprae* could be useful in third-world countries, where these pathogens are infecting people (Matthews and Kricka, 1988).

DNA-based techniques are going to be increasingly important in forensic analysis and diagnosis of infectious diseases and malignancies. A new procedure called polymerase chain reaction not only adds to the speed of analysis, but the amplification step allows the use of a small amount of materials, such as chorionic villus samples from pregnant women (Saiki *et al.*, 1988). This new technique amplifies *in vitro* a target DNA sequence at least 100,000 fold in a few hours and utilizes the thermostable enzyme Taq polymerase (Mullis, 1987). The automation of the polymerase chain reaction and an automated DNA sequencer would increase the clinical applications of this technique by additional orders of magnitude.

XII. Inherited Diseases

Recombinant DNA technology has begun to unlock the secrets of inherited diseases. The molecular basis for many such diseases will be unraveled in the near future (Table V). With the new techniques of genetic engineering, what was thought impossible a few years ago is now possible. Of 3000 different genetic diseases, specific DNA probes for routine screening of inherited diseases are quickly becoming available. Genes implicated in cancers of various types have been discovered, and many human genes have already been sequenced (Watson et al., 1983; Bishop, 1987). Loss of a part of chromosome 5 has been shown in human colon cancer tumors, the second most common cancer in the United States. Furthermore, a small region of chromosome 5 seems to determine which members of a susceptible family will develop a relatively rare form of colon cancer (Bodmer, 1987). Many diseases, such as sickle-cell and other anemias, a congenital form of emphysema, Huntington's disease, and Duchenne's muscular dystrophy, can be detected by using DNA probes and the restriction enzyme pattern of DNA (Caskey, 1987). By the end of this century, the human genome will have a restriction map and most of it will be sequenced.

Over 300 cloned human DNA fragments are available from the American Type Culture Collection for distribution to scientists. Included are clones mapping for each of the human chromosomes and chromosome-specific libraries. EcoRI and HindIII total digest libraries, as well as a partial digest from flow-sorted human chromosomes, have been constructed for each human chromosome by the Los Alamos and Lawrence Livermore National Laboratories.

Actual replacement of defective genes in the sperm or egg may be possible. The human gene for growth hormone has already been implanted into the mouse embryo to produce mice 80% larger than normal. The structural gene coding for human growth hormone was fused to the regulatory region of the mouse gene for metallothionein-1 and microinjected into the fertilized eggs of mice. Twenty-three mice (70%) stably incorporated fusion genes, had a higher concentration of human growth hormone in their serum, and grew significantly larger than control mice. Cadmium or zinc, inducers of the metallothionein gene, further induced synthesis of human growth hormone (Palmiter et al., 1983). Harmless viruses may be used to carry the necessary gene into the patient's cells. The problem of gene regulation still has to be solved. Activating inactive genes is another avenue that is being used to research sickle-cell anemia and thalassemia. Administration of the anti-tumor drug 5-azacytidine reactivated the fetal gene to produce sufficient normal hemoglobin to improve health.

Homologous recombination between DNA sequences residing in the chromosome and newly introduced DNA sequences make it possible to routinely alter genes in the genome of yeast (Botstein and Fink, 1988). This may be possible with higher forms of life in the near future (Doetschman et al., 1987). Song et al. (1987) used homologous recombination to introduce a DNA sequence into the chromosomal β-globin locus of mice. Mutant neomycin resistance genes in the mice chromosome were corrected with the plasmid DNA carrying a different mutation in the neo gene. They derived recipient mouse cell lines with a resident mutant neo gene that has an amber mutation causing premature termination of translation. This amber mutation in the initial portion of the coding region of neo was corrected by microinjecting a mutant neo that has a deletion removing the last 52 amino acids from the protein. At a frequency of 1 per 1000 infections, neo-resistant cells were obtained, half of which had a normal neo gene with the amber mutation removed. Surprisingly, in the rest of the neo-resistant cell lines, the amber mutation was compensated for by another mutation resulting from incorrect repair of the heteroduplex formed between the introduced and the resident DNA sequence. Implications of implantation of genetically engineered fibroblasts into mice and its role in gene therapy offers attractive possibilities for treating many diseases (Selden et al., 1987).

The California Institute of Technology (Pasadena) researchers have used gene transfer techniques to cure the neurological disease of uncontrollable shivering, convulsions, and early death in a sick strain of mice. This ailment is due to the destruction of white fatty insulation, myelin, on nerve cells and also occurs in humans who suffer from multiple sclerosis and Lou Gehrig's disease. Two hundred copies of the gene that determines the basic myelin protein were injected into fresh mouse embryos, which were then implanted in female mice. Infant mice, in whose chromosomes duplicate copies of the gene were incorporated, did not develop the disease. This approach altered the genetic make-up of the offspring. Genes can also be transferred into cells where they are usually expressed. Diseases such as sickle-cell anemia, Tay–Sachs disease, and cystic fibrosis are candidates for this approach, which will cure the recipient but still allow the disease to be passed on to its progeny.

XIII. Embryo Transfer and Animal Husbandry

Embryo transfer technology has tremendous potential for the enhancement of animal reproduction. Embryos collected from donor females can be stored by cryopreservation and transferred to recipient

TABLE V
CHROMOSOMAL LOCALIZATION AND GENE ABNORMALITIES IN SELECTED NEUROLOGICAL DISEASES[a]

Genetic classification and disease	Chromosome	Gene defect	Comments on genetic heterogeneity
Autosomal dominant			
Huntington's disease	4p16	Unknown	None demonstrated in 50 pedigrees
Myotonic dystrophy	19 Centromeres	Unknown	None demonstrated
Familial Alzheimer's disease	21q21	Unknown	Not adequately studied
Familial amyloidotic polyneuropathy	18q11.2–q12.1	Single-base pair substitution in mRNA for transthyretin	Allelic heterogeneity
Manic–depressive illness	11p, Xp	Unknown	Evidence for nonallelic heterogeneity
Spinocerebellar atrophy	6	Unknown	Unknown
Von Recklinghausen's neurofibromatosis	17	Unknown	None demonstrated in >25 pedigrees
Bilateral acoustic neurofibromatosis	22	Unknown	Unknown
Charcot–Marie–Tooth disease (type 1)	1q2	Unknown	Unknown

X-linked recessive			
Duchenne's dystrophy	Xp21	Deletions in 5–10%	Multiallelic heterogeneity
Adrenoleukodystrophy	Xq27–q28	Unknown	Unknown
Lesch–Nyhan syndrome	Xq27	HPRT deficiency, many variations	Multiallelic heterogeneity
Pelizaeus–Merzbacher disease	Xq21–q22	Defect in myelin proteolipid protein	Unknown
Autosomal recessive			
Gaucher's disease	1q21	Amino acid substitution in glucocerebrosidase	Allelic heterogeneity
G_{M1} gangliosidosis	3pl	Partially characterized	Unknown
G_{M2} gangliosidosis			
Tay–Sachs disease (type 1)	15q22–q25	Mutation in gene encoding α-chain of hexosaminidase	Allelic heterogeneity
Sandhoff disease (type 2)	5q13	Mutation in gene encoding β-chain of hexosaminidase	Allelic heterogeneity
Wilson's disease	13q14.11	Unknown	Unknown
Recessive with germinal chromosomal deletion			
Central neurofibromatosis	22q11–q13	Unknown	Unknown
Retinoblastoma	13q14	Partially characterized	Allelic heterogeneity
Meningioma	22q12.3–qter	Unknown	Unknown

[a] Reproduced with permission from Martin (1987).

females, which serve as surrogate mothers. Embryo transfer technology is now possible with nonsurgical procedures in the cow, and the potential of this rapidly developing science will continue to increase with many practical applications (Mapletoft, 1984). Sperm injection, nuclear implantation, and manipulation of individual genes and chromosomes will be useful in animal breeding. Transgenic animals generated by microinjection of foreign genes into fertilized eggs represent one of the open new avenues to exploiting these techniques for basic as well as applied studies. The growth hormone gene, injected into pronuclei of one-cell fertilized ova, has already been integrated into genomes, and has been expressed and transmitted to the progeny of mice, sheep, fish, and several other animals (Jaenisch, 1988). Introduction of the growth hormone gene in marine life will allow quick exploitation of the ocean for producing food (MacLean et al., 1987). It may be possible to add genes for anatomical and production traits or disease resistance directly to the embryo from any source, since DNA transfer is not species specific.

Goats, sheep, and chickens are good subjects for transgenetic engineering to improve the value of their milk and eggs. Integrated Genetics, Inc., in Framingham, Massachusetts, has expressed the tissue plasminogen activator gene in the mammary glands of mice (Gordon et al., 1987). In this way, milk can be made into a source of human proteins of therapeutic value. The c-DNAs for human blood clotting factor IX and α-antitrypsin have been expressed in sheep from the sheep β-lactoglobulin gene promoter, and many more foreign genes will be introduced in these animals in the near future.

XIV. Future Prospects

The name of the game today is recombinant DNA and the pace of progress is fast. The technology is so powerful that it is allowing manipulation of the genome in ways hard to imagine only a decade ago. Several products of recombinant DNA, including human insulin, growth hormone, and hepatitis B vaccine, are already on the market and many are in the pipeline.

The short-term outlook for biotechnology applications to improve the quantity and quality of the world's food supply is good. The easy availability of technology and, especially, key genes to developing countries could help fight world hunger and malnutrition. The most progress has been made in isolating single genes from nature and introducing them into systems in which they make a positive impact. These isolated, well-defined DNA sequences that determine resistance to disease, herbicides, insecticides, and pesticides should be introduced into crop varieties that provide foods for third-world populations. Similarly milk, meat, and wool production may be improved by the transplanting of growth hormone genes as well as by embryo transfer in domestic and sea animals. If patent laws prohibit use of key DNA sequences and useful genes by third-world countries, then the gap would widen even further. It is, therefore, important that the nuts and bolts of the technology and key genes be made available for introduction into the genomes of the varieties of plants and animals that third-world populations need for their survival.

In order to be successful and competitive in biotechnology, appropriate infrastructure is required. It is essential that the people involved have the necessary knowledge and training in the various aspects of this multifaceted discipline. To this end, the curricula in a number of United States colleges have been designed to provide not only the underlying basic knowledge in general and cell biology, microbiology, immunology, general and organic chemistry, biochemistry, enzymology, and other areas, but also in specialized topics. These include genetics, genetic engineering, prokaryotic and eukaryotic molecular genetics with recombinant DNA laboratory training, nucleic acid sequencing and hybridization, animal and plant tissue culture, monoclonal antibody technology, hybridoma techniques, peptide and protein synthesis, industrial microbiology, fermentation scale-up, computer science, and industrial management. The purpose is to transfer this knowledge to encourage further innovation into the biotechnological setting.

Efforts in other countries are also aimed at expanded education in biotechnology in an effort to keep abreast of this rapidly advancing field (Carter, 1987).

> In modern times development of science and technology is one of the most important factors in determining national power. Science and technology which are driving forces in the cultural development of mankind have greatly contributed toward rapid economic growth and human welfare. (Korean President Park Chung Hee, 1966)

References

Abdel-Meguid, S. S., Shieh, M., Smith, W. W., Dayringer, H. E., Bioland, B. N., and Bentle, L. A. (1987). *Proc. Natl. Acad. Sci., U.S.A.* **84**, 6434–6437.
Barton, K. A., Whitely, M. R., and Yang, N. Y. (1987). *Plant Physiol.* **85**, 1103.
Basisan, N., Armitage, P., Scott, R. J., and Draper, J. (1987). *Plant Cell Rep.* **6**, 396–398.

Beachy, R. N., Stark, D. M., Deom, C. M., Oliver, M. J., and Fraley, R. T. (1987). In "Tailoring Genes for Crop Improvement" (G. Bruening, J. Harada, and T. Kosuge, eds.). Plenum, New York.

Beachy, R. N., Nelson, R. S., Register, J., Fraley, R. T., and Tumer, N. (1988). In "Genetic Improvement of Agriculturally Important Crops: Progress and Issues" (R. T. Fraley, N. M. Frey, and J. Schell, eds.), pp. 47–53. Cold Spring Harbor Lab., Cold Spring Harbor, New York.

Beckman, J. S., and Soller, M. (1987). Biotechnology **5**, 573–576.

Biemann, K., and Scoble, H. A. (1987). Science **237**, 992–998.

Bishop, J. M. (1987). Science **235**, 305–311.

Bodmer, W. F. (1987). Nature (London) **328**, 614–616.

Botstein, D., and Fink, G. R. (1988). Science **240**, 1439–1442.

Boynton, J. E., Gillham, N. W., Harris, E. H., Hosler, J. P., Johnson, A. M., Jones, A. R., Anderson, B. L., Robertson, D., Klein, T. M., Shark, K. B., and Sanford, J. C. (1988). Science **240**, 1534–1538.

Brock, T. D., ed. (1986). "Thermophiles." Wiley (Interscience), New York.

Buchanan-Wallaston, V., Passiatore, J. E., and Cannon (1987). Nature (London) **328**, 172.

Burke, D. T., Carle, G. F., and Olson, M. V. (1987). Science **236**, 806–812.

Bytebier, B., Deboeck, F., Greve, H. D., Montagu, M. V., and Hernalsteens, J. P. (1987). Proc. Natl. Acad. Sci. U.S.A. **84**, 5345–5348.

Cantoral, J. M., Diez, B., Barredo, J. L., Alvarez, E., and Martin, J. F. (1987). Bio/Technology **5**, 494–497.

Carle, G. F., and Olson, M. V. (1984). Nucleic Acids Res. **12**, 5647–5664.

Carter, P., and Wells, J. A. (1987). Science **237**, 394–399.

Carter, T. H. (1987). Bio/Technology **5**, 347–349.

Caskey, C. T. (1987). Science **236**, 1223.

Certa, V. (1988). Science **240**, 1036–1038.

Chaleff, R. S., and Ray, T. B. (1984). Science **223**, 1148.

Chirikjan, J. G. (1987). "Application of Nucleic Acid Probe Technology in Medicine." Am. Elsevier, New York.

Christou, P., Murphy, J. E., and Swain, W. F. (1987). Proc. Natl. Acad. Sci. U.S.A. **84**, 3962–3966.

Chu, G., Vollrath, D., and Davis, R. W. (1987). Science **234**, 1582.

Cioppa, G. D., Bauer, S. C., Taylor, M. L., Rochester, D. E., Klein, B. K., Shah, D. M., Fraley, R. T., and Kishore, G. M. (1987). Bio/Technology **5**, 579–584.

Cocking, E. C., and Davey (1987). Science **236**, 1259.

Collinge, M. (1986). Trends Biotechnol. **4**, 209–212.

Comai, L., and Stalker, D. (1986). Oxford Surv. Plant Mol. Cell Biol. **3**, 456–480.

Cooper, D. G., and Paddock, D. A. (1984). Appl. Environ. Microbiol. **47**, 173–176.

Cram, D. S., Sherf, B. A., Libby, R. T., Mattaliano, R. J., Ramachandran, K. L., and Reeve, J. N. (1987). Proc. Natl. Acad. Sci. U.S.A. **84**, 3992–3996.

Cregg, J. M. (1987). Bio/Technology **5**, 479–485.

Cullen, D., Gray, G. L., Wilson, L., Hayenga, K. J., Lamsa, M. H., Rey, M. W., Norton, S., and Berka, R. M. (1987). Bio/Technology **5**, 369–376.

D'Amore, T., and Stewart, G. G. (1987). Enzyme Microb. Technol. **9**, 322–326.

Dandekar, A. M., Gupta, P. K., Durzan, D. J., and Knauf, V. (1987). Bio/Technology **5**, 587–326.

David, C., Chilton, M. D., and Tempe, J. (1984). Bio/Technology **2**, 73–78.

Davies, P. A. (1986). Theor. Appl. Genet. **72**, 644.

DeBlock, M., Botterman, J., Vandewide, M., Docky, J., Thoen, C., Gosselle, V., Movva, N., Thompson, C., Montagu, M., and Leemans, J. (1987). *EMBO J.* **6,** 2513–2518.
Deom, C. M., Oliver, M. J., and Beachy, R. N. (1987). *Science* **237,** 389–393.
Doetschmann, T., Gregg, R. G., Malda, N., Hooper, M. L., Melton, D. W., Thompson, S., and Smithies, O. (1987). *Nature (London)* **330,** 576–578.
Eaton, R. W., and Timmis, K. N. (1986a). *J. Bacteriol.* **168,** 123–131.
Eaton, R. W., and Timmis, K. N. (1986b). *J. Bacteriol.* **168,** 428–430.
Ecker, R. (1986). *Mol. Gen. Genet.* **205,** 14–24.
Eilert, V., DeLuca, V., Kurtz, W. G. W., and Constabel, F. (1987). *Plant Cell Rep.* **6,** 271–275.
Everett, N. P., Robinson, K. E. P., and Mascarenhas, D. (1987). *Bio/Technology* **5,** 1201–1204.
Fangman, W. L. (1978). *Nucleic Acids Res.* **5,** 653–668.
Fontanel, A., and Tabata, M. (1988). *Plant Cell Rep.* **7,** 206.
Fraley, R. T., Frey, N. M., and Schell, J., eds. (1988). "Genetic Improvement of Agriculturally Important Crops: Progress and Issues." Cold Spring Harbor Lab., Cold Spring Harbor, New York.
Frier, L. S., Merlo, D., Zinnen, T., Burhop, L., Hill, K., Krahm, K., Jarvis, J., Nelson, S., and Halk, E. (1987). *EMBO J.* **6,** 1845–1851.
Fromm, M. E., Taylor, L. P., and Walbot, V. (1986). *Nature (London)* **319,** 791–793.
Fry, J., Barnason, A., and Horsch, R. B. (1987). *Plant Cell Rep.* **6,** 321–325.
Fukui, S., and Tanaka, A. (1980). *Adv. Biochem. Eng.* **17,** 1–95.
Gardiner, K., Laas, W., and Patterson, D. (1986). *Somatic Cell Mol. Genet.* **12,** 185–195.
Gellenbeck, K. W., and Chapman, D. J. (1983). *Endeavour* **7,** 31–38.
Gemmill, R. M., Coyle-Morris, J. F., McPeek, F. D., Ware-Uribe, L., and Hecht, F. (1987). *Gene Anal. Technol.* **4,** 119–121.
Gerlach, W. L., Llewellyn, D., and Haseloff, J. (1987). *Nature (London)* **328,** 802–805.
Germino, J., and Bastia, D. (1984). *Proc. Natl. Acad. Sci. U.S.A.* **81,** 4692–4696.
Gheysen, G., Montagu, J., and Zambryski, P. (1987). *Proc. Natl. Acad. Sci.* **84,** 6169–6173.
Gish, G., and Eckstein, F. (1988). *Science* **240,** 1520–1522.
Goldberg, R. B. (1988). *Science* **240,** 1460–1467.
Goodman, R., Hauptli, H., Crossway, A., and Knauf, V. C. (1987). *Science* **236,** 48–54.
Gordon, K., Lac, E., Vitale, J. A., Smith, A. E., Westphal, H., and Henninghausen (1987). *Bio/Technology* **5,** 1183–1187.
Gray, G. L., McKeown, K. A., Jones, A. J. S., Seeburg, P. H., and Heynekes, H. L. (1984). *Bio/Technology* **2,** 161–165.
Gray, J. W., Dean, P. N., Fuscoe, J. C., Peters, D. C., Trask, B. J., Engh, G. J., and Dilla, M. A. (1987). *Science* **238,** 323–330.
Green, P. (1986). *Annu. Rev. Biochem.* **55,** 569–597.
Gupta, P. K., and Durzan, D. J. (1987). *Bio/Technology* **5,** 147–151.
Haggin, J. (1988). *Chem. Eng. News* July 11, p. 30.
Haissig, B. E., Nelson, N. D., and Kidd, G. E. (1987). *Bio/Technology* **5,** 52–56.
Hallewell, R. A. (1987). *Bio/Technology* **5,** 363–366.
Harrison, B. D., Mayo, M. A., and Baulcombe, D. C. (1987). *Nature (London)* **328,** 799–802.
Herrington, D. A. (1987). *Nature (London)* **328,** 257.
Hewick, R. M., Hunkapillar, M. W., Hood, L. E., and Dryer, W. J. (1981). *J. Biol. Chem.* **256,** 7990–7995.
Hilder, V. A., Gatehouse, B. D., Sheurman, S., and Barker (1987). *Nature (London)* **330,** 160.

Hogle, J. M. (1988). *Nature (London)* **332**, 13.
Hommel, R., Stuwer, O., Stuber, W., Haferburg, D., and Kleber, H. P. (1987). *Appl. Microbiol. Biotechnol.* **26**, 199–205.
Husic, D. W., Husic, H. D., and Tolbert, N. E. (1987). *CRC Crit. Rev. Plant Sci.* **5**, 45–100.
Husson, R. N., and Young, R. A. (1987). *Proc. Natl. Acad. Sci. U.S.A.* **84**, 1679–1683.
Ide, J. A., Park, R. J., Leppik, R. A., and Dunn, N. W. (1987). *Appl. Microbiol. Biotechnol.* **26**, 234–236.
Jaenisch, R. (1988). *Science* **240**, 1468–1474.
Jasin, M., and Berg, P., *Genes and Dev.* **2**, 1353.
Jasin, M., Villiers, J. D., Weber, F., and Schaffner, W. (1985). *Cell* **43**, 695.
Johnson, S. A., Anziano, P. Q., Shark, K., Sanford, J. C., and Butow, R. A. (1988). *Science* **240**, 1538–1541.
Kaback, M. R. (1987). *Biochemistry* **26**, 2071–2076.
Kaiser, C. A., Preuss, D., Grisafi, P., and Botstein, D. (1987). *Science* **235**, 312–317.
Kawamori, M., Hashimoto, Y., Katsumata, R., Okachi, R., and Takayama, K. (1983). *Agric. Biol. Chem.* **47**, 2503–2509.
Kawashima, H., Nakahara, T., Oogaki, M., and Tabuchi, T. (1983). *J. Ferment. Technol.* **61**, 143–149.
Kitano, K., Nakao, M., Itoh, Y., and Fujisawa, Y. (1987). *Bio/Technology* **5**, 281–288.
Klass, D. L. (1984). *Science* **223**, 1021–1029.
Klausner, A. (1987). *Bio/Technology* **5**, 551–556.
Knecht, D. A., and Loomis, W. F. (1987). *Science* **236**, 1081–1086.
Knowles, J. R. (1987). *Science* **236**, 1252–1258.
Koncz, C., and Schell, J. (1986). *Mol. Gen. Genet.* **204**, 383–393.
Koshland, D. E. (1988). *Science* **240**, 1385.
Kretschmer, A., Bock, H., and Wagner, F. (1982). *Appl. Environ. Microbiol.* **44**, 864–870.
Kudrika, D. T. (1986). *Can. J. Genet. Cytol.* **28**, 808–813.
Kuliopulos, A., Shortle, D., and Talalay, P. (1987). *Proc. Natl. Acad. Sci. U.S.A.* **84**, 8893–8897.
Lambing, J. L., Foster, L. C., Kearney, D. J., Barnes, M. R., Tran, T. T., and Yang, B. (1988). *Biochem. Biophys. Res. Commun.* **151**, 693–700.
Lefebure, D. D., Miki, B. L., and Laliberte, J. (1988). *Bio/Technology* **5**, 1053–1056.
Lennarz, W. J. (1987). *Biochemistry* **17**, 7205–7210.
Lozanne, A. D., and Spuditch, J. A. (1987). *Science* **236**, 1086–1091.
McCullough, J. E. (1983). *Bio/Technology* **1**, 879–892.
MacLean, N., Pewnman, D., and Zhu, Z. (1987). *Bio/Technology* **5**, 257–261.
McPeek, F. D., Coyle-Morris, J. F., and Gemmill, R. M. (1986). *Anal. Biochem.* **156**, 274–285.
Malik, V. S. (1979). *Adv. Genet.* **20**, 37–114.
Malik, V. S. (1981). *Adv. Appl. Microbiol.* **27**, 1–84.
Malik, V. S. (1982a). *Adv. Appl. Microbiol.* **28**, 27–115.
Malik, V. S. (1982b). *Process Biochem.* March, April 1982, 38–42.
Malik, V. S. (1982c). *Z. Allg. Mikrobiol.* **22**, 261–266.
Malik, V. S. (1986). In "Biotechnology" (H. J. Rehm and G. Reed, eds.), Vol. 4, pp. 39–68. Springer-Verlag, New York.
Mansour, S. L., Thomas, K. R., and Capecchi, M. R. (1988). *Nature (London)* **336**, 348.
Mapletoft, R. J. (1984). *Bio/Technology* **2**, 149–160.
Marcotrigiano, M. (1986). *Herb Spice Med. Dig.* **4**,(4), 1–8.
Martin, J. B. (1987). *Science* **238**, 765–772.
Martin, J. F., and Gil, J. A. (1984). *Bio/Technology* **2**, 63–71.

Masui, Y., Mizuno, T., and Inouye, M. (1984). Bio/Technology **2**, 81–85.
Matthews, J. A., and Kricka, L. J. (1988). Anal. Biochem. **169**, 1–25.
Mazur, B. J., Chui, C., and Smith, J. K. (1987). Plant Physiol. **85**, 1110.
Mermod, N., Lehrbach, P. R., Don, R. H., and Timmis, K. N. (1986). In "The Biology of Pseudomonas" (J. R. Sokatch, ed.), Vol. 10. Academic Press, Orlando, Florida.
Metting, B., and Pyne, J. W. (1986). Enzyme Microb. Technol. **8**, 386–394.
Meyer, P., Heidmann, I., Forkmann, G., and Saedler, H. (1987). Nature (London) **330**, 677–678.
Michiels, F., Burmeister, M., and Lehrach, H. (1987). Science **236**, 1305.
Moores, J. C. (1987). Anal. Biochem. **163**, 1–9.
Morreale, A., Murphy, K. P., Cera, E. D., Fall, R., De Vries, A. L., and Gill, S. J. (1988). Nature (London) **333**, 782–783.
Mullens, I. A., and Dalton, M. (1987). Bio/Technology **5**, 490–493.
Mullis, K. B. (1987). U.S. Patent 4,683,202.
Murakami, T. (1983). J. Antibiot. **36**(10), 1365–1368.
Nevins, D., ed. (1987). "Tomato Biotechnology." Alan R. Liss, New York.
Niwa, O., Matsumoto, T., and Yanagida, M. (1986). Mol. Gen. Genet. **203**, 397–405.
Nuell, M. J., Fang, G., Kenigsberg, P., and Hager, L. P. (1988). J. Bacteriol. **170**, 1007–1016.
Obukowicz, M. G., Perlak, F. J., Bolten, S. L., Kretzmer, K., Mayer, E. J., and Watrud, L. S. (1987). Gene **51**, 91–96.
Palmiter, R. D., Norstedt, G., Gelinas, R. E., Hammer, R. E., and Brinster, R. L. (1983). Science **222**, 809–814.
Pantoliano, M. W., Ladner, R. C., Bryan, P. N., Rollence, M. L., Wood, J. F., and Poulos, T. L. (1987). Biochemistry **26**, 2077–2082.
Paoletti, E., Lipinskas, B. R., Samsonoff, C., Mercer, S., and Paniceli, D. (1984). Proc. Natl. Acad. Sci. U.S.A. **81**, 1983–1987.
Pelletier, C. (1986). Oxford Surv. Plant. Mol. Cell Biol. **3**, 96–114.
Pharmacia Analects (1988). **16**, 6–8.
Pichersky, E., Bernaztsky, R., Tanksley, S. D., Malik, V. S., and Cashmore, A. R. (1987a). In "Tomato Biotechnology" (D. Nevins, ed.), pp. 229–238. Alan R. Liss, New York.
Pichersky, E., Hoffman, N. E., Malik, V. S., Bernatzky, R., Tanksley, S. D., Szabo, L., and Cashmore, A. R. (1987b). Plant Mol. Biol. **9**, 109.
Prober, J. M., Trainor, G. L., Dam, R. J., Hobbs, F. W., Robertson, C. W., Zagursky, R. J., Cocuzza, A. J., Jensen, M. A., and Baumeister, K. (1987). Science **238**, 336–341.
Ramos, J. L., Wasserfallen, A., Rosi, K., and Timmis, K. N. (1987). Science **235**, 593–596.
Rangaswami, S. R. S. (1987). Int. Rice Res. Inst. Newsl. **12**(1), 15–20.
Reinikainen, P., Lahde, M., Karp, M., Suominen, I., Markkanen, P., and Mantsala, P. (1988). Biotechnol. Lett. **10**, 149–154.
Ristau, E., and Wagner, F. (1983). Biotechnol. Lett. **5**, 95–100.
Robinson, A., and Austen, B. (1987). Biochem. J. **246**, 294–261.
Rochaix, J. D. (1987). Microbiol. Rev. **46**, 13–40.
Rochaix, J. D., and Erickson, J. (1988). Microbiol. Rev. **13**, 56–59.
Rojo, F., Pieper, D. H., Engesser, K. H., Knackmuss, H. J., and Timmis, K. N. (1987). Science **238**, 1395–1398.
Rothstein, S. J., Dimaio, S. M., and Rice, D. (1988). In "Genetic Improvement of Agriculturally Important Crops: Progress and Issues" (R. T. Fraley, N. M. Frey, and J. Schell, eds.), pp. 41–46. Cold Spring Harbor Lab., Cold Spring Harbor, New York.
Ruy, Y. W., and Ruy, D. D. Y. (1987). Enzyme Microb. Technol. **9**, 339–344.
Sabour, M. (1986). Plant Breed. **97**, 324.

Saiki, R. K., Gelfand, D. H., Stoffel, S., Scharf, S. J., Higuchi, R., Horn, G. T., Mullis, K. B., and Erlich, H. A. (1988). *Science* **239**, 486–494.
Samson, S. M. (1987). *Bio/Technology* **5**, 1207–1214.
Sassenfeld, H. M., and Brewer, S. J. (1984). *Bio/Technology* **2**, 76–81.
Schell, J. S. (1987). *Science* **237**, 1176–1182.
Schnell, D. J., Pfannenstiel, M. A., and Nickerson, K. W. (1984). *Science* **223**, 1191–1193.
Schwartz, D. C. (1983). *Cold Spring Harbor Symp. Quant. Biol.* **47**, 189–195.
Schwartz, D. C., and Cantor, C. R. (1984). *Cell (Cambridge, Mass.)* **337**, 67–75.
Scopes, R. K. (1987). "Protein Purification," 2nd ed. Springer-Verlag, New York.
Scott, S. D. (1987). *Enzyme Microb. Technol.* **9**, 66–73.
Selden, R. F., Skoskiewitz, M. J., Howie, K. B., Russell, P. S., and Goodman, H. M. (1987). *Science* **236**, 714–718.
Sengupta-Gopalan, C., Reichert, N. A., Barker, R. F., Hall, T. C., and Kemp, J. D. (1985). *Proc. Natl. Acad. Sci. U.S.A.* **82**, 3320–3324.
Seno, M., Sasada, R., Iwani, M., Sudo, K., Kurokawa, T., Ito, K., and Igarashi, K. (1988). *Biochem. Biophys. Res. Commun.* **151**, 701–708.
Sgarrella, F., and Szulmajster, J. (1987). *Biochem. Biophys. Res. Commun.* **143**, 901–905.
Sheerman, S., and Bevan, M. W. (1988). *Plant Cell Rep.* **7**, 13–16.
Singh, H. S., and Bhardwaj, T. R. (1986). *Indian J. Chem., Sect. B* **25B**, 989–998.
Smith, A., and Berg, P. (1984). *Cold Spring Harbor Symp. Quant. Biol.* **49**, 171–175.
Smith, C. L., Econome, J. G., Schutt, A., Klco, S., and Cantor, C. R. (1987). *Science* **236**, 1448–1453.
Smithies, O. (1985). *Nature (London)* **317**, 230.
Song, K., Schwartz, F., Malda, N., Smithies, O., and Kucherlapati, R. (1987). *Proc. Natl. Acad. Sci. U.S.A.* **84**, 6820–6824.
Summers, A. O. (1985). *Trends Biotechnol.* **3**, 122–124.
Trevors, J. T. (1987). *Enzyme Microb. Technol.* **9**, 331–334.
Tumer, N. E., Connell, K. M., Nelson, R. S., Sanders, P. R., Beachy, R. N., Fraley, R. T., and Shaw, D. M. (1987). *EMBO J.* **6**, 1181–1188.
Umbeck, P. et al. (1987). *Bio/Technology* **5**, 263–267.
Umesono, K., Hiraoka, Y., Toda, T., and Yanagida, M. (1983). *Curr. Genet.* **17**, 123–128.
Urdea, M. S., Merryweather, J. P., Mullenbach, G. T., Coit, D., Heberlein, V., Valenzuela, P., and Barr, P. (1983). *Proc. Natl. Acad. Sci. U.S.A.* **80**, 7461–7465.
Van der Krol, A. R., Lenting, P. E., Veenstra, J., Meer, I. M., Koer, R. E., Gerats, G. M., Mol, J. N. M., and Stuitje, A. R. (1988). *Nature (London)* **333**, 866–869.
Van der Ploeg, L. H. T. et al. (1984). *Cell (Cambridge, Mass.)* **37**, 77–84.
Vehmaanpera, J., Nybergh, P. M. A., Tanner, R., Pohjonen, E., Bergelin, R., and Korkola, M. (1987). *Microb. Technol.* **9**, 547–549.
Watson, J. D., Tooze, J., and Kurtz, D. T. (1983). "Recombinant DNA: A Short Course." Freeman, New York.
White, R., and Caskey, C. T. (1988). *Science* **240**, 1483–1488.
Wilson, T. (1984). *Bio/Technology* **2**, 29–39.

INDEX

A

Acetate, secondary metabolism and, 17
Acetic acid
　biotechnology and, 268
　monosaccharides and, 147
Acetobacter, monosaccharides and, 156, 171
Acetobacter xylinum, monosaccharides and, 150, 151
Acetyl coenzyme A
　gibberellins and, 52, 79, 81, 90
　secondary metabolism and, 6
Acquired immune deficiency syndrome (AIDS)
　biotechnology and, 290, 295
　substances from fungi and, 183, 202
Actinomycetes
　gibberellins and, 49, 51
　substances from fungi and, 184, 202
ADP, monosaccharides and, 149
Aeration, gibberellins and
　economics, 117
　submerged fermentation, 62, 63, 79, 88
Aflatoxin, substances from fungi and, 211, 229
Agitation, gibberellins and
　economics, 117
　submerged fermentation, 62, 63, 79, 82
Aldehydes, monosaccharides and, 145
Alditols, monosaccharides and, 141
　applications, 160–166
　fermentation, 157
　mechanisms, 144
　structure, 150–153
Aldolase, biotechnology and, 294
Aldonate, monosaccharides and
　applications, 169–174
　structure, 153, 154
Aldonic acids, monosaccharides and, 141
　applications, 168, 172
　fermentation, 156, 157
　mechanisms, 146
　nature, 149
　structure, 152–155
Aldoses, monosaccharides and
　applications, 167–169
　fermentation, 156, 157
　structure, 152, 153
Aleurone tissues, gibberellins and, 39, 40, 44
Algae
　biotechnology and, 266, 267, 285
　gibberellins and, 41, 42
Amino acids
　biotechnology and
　　crop improvement, 286, 287, 289
　　human protein, 290
　　industrial organism, 266
　　inherited diseases, 297
　　technology, 271, 274
　　vaccines, 294
　gibberellins and, 93, 114
　secondary metabolism and, 17, 18, 22
　substances from fungi and, 235, 243
6-Aminocephalosporanic acid, biotechnology and, 280
7-Aminocephalosporanic acid, biotechnology and, 280
Amoxicillin, biotechnology and, 280
Amyloglucosidase, gibberellins and, 121
Anguidine, substances from fungi and, 229
Anti-sense RNA, biotechnology and, 272–274, 288

Antibiotics
  biotechnology and, 264, 280, 281
  gibberellins and, 88, 101
  secondary metabolism and, 5, 7–9, 12
  substances from fungi and
    ascomycetes, 242, 243
    basidiomycetes, 238, 240, 241
    fungi imperfecti, 230–232, 234, 236
    history, 184
    phycomycetes, 244
    production, 186–206
    screening, 185
Antibodies, biotechnology and, 290, 292, 294
Antigens, biotechnology and, 290, 292, 294, 295
Antitumor substances from fungi, see Fungi, substances from
Antiviral substances from fungi, see Fungi, substances from
Aphidicolin, substances from fungi and, 232
D-Arabinitol, monosaccharides and, 158, 159, 161, 162
Aranoflavins, substances from fungi and, 242
Aranotins, substances from fungi and, 230, 243
Ascochlorin, substances from fungi and, 233
Ascomycetes, substances from fungi and, 202, 206, 236–241
  antibiotics, 186, 202
  screening, 186
L-Ascorbic acid, monosaccharides and, 170
Asterriquinone, substances from fungi and, 235
ATP, secondary metabolism and, 22
Auxins, gibberellins and, 33

B

Bacteria
  biotechnology and
    crop improvement, 287–289
    human protein, 291
    technology, 272
    vaccines, 292
  gibberellins and, 49, 51, 101, 102
  monosaccharides and
    applications, 161, 162, 166, 169–172, 174
    fermentation, 155–157
    mechanisms, 142, 144–146
    nature, 147–150
    structure, 150–155
  secondary metabolism and, 2, 5, 11, 22
  substances from fungi and, 202, 230
Bakanae disease, gibberellins and, 32, 48, 49, 90
Barley aleurone bioassay, gibberellins and, 113
Basidiomycetes, substances from fungi and, 202, 206, 236–241
  antibiotics, 186, 202
  screening, 186
Bikaverins
  gibberellins and
    solid-state fermentation, 108
    submerged fermentation, 78, 79, 87, 88
  substances from fungi and, 235
Bioassays, gibberellins and, 111–113
Biosynthesis
  biotechnology and, 280, 284
  gibberellins and
    analytical methods, 112
    pathways, 51–57
    solid-state fermentation, 105, 107
    submerged fermentation, 62, 70, 74, 90, 91
  secondary metabolism and, 5, 6, 11, 16
Biotechnology, 263, 264, 300, 301
  antibiotics, 280, 281
  crop improvement
    plants, 282–289
    tissue culture, 281
  embryo transfer, 297, 300
  energy, 277, 280
  enzymes, 276–279
  gibberellins and, 110
  human protein, 289, 290
    growth hormone, 291
    insulin, 291
    interferon, 291
    interleukin-2, 292
    recombinant DNA, 293
    renin, 292
    tissue plasminogen activator, 291, 292
  industrial organism, 264–269
  inherited diseases, 296–299

INDEX

monoclonal antibodies, 294, 295
pollutants, 275
technology
  protein chemistry, 274, 275
  recombinant DNA, 269-271
  reverse genetics, 271-274
vaccines, 292, 294
Brassinosteroids, gibberellins and, 108-111, 122
Bredinin, substances from fungi and, 234, 243

C

Calcium
  gibberellins and, 64, 73
  monosaccharides and, 167, 170, 171, 173
Calgene, biotechnology and, 287
Calvacin, substances from fungi and, 241
Calvatic acid, substances from fungi and, 241
Carbohydrate
  biotechnology and, 266
  gibberellins and
    economics, 117
    growth phases, 76-79
    kinetic studies, 82, 83
    submerged fermentation, 64, 70, 87
  secondary metabolism and, 4, 6
  substances from fungi and, 242
Carbon
  biotechnology and, 285
  gibberellins and
    biosynthesis pathways, 55
    chemistry, 36
    economics, 119
    growth phases, 77, 79
    immobilized whole cells, 99, 100
    liquid surface fermentation, 58
    mode of action, 43
    process, 85-87
    regulation, 79-81
    solid-state fermentation, 100
    submerged fermentation, 64-67, 71
  monosaccharides and
    applications, 165, 167, 172
    fermentation, 157
    mechanisms, 144
    structure, 150, 152-154

Carcinoma, substances from fungi and
  ascomycetes, 242
  basidiomycetes, 237, 238, 241
  fungi imperfecti, 229, 232-235
Catalysts, biotechnology and, 276-279
Cephalosporins, biotechnology and, 280
Cereal, gibberellins and, 39, 40
Chitinase, biotechnology and, 288
Chlamydocin, substances from fungi and, 234
Chloramphenicol, biotechnology and, 280
Chromosomes, biotechnology and
  crop improvement, 281
  inherited diseases, 296-299
  monoclonal antibodies, 296
  technology, 269-271
Chymopapain, biotechnology and, 277
Citrate, secondary metabolism and, 11, 12
Citric acid
  biotechnology and, 294
  gibberellins and, 120, 121
  monosaccharides and, 172
  secondary metabolism and, 2, 4
Clones, biotechnology and
  crop improvement, 281, 286
  enzymes, 277
  human protein, 289-291
  technology, 270, 274
  vaccines, 292, 294
Collagenase, biotechnology and, 277
Continuous culture, gibberellins and, 87, 88
Continuous submerged fermentation, gibberellins and, 118
Cordycepin, substances from fungi and, 243
Coriolins, substances from fungi and, 239
Cucumber hypocotyl bioassay, gibberellins and, 112
Cyanein, substances from fungi and, 233
Cyclopin, substances from fungi and, 211
Cylindrochlorin, substances from fungi and, 236
Cysteine, biotechnology and, 274
Cystic fibrosis, biotechnology and, 297
Cytochalasins, substances from fungi and, 232
Cytochromes
  gibberellins and, 55, 57
  monosaccharides and, 145, 146, 148
  secondary metabolism and, 10

Cytoplasm
  biotechnology and, 268, 274
  monosaccharides and
    applications, 166, 173
    mechanisms, 142, 143
    nature, 148
    structure, 152
Cytosol, monosaccharides and
  fermentation, 160
  mechanisms, 142–145
  nature, 149
  structure, 152

D

Dactylarin, substances from fungi and, 234
Decomposition, gibberellins and, 38
Dehydrogenases of monosaccharides, see Monosaccharides
Deoxyalditols, monosaccharides and, 151, 164
Deoxyaldoses, monosaccharides and, 169
Deoxyaminoalditols, monosaccharides and, 165
Deoxyhalogenalditols, monosaccharides and, 165
Differentiation
  monosaccharides and, 149
  secondary metabolism and, 16
Dihydroxyacetone, monosaccharides and, 142, 161
Dimethyl-allyl pyrophosphate (DMAPP), gibberellins and, 52, 53
Dinitrophenol, monosaccharides and, 158, 160
Disulfide bonds, biotechnology and, 268
DNA
  biotechnology and, 264, 300, 301
    crop improvement, 282, 284–286
    embryo transfer, 300
    human protein, 289, 290, 292
    industrial organism, 265, 268, 269
    inherited diseases, 296, 297
    monoclonal antibodies, 295
    pollutants, 275
    technology, 269–271, 273
  gibberellins and, 40, 78, 87
  secondary metabolism and, 23
  substances from fungi and, 185, 229, 233, 235

DNA polymerase, biotechnology and, 281
Downstream processing, gibberellins and, 122
  economics, 117, 120
  solid-state fermentation, 108
  submerged fermentation, 95–98
Drosophila, biotechnology and, 269
Dry moldy bran, gibberellins and
  analytical methods, 112, 113
  economics, 119–121
  solid-state fermentation, 103–111
Duclauxin, substances from fungi and, 234
Dwarf maize bioassay, gibberellins and, 112
Dwarf pea bioassay, gibberellins and, 112

E

Ehrlich ascites cells, substances from fungi and
  ascomycetes, 243
  basidiomycetes, 239–241
  fungi imperfecti, 229, 231, 232
Electron transport system, monosaccharides and
  mechanisms, 144–147
  nature, 147–149
Embryo transfer, biotechnology and, 297, 300, 301
Ent-kaurene, gibberellins and, 44, 54, 55, 57
Enzymes, biotechnology and, 276–279
Epidermal growth factor, biotechnology and, 290
Epidithiopiperazinediones, substances from fungi and, 230
Erythritol, monosaccharides and, 161
Escherichia coli, biotechnology and
  crop improvement, 285, 286
  enzymes, 277
  human protein, 290–292
  industrial organism, 268, 269
  vaccines, 294
Esters
  gibberellins and, 39, 74
  substances from fungi and, 229, 241
Ethanol, biotechnology and, 277, 280
Eubacteria, substances from fungi and, 184

INDEX

Eukaryotes
  biotechnology and, 268, 271, 301
  secondary metabolism and, 10
Extraction, gibberellins and, 109, 110

F

Farnesyl pyrophosphate, gibberellins and, 53, 54
Fatty acids, substances from fungi and, 231
Fermentation, see also specific fermentation
  biotechnology and, 264, 301
    crop improvement, 287, 289
    energy, 277, 280
    human protein, 291
    industrial organism, 266–268
  gibberellins and, 31, 121
    analytical methods, 113, 114
    chemistry, 38
    economics, 116
    history, 33, 34
    mode of action, 40
    routes, 47, 48
  monosaccharides and, 141, 142, 155, 156
    applications, 163, 164, 166–168, 170–172, 174
    enzyme inhibitors, 158–160
    individual groups, 156, 157
    mechanisms, 144
    nature, 147
    structure, 152, 153
    technique, 157
  secondary metabolism and, 12
  substances from fungi and, 229, 235, 236, 241
Fibroblasts
  biotechnology and, 297
  substances from fungi and, 229, 230, 234, 235
Filaments
  gibberellins and, 87, 102
  secondary metabolism and, 9
Flammulin, substances from fungi and, 239, 240
Fluoride, monosaccharides and, 158–160
Fluorometric assay, gibberellins and, 114, 115

Foam control, gibberellins and, 89, 90
Fomecin, substances from fungi and, 240
Fructose
  biotechnology and, 264
  monosaccharides and, 165–168
D-Fructose, monosaccharides and
  applications, 163, 165–168
  fermentation, 160
  structure, 152
Fujic acid, gibberellins and, 108
Fumagillin, substances from fungi and, 231
Fungi
  biotechnology and, 288
  gibberellins and
    biosynthesis pathways, 52
    history, 32
    liquid surface fermentation, 58
    microorganisms, 49, 50
    solid-state fermentation, 102, 108
    submerged fermentation, 60, 90
  monosaccharides and, 141, 146, 155, 167
  secondary metabolism and, 4–6, 10, 14, 22
Fungi, substances from, 183, 184
  antitumor-antiviral substances, 202, 206–211
  ascomycetes, 226–228, 241–243
  basidiomycetes, 236–241
  fungi imperfecti, 206, 211–226, 229–236
  phycomycetes, 244
  history, 184, 185
  screening, 185, 186
Fungi imperfecti, 202, 206, 211, 226, 229–236, 243, 244
  antibiotics, 202, 206
  screening, 186
Funiculosin, substances from fungi and, 235
Furanose, monosaccharides and, 152
Fusarenon X, substances from fungi and, 229
Fusaric acid, gibberellins and, 92
Fusarin, gibberellins and, 58
Fusarium
  gibberellins and, 33, 40, 73, 92
  substances from fungi and, 229, 235
Fusarium moniliforme, gibberellins and
  biosynthesis pathways, 56

Fusarium moniliforme, gibberellins and (*cont.*)
  chemistry, 35
  concomitant products, 92, 93
  history, 32
  immobilized whole cells, 99
  mathematic model, 83
  microorganisms, 49, 50
  process operation, 87, 89, 90
  solid-state fermentation, 103
  submerged fermentation, 62–64, 70, 71, 74, 75
Fusidic acid, substances from fungi and, 202

G

GA_{12}-aldehyde, gibberellins and, 55–57
Galactose, substances from fungi and, 237
D-Galactose, monosaccharides and, 169
Gallic acid, secondary metabolism and, 2, 4, 15
Gas chromatography, gibberellins and, 116
Geranyl geranyl pyrophosphate, gibberellins and, 53, 54
Geranyl pyrophosphate, gibberellins and, 53
Gibberella fujikuroi, 122
  biosynthesis pathways, 51, 52, 56, 57
  chemistry, 35, 36, 39
  history, 32
  immobilized whole cells, 99
  liquid surface fermentation, 58
  microorganisms, 49, 50
  routes, 48
  solid-state fermentation, 103–111
  submerged fermentation, 60, 62–64, 70, 71, 74
    concomitant products, 92, 93
    growth phases, 76, 78
    kinetic studies, 81
    process operation, 86–90
  substances from fungi and, 235
Gibberellic acid, 30, 31, 121, 122
  analytical methods, 112, 114–116
  biosynthesis pathways, 52
  chemistry, 36–38
  economics, 117–121
  immobilized whole cells, 98–100
  microorganisms, 49
  routes, 45, 47–49
  solid-state fermentation, 103–111
  submerged fermentation, 60–64
    concomitant products, 91, 92
    downstream processing, 95
    kinetic studies, 82
    mathematic models, 83, 84
    mutation, 64, 66, 67, 70, 73–75
    process, 84–89
    regulation, 80, 81
Gibberellins, 30, 31, 121, 122
  analytical methods, 111
    bioassays, 111–113
    instrumentation, 113–116
  biosynthesis pathways, 51, 52
    ent-kaurene, 54, 55
    GA_{12}-aldehyde, 55–57
    inhibitors, 57
    isopentenyl pyrophosphate, 52, 53
    terpenes, 52–54
  chemistry, 34–36
    esters, 39
    gibberellic acid, 36–38
    glycosides, 38
  economics, 116
    comparative, 120, 121
    continuous submerged fermentation, 118
    immobilized whole cells, 119
    liquid surface fermentation, 117
    solid-state fermentation, 119, 120
    submerged batch fermentation, 117, 118
  history, 31–34
  immobilized whole cells, 98–100
  liquid surface fermentation, 57–59
  microorganisms, 49–51
  mode of action
    algae, 41, 42
    animals, 41–43
    mechanism, 43
    microorganisms, 40, 41
    plant tissues, 39, 40
    structure, 43, 44
  routes
    chemical synthesis, 45, 47
    fermentation, 48, 49
    plants, 47, 48
  solid-state fermentation
    concomitant products, 107, 108
    downstream processing, 108–111
    fed-batch process, 106

INDEX

growth pattern, 107
large-scale trial, 105, 106
nutrition, 104, 105
physical factors, 104
potential, 100-103
submerged fermentation
concomitant products, 90-94
downstream processing, 95-98
growth phases, 75-79
kinetic studies, 81-83
mathematic models, 83, 84
nutrition, 64-75
physical factors, 61-64
process operation, 84-90
regulation, 79-81
technique, 59, 60
uses, 45-47
Gliocladic acid, substances from fungi and, 235, 243
Gliotoxins, substances from fungi and, 230, 243
Glucan, substances from fungi and, 238, 239, 242
D-Glucan, substances from fungi and, 237
Glucitol, monosaccharides and, 163
D-Glucitol, monosaccharides and
applications, 162-164, 166, 167, 170
fermentation, 158
Glucoamylase, biotechnology and, 269
Gluconate, monosaccharides and, 142
applications, 167, 171, 173
fermentation, 157
structure, 152-155
D-Gluconate, monosaccharides and
applications, 168-170, 174
fermentation, 155, 156, 158, 160
mechanisms, 143, 145, 146
structure, 153-155
D-Gluconic acid, monosaccharides and, 147, 156, 167
Gluconobacter, monosaccharides and, 156, 172
Gluconobacter oxydans, monosaccharides and
applications, 161-165, 167-169, 172, 173
fermentation, 156-158, 160
mechanisms, 144, 145
nature, 147, 149, 150
structure, 150-153
Gluconobacter suboxydans, monosaccharides and, 145, 165, 172

Glucose
biotechnology and, 264
gibberellins and
chemistry, 38
concomitant products, 90
growth phases, 75-78
immobilized whole cells, 99
kinetic studies, 82, 83
liquid surface fermentation, 58
process, 84-86
regulation, 79, 80
solid-state fermentation, 106, 107
submerged fermentation, 60, 61, 63-66, 71
monosaccharides and, 168, 171-174
secondary metabolism and, 10
substances from fungi and
ascomycetes, 243
basidiomycetes, 237, 239, 241
fungi imperfecti, 234, 235
D-Glucose
monosaccharides and
applications, 167, 168, 170, 171, 174
fermentation, 155, 156, 158, 160
mechanisms, 143, 146, 147
structure, 155
substances from fungi and, 236-238
Glucose oxidase, monosaccharides and, 158, 167
Glucosides, gibberellins and, 38, 39, 44
D-Glucuronate, monosaccharides and, 155
Glycerol
gibberellins and
liquid surface fermentation, 58
submerged fermentation, 61, 65, 66, 79, 80
monosaccharides and, 161
substances from fungi and, 234
Glycoprotein
biotechnology and, 286
substances from fungi and, 231
Glycosides, gibberellins and, 38
Glycosylation, biotechnology and, 268, 286
Griseofulvin, substances from fungi and, 202
Growth factors, gibberellins and, 73
Growth hormone, biotechnology and, 264, 300, 301
human protein, 291
industrial organism, 268
inherited diseases, 296

H

Hadacidin, substances from fungi and, 235
Hepatitis, biotechnology and, 290, 294
Hepatitis B virus, biotechnology and, 268, 300
Heptitols, monosaccharides and, 164
Hexitols, monosaccharides and, 165
Homologous recombination,
  biotechnology and, 271, 272
  crop improvement, 285
  inherited diseases, 297
Hormones
  gibberellins and, 30, 43
  secondary metabolism and, 14
Human growth hormone, biotechnology and, 268
Huntington's chorea, biotechnology and, 295, 296
Hyaluronic acid, biotechnology and, 264
Hybridization, biotechnology and, 301
  antibiotics, 280
  crop improvement, 281–283
  technology, 270
Hybridomas, biotechnology and, 264, 294, 301
Hydrogen, monosaccharides and, 143, 146, 148, 174
Hydrolysis
  biotechnology and, 264, 280, 288
  gibberellins and, 38, 39, 106, 107
  monosaccharides and, 152, 169
  substances from fungi and, 232, 240, 242
Hydroxylation, gibberellins and, 43, 44, 56, 57
Hydroxymethylglutaryl CoA, gibberellins and, 52
14-α-Hydroxywithaferin A, substances from fungi and, 244

I

Idiophase, gibberellins and, 78, 79, 87
L-Idonate, monosaccharides and, 170
Illudins, substances from fungi and, 239
Immobilized whole cells, gibberellins and, 98–100, 119
Infrared spectra, gibberellins and, 108, 115, 116

Inoculum
  gibberellins and
    solid-state fermentation, 101–104
    submerged fermentation, 88, 89
  monosaccharides and, 161
Insulin, biotechnology and, 264, 268, 291, 300
Interferon
  biotechnology and, 264, 268, 290, 291
  substances from fungi and, 211
Interleukin-2, biotechnology and, 290, 292
Isopentenyl pyrophosphate, gibberellins and, 52, 53

K

Ketoaldonic acids, monosaccharides and, 141, 152, 156, 157, 169
Ketofermentations, monosaccharides and, 141, 142, 152
Ketogenic cells, monosaccharides and, 144
Ketogluconates, monosaccharides and, 142, 145, 154, 158
Ketoses, monosaccharides and, 141
  applications, 165–167
  structure, 150

L

β-Lactam antibiotics, biotechnology and, 280, 281
Lactones
  monosaccharides and, 141, 167
  substances from fungi and, 244
Lactose
  gibberellins and
    immobilized whole cells, 99
    production, 89
    submerged fermentation, 65, 66, 79, 80, 83
  substances from fungi and, 235
Lentinan, substances from fungi and, 237, 242
Lettuce hypocotyl bioassay, gibberellins and, 112, 113
Leucinostatin, substances from fungi and, 231

INDEX 315

Leukemia
  biotechnology and, 291
  substances from fungi and
    basidiomycetes, 236, 238, 239, 241
    fungi imperfecti, 229, 232, 233, 235
    phycomycetes, 244
Lipids
  biotechnology and, 264, 266
  gibberellins and, 110
  secondary metabolism and, 20
Liquid surface fermentation, gibberellins
    and, 57-59
  economics, 117, 120, 121
  routes, 48
  submerged fermentation, 60
Liver, substances from fungi and, 237
Luxury metabolite, secondary metabolism
    and, 15
Lymphokines, biotechnology and, 264
L-Lysine-oxidase, substances from fungi
    and, 233

M

Magnesium
  gibberellins and, 70, 76, 82, 83
  monosaccharides and, 160
Malate, secondary metabolism and, 11
Malic acid, secondary metabolism and, 2
Manganese, biotechnology and, 269
Mannitol
  monosaccharides and, 165
  secondary metabolism and, 4
D-Mannitol, monosaccharides and, 165, 166, 171
D-Mannonate, monosaccharides and, 172, 173
Mannose, substances from fungi and, 237
Meiosis, biotechnology and, 285
Melinacidins, substances from fungi
    and, 230
Merulinic acids, substances from fungi
    and, 241
Methane, biotechnology and, 267
Methanol, biotechnology and, 269
Methionine, biotechnology and
  crop improvement, 286
  human protein, 291
  industrial organism, 268, 269
  technology, 274

Mevalonic acid, gibberellins and, 52, 74, 81
Microbial dehydrogenases of
  monosaccharides, see
  Monosaccharides
Microbial fermentation, gibberellins and, 48, 49
Microbial secondary metabolism, 4-11, 23, 24
  history, 1-4
  naming, 11
    function, 13-16
    growth phase, 12, 13
    overproduction, 11, 12
  nomenclature
    alternatives, 16-18
    definitions, 22, 23
    semantics, 18-22
Microorganisms
  biotechnology and, 263, 264, 293
  gibberellins and
    chemistry, 34-36
    mode of action, 40, 41
    production, 49-51
    solid-state fermentation, 100-103
    submerged fermentation, 81, 93
  secondary metabolism and, 17
  substances from fungi and, 184, 231
Mineral salts, gibberellins and, 70, 72
Mitochondria
  biotechnology and, 271
  gibberellins and, 92
  monosaccharides and, 145, 148, 149
  secondary metabolism and, 10
Monoclonal antibodies, biotechnology
    and, 294, 295, 301
Monoglycerides, substances from fungi
    and, 231, 240
Monosaccharides, microbial
    dehydrogenases of, 141, 142
  applications
    alditols, 160-166
    aldonates, 169-174
    aldoses, 167-169
    ketoses, 166, 167
  fermentation, 155, 156
    enzyme inhibitors, 158-160
    individual groups, 156, 157
    technique, 157
  mechanisms
    cytosol, 144, 145

Monosaccharides, microbial
  dehydrogenases of, mechanisms (cont.)
    electron transport system, 145–147
    membrane, 143, 144
  nature, 147, 148
    electron transport system, 148, 149
    Gluconobacter oxydans, 149, 150
    respiration, 149
  structure, 155
    alditols, 150–152
    aldonic acids, 153–155
    aldoses, 152, 153
mRNA, biotechnology and, 272
Mucoprotein, substances from fungi
  and, 240
Multiple-contact countercurrent leaching,
  gibberellins and, 109
Mutation
  biotechnology and
    crop improvement, 281, 282, 285,
      286, 288
    enzymes, 276
    industrial organism, 267, 268
    inherited diseases, 297
    pollutants, 275
    technology, 269, 271, 273–275
  gibberellins and, 52, 56, 112
  secondary metabolism and, 9
  substances from fungi and, 242
Mycotoxins, substances from fungi and,
  211, 219
Mycoviruses, substances from fungi and,
  206, 211
Myelin, biotechnology and, 297
Myxomycetes, substances from fungi
  and, 186

N

NAD, monosaccharides and, 142, 144
NADP, monosaccharides and, 142, 144, 145
NADPHDH, monosaccharides and, 145
National Cancer Institute (NCI),
  substances from fungi and, 183, 184
Nebularine, substances from fungi
  and, 240
Neoplasia, substances from fungi
  and, 202
Neoschizophyllan, substances from fungi
  and, 237

Nitrogen
  biotechnology and, 266, 268, 285, 289
  gibberellins and
    biosynthesis pathways, 57
    economics, 117
    growth phases, 75–79
    kinetic studies, 82, 83
    production, 85–87
    solid-state fermentation, 100, 105
    submerged fermentation, 61, 63, 64,
      66–71
  monosaccharides and, 149, 167
  secondary metabolism and, 6, 10
Nuclear magnetic resonance, gibberellins
  and, 39, 111
Nucleic acid, substances from fungi
  and, 232
Nucleotides
  biotechnology and, 271
  secondary metabolism and, 20, 22

O

Octitols, monosaccharides and, 164
Oncogenes, biotechnology and, 270
Optimization, gibberellins and, 74, 75,
  103, 119
Oryzachlorin, substances from fungi
  and, 230
Ovaries
  biotechnology and, 291
  substances from fungi and, 233
Overproduction, secondary metabolism
  and, 11, 12
Oxalic acid, secondary metabolism
  and, 6
Oxidation
  monosaccharides and
    applications, 164, 165, 167–169, 172
    fermentation, 155, 156, 158
    mechanisms, 145, 146
    nature, 147–149
    structure, 155
  substances from fungi and, 237
Oxygen
  biotechnology and, 285
  gibberellins and
    biosynthesis pathways, 57
    concomitant products, 92
    economics, 117

growth phases, 76, 79
immobilized whole cells, 99
kinetic studies, 82
submerged fermentation, 62, 63
monosaccharides and
  applications, 163, 164, 168, 172
  fermentation, 155, 156, 158
  mechanisms, 145, 146
  nature, 147–149
  structure, 155

P

Pachyman, substances from fungi and, 239
Penicillin
  biotechnology and, 264, 280
  gibberellins and, 33
  secondary metabolism and, 4, 5, 19
  substances from fungi and, 184
Penicillium
  secondary metabolism and, 8, 9
  substances from fungi and, 230–235, 244
Pentitols, monosaccharides and, 165
Peptides, biotechnology and, 204, 280, 295, 301
Peroxisomes, monosaccharides and, 142
pH
  gibberellins and
    concomitant products, 92
    growth phases, 77, 78
    history, 32
    immobilized whole cells, 99
    kinetic studies, 82
    liquid surface fermentation, 58
    production, 84
    routes, 48
    solid-state fermentation, 104
    submerged fermentation, 60, 61, 70
  monosaccharides and, 161, 167, 171, 173
Phenotype
  biotechnology and, 271, 273, 274, 282
  secondary metabolism and, 16
Phenylalanine ammonia-lysine,
  substances from fungi and, 233
Phosphinothricin, biotechnology
  and, 288
Phosphorus, gibberellins and, 70, 76
Phosphorylation
  gibberellins and, 92
  monosaccharides and, 142, 144, 149, 160

Phycomycetes, substances from fungi
  and, 202, 206, 244
  antibiotics, 186, 202
  screening, 186
Plant growth hormones, gibberellins and, 30, 31
Plasma, secondary metabolism and, 3
Plasmids, biotechnology and
  crop improvement, 282, 284, 285
  inherited diseases, 297
  technology, 271
Plasminogen activator, biotechnology
  and, 268, 290–292
Plastids, gibberellins and, 52
Pleuromutilin, substances from fungi
  and, 240
Pollutants, biotechnology and, 275
Polyadenylation, biotechnology and, 286
Polyarginine, biotechnology and, 290
Polyketides, gibberellins and, 80, 108, 122
Polymerization, secondary metabolism
  and, 6
Polypeptides, substances from fungi
  and, 234
Polysaccharides
  biotechnology and, 266, 277
  secondary metabolism and, 10
  substances from fungi and, 236–239
Polysaccharopeptide, substances from
  fungi and, 236–239, 241, 242
Poricin, substances from fungi and, 240
Potassium, gibberellins and, 70, 76, 77, 84
Potato dextrose agar, gibberellins and, 88, 89
Precursors, gibberellins and, 73–75
Preprosubtilisin, biotechnology and, 274
Prokaryotes
  biotechnology and, 268, 271, 301
  secondary metabolism and, 10
Prorennin, biotechnology and, 277
Protein
  biotechnology and, 264, 301
    crop improvement, 285–289
    embryo transfer, 300
    enzymes, 277
    human, 289–293
    industrial organism, 266, 269
    inherited diseases, 297
    monoclonal antibodies, 295
    pollutants, 275

Protein, biotechnology and (cont.)
  technology, 271, 274, 275
  vaccines, 294
 gibberellins and
  analytical methods, 114
  mode of action, 43
  solid-state fermentation, 101, 110
  submerged fermentation, 78
 monosaccharides and, 146, 171
 substances from fungi and
  ascomycetes, 242
  basidiomycetes, 237–240
  fungi imperfecti, 229, 232
  history, 185
Proteolysis, biotechnology and, 290
PSX-1, substances from fungi and, 234
Pterin deaminases, substances from fungi and, 232, 233
Pterine deaminase, substances from fungi and, 244
Purification
 biotechnology and, 274
 gibberellins and, 110, 111, 114, 119
Pyocyanin, monosaccharides and, 149, 156
Pyranose, monosaccharides and, 152

R

Ramycin, substances from fungi and, 202
Recombinant DNA, biotechnology and, 300, 301
 crop improvement, 282, 284
 human protein, 293
 industrial organism, 268, 269
 pollutants, 275
 technology, 269–271
Renin, biotechnology and, 264, 292
Rennin, biotechnology and, 277
Replication, biotechnology and, 270, 281
Respiratory systems, monosaccharides and, 147–149
Reverse genetics, biotechnology and, 271–274
Rhizopin, substances from fungi and, 244
Rhizoxin, substances from fungi and, 244
Ribitol, monosaccharides and, 162
Riboflavin, secondary metabolism and, 11, 12

RNA
 biotechnology and
  crop improvement, 285, 288
  monoclonal antibodies, 295
  technology, 271–274
 gibberellins and, 40, 78, 79, 87
 substances from fungi and, 230, 232, 233, 235
Roridins, substances from fungi and, 229

S

Schizophyllan, substances from fungi and, 237
Scleroglucan, substances from fungi and, 242
Secondary metabolism, microbial, *see* Microbial secondary metabolism
Secondary metabolites
 gibberellins and
  downstream processing, 95
  production, 87
  submerged fermentation, 74, 78, 79, 81
 substances from fungi and, 185, 229, 230
Shikimate, secondary metabolism and, 11
Shunt metabolite, secondary metabolism and, 6, 11
Sickle-cell anemia, biotechnology and, 297
Sitosterol, biotechnology and, 264
Sodium, monosaccharides and, 167
Soil, substances from fungi and, 185, 202
Solid-state fermentation, gibberellins and, 122
 analytical methods, 114, 115
 concomitant products, 107, 108
 downstream processing, 108–111
 economics, 119–121
 fed-batch process, 106
 growth pattern, 107
 history, 34
 large-scale trial, 105, 106
 nutrition, 104, 105
 physical factors, 104
 potential, 100–103
Somatostatin, biotechnology and, 268
Sorbose, monosaccharides and, 142
 applications, 161–164, 167, 171

fermentation, 160
structure, 151, 153
L-Sorbose, monosaccharides and
  applications, 162–164, 167
  fermentation, 158, 160
Spectrofluorodensitometry, gibberellins
  and, 115, 122
Spectrophotometry, gibberellins and, 114
Spermidine, secondary metabolism
  and, 15
Spleen, substances from fungi and, 237
Sporidesmins, substances from fungi
  and, 243
Stemlon, substances from fungi and, 236
Sterigmatocystins, substances from fungi
  and, 229
Steroids, biotechnology and, 264,
  286, 287
Sterols, biotechnology and, 266, 286, 287
Streptomycin, substances from fungi
  and, 184
Submerged batch fermentation,
  gibberellins and, 117, 118
Submerged fermentation, gibberellins
  and, 31, 122
  analytical methods, 114, 115
  concomitant products, 90–94
  downstream processing, 95–98
  economics, 117, 118, 120
  growth phases, 75–79
  history, 34
  kinetic studies, 81–83
  liquid surface fermentation, 58
  mathematic models, 83, 84
  nutrition, 64–75
  physical factors, 61–64
  process operation, 84–90
  regulation, 79–81
  routes, 48, 49
  solid-state fermentation, 101, 105,
    107–109
  technique, 59, 60
Subtilisin, biotechnology and, 274
Sucrose, gibberellins and
  liquid surface fermentation, 58
  submerged fermentation, 60, 64–67,
    80, 85
Sulfonylurea, biotechnology and, 287
Supercritical fluid extraction,
  gibberellins and, 110

T

Tan-ginbozu dwarf microdrop bioassay,
  gibberellins and, 112
Tartaric acid, monosaccharides and, 161
Tay-Sach's disease, biotechnology and, 297
Temperature, gibberellins and
  solid-state fermentation, 102, 109, 110
  submerged fermentation, 61, 62, 81–83, 87
Tenuazonic acid, substances from fungi
  and, 233
Terpenes
  biotechnology and, 264
  gibberellins and, 52–54
Terpenoids, gibberellins and, 52–54
Thaumatin, biotechnology and, 287
Thin-layer chromatography, gibberellins
  and, 111, 115
Tissue plasminogen activator,
  biotechnology and, 268, 291, 292
Tissue specificity
  biotechnology and, 274
  secondary metabolism and, 8
Toxic pollutants, biotechnology and, 275
Trace elements, gibberellins and, 73
Trichothecenes, substances from fungi
  and, 229
Tricothecin, substances from fungi
  and, 229
Trihydroxytetralones, substances from
  fungi and, 231
Triornicin, substances from fungi and, 232
Triterpene acids, substances from fungi
  and, 241
Trophophase
  gibberellins and, 78, 81, 87
  substances from fungi and, 185
Tumor
  biotechnology and, 282, 295, 296
  substances from fungi and, 183
    ascomycetes, 242, 243
    basidiomycetes, 236–241
    fungi imperfecti, 229, 231–235
Tumor inhibitors, substances from fungi
  and, 240, 241

U

Ultraviolet spectra, gibberellins and,
  108, 115

V

Vaccine, biotechnology and, 264, 292, 294, 300
Vaccinia virus, substances from fungi and, 234, 235
Vermiculine, substances from fungi and, 232
Verrucarins, substances from fungi and, 229
Vesiculogen, substances from fungi and, 243
Virus
  biotechnology and
    crop improvement, 288, 289
    human protein, 292
    inherited diseases, 296
    technology, 271, 274
    vaccines, 294
  substances from fungi and
    basidiomycetes, 236, 240
    fungi imperfecti, 229, 230, 232, 233, 235
Vitamin C
  gibberellins and, 73
  monosaccharides and, 142, 163, 170–174

W

Wheat bran
  gibberellins and, 103–107, 109

substances from fungi and, 232, 244

X

D-Xylonate, monosaccharides and, 169
D-Xylulose, monosaccharides and, 158, 159, 161

Y

Yeast
  biotechnology and
    crop improvement, 285
    energy, 277
    human protein, 290–292
    industrial organism, 266, 269
    inherited diseases, 297
    technology, 269–271
    vaccines, 292
  gibberellins and, 49, 51, 102

Z

Zymosan, substances from fungi and, 242

CONTENTS OF PREVIOUS VOLUMES

Volume 15

Medical Applications of Microbial Enzymes
Irwin W. Sizer

Immobilized Enzymes
K. L. Smiley and G. W. Strandberg

Microbial Rennets
Joseph L. Sardinas

Volatile Aroma Components of Wines and Other Fermented Beverages
A. Dinsmoor Webb and Carlos J. Muller

Correlative Microbiological Assays
Ladislav J. Haňka

Insect Tissue Culture
W. F. Hink

Metabolites from Animal and Plant Cell Culture
Irving S. Johnson and George B. Boder

Structure–Activity Relationships in Coumermycins
John C. Godfrey and Kenneth E. Price

Chloramphenicol
Vedpal S. Malik

Microbial Utilization of Methanol
Charles L. Cooney and David W. Levine

Modeling of Growth Processes with Two Liquid Phases: A Review of Drop Phenomena, Mixing and Growth
P. S. Shah, L. T. Fan, I. C. Kao, and L. R. Erickson

Microbiology and Fermentations in the Prairie Regional Laboratory of the National Research Council of Canada 1946–1971
R. H. Haskins

AUTHOR INDEX—SUBJECT INDEX

Volume 16

Public Health Significance of Feeding Low Levels of Antibiotics to Animals
Thomas H. Jukes

Intestinal Microbial Flora of the Pig
R. Kenworthy

Antimycin A., a Piscicidal Antibiotic
Robert E. Lennon and Claude Vézina

Ochratoxins
Kenneth L. Applegate and John R. Chipley

Cultivation of Animal Cells in Chemically Defined Media, A Review
Kiyoshi Higuchi

Genetic and Phenetic Classification of Bacteria
R. R. Colwell

Mutation and the Production of Secondary Metabolites
Arnold L. Demain

Structure–Activity Relationships in the Actinomycins
Johannes Meienhofer and Eric Atherton

Development of Applied Microbiology at the University of Wisconsin
William B. Sarles

AUTHOR INDEX—SUBJECT INDEX

Volume 17

Education and Training in Applied Microbiology
  Wayne W. Umbreit

Antimetabolites from Microorganisms
  David L. Pruess and James P. Scannell

Lipid Composition as a Guide to the Classification of Bacteria
  Norman Shaw

Fungal Sterols and the Mode of Action of the Polyene Antibiotics
  J. M. T. Hamilton-Miller

Methods of Numerical Taxonomy for Various Genera of Yeasts
  I. Campbell

Microbiology and Biochemistry of Soy Sauce Fermentation
  F. M. Young and B. J. B. Wood

Contemporary Thoughts on Aspects of Applied Microbiology
  P. S. S. Dawson and K. L. Phillips

Some Thoughts on the Microbiological Aspects of Brewing and Other Industries Utilizing Yeast
  G. G. Stewart

Linear Alkylbenzene Sulfonate: Biodegradation and Aquatic Interactions
  William E. Gledhill

The Story of the American Type Culture Collection—Its History and Development (1899–1973)
  William A. Clark and Dorothy H. Geary

Microbial Penicillin Acylases
  E. J. Vandamme and J. P. Voets

SUBJECT INDEX

Volume 18

Microbial Foundation of Environmental Pollutants
  Martin Alexander

Microbial Transformation of Pesticides
  Jean-Marc Bollag

Taxonomic Criteria for Mycobacteria and Nocardiae
  S. G. Bradley and J. S. Bond

Effect of Structural Modifications on the Biological Properties of Aminoglycoside Antibiotics Containing 2-Deoxystreptamine
  Kenneth E. Price, John C. Godfrey, and Hiroshi Kawaguchi

Recent Developments of Antibiotics Research and Classification of Antibiotics According to Chemical Structure
  János Bérdy

SUBJECT INDEX

Volume 19

Culture Collections and Patent Depositions
  T. G. Pridham and C. W. Hesseltine

Production of the Same Antibiotics by Members of Different Genera of Microorganisms
  Hubert A. Lechevalier

Antibiotic-Producing Fungi: Current Status of Nomenclature
  C. W. Hesseltine and J. J. Ellis

Significance of Nucleic Acid Hybridization to Systematics of Actinomycetes
  S. G. Bradley

Current Status of Nomenclature of Antibiotic-Producing Bacteria
  Erwin F. Lessel

Microorganisms in Patent Disclosures
  Irving Marcus

Microbiological Control of Plant Pathogens
  Y. Henis and I. Chet

Microbiology of Municipal Solid Waste Composting
  Melvin S. Finstein and Merry L. Morris

Nitrification and Dentrification Processes Related to Waste Water Treatment
D. D. Focht and A. C. Chang

The Fermentation Pilot Plant and Its Aims
D. J. D. Hockenhull

The Microbial Production of Nucleic Acid-Related Compounds
Koichi Ogata

Synthesis of L-Tyrosine-Related Amino Acids by β-Tyrosinase
Hideaki Yamada and Hidehiko Kumagai

Effects of Toxicants on the Morphology and Fine Structure of Fungi
Donald V. Richmond

SUBJECT INDEX

Volume 20

The Current Status of Pertussis Vaccine: An Overview
Charles R. Manclark

Biologically Active Components and Properties of Bordetella pertussis
Stephen I. Morse

Role of the Genetics and Physiology of Bordetella pertussis in the Production of Vaccine and the Study of Host–Party Relationships in Pertussis
Charlotte Parker

Problems Associated with the Development and Clinical Testing of an Improved Pertussis Vaccine
George R. Anderson

Problems Associated with the Control Testing of Pertussis Vaccine
Jack Cameron

Vinegar: Its History and Development
Hubert A. Conner and Rudolph J. Allgeier

Microbial Rennets
M. Sternberg

Biosynthesis of Cephalosporins
Toshihiko Kanzaki and Yukio Fujisawa

Preparation of Pharmaceutical Compounds by Immobilized Enzymes and Cells
Bernard J. Abbott

Cytotoxic and Antitumor Antibiotics Produced by Microorganisms
J. Fuska and B. Proksa

SUBJECT INDEX

Volume 21

Production of Polyene Macrolide Antibiotics
Juan F. Martin and Lloyd E. McDaniel

Use of Antibiotics in Agriculture
Tomomasa Misato, Keido Ko, and Isamu Yamaguchi

Enzymes Involved in β-Lactam Antibiotic Biosynthesis
E. J. Vandamme

Information Control in Fermentation Development
D. J. D. Hockenhull

Single-Cell Protein Production by Photosynthetic Bacteria
R. H. Shipman, L. T. Fan, and I. C. Kao

Environmental Transformation of Alkylated and Inorganic Forms of Certain Metals
Jitendra Saxena and Philip H. Howard

Bacterial Neuraminidase and Altered Immunological Behavior of Treated Mammalian Cells
Prasanta K. Ray

Pharmacologically Active Compounds from Microbial Origin
Hewitt W. Matthews and Barbara Fritche Wade

SUBJECT INDEX

Volume 22

Transformation of Organic Compounds by Immobilized Microbial Cells
Ichiro Chibata and Tetsuya Tosa

Microbial Cleavage of Sterol Side Chains
Christoph K. A. Martin

Zearalenone and Some Derivatives: Production and Biological Activities
P. H. Hidy, R. S. Baldwin, R. L. Greasham, C. L. Keith, and J. R. McMullen

Mode of Action of Mycotoxins and Related Compounds
F. S. Chu

Some Aspects of the Microbial Production of Biotin
Yoshikazu Izumi and Koichi Ogata

Polyether Antibiotics: Versatile Carboxylic Acid Inophores Produced by Streptomytes
J. W. Westley

The Microbiology of Aquatic Oil Spills
R. Bartha and R. M. Atlas

Comparative Technical and Economic Aspects of Single-Cell Protein Processes
John H. Litchfield

SUBJECT INDEX

Volume 23

Biology of Bacillus popilliae
Lee A. Bulla, Jr., Ralph N. Costilow, and Eugene S. Sharpe

Production of Microbial Polysaccharides
M. E. Slodki and M. C. Cadmus

Effects of Cadmium on the Biota: Influence of Environmental Factors
H. Babich and C. Stotzky

Microbial Utilization of Straw (A Review)
Youn W. Han

The Slow-Growing Pigmented Water Bacteria: Problems and Sources
Lloyd G. Herman

The Biodegration of Polyethylene Glycols
Donald P. Cox

Introduction to Injury and Repair of Microbial Cells
F. F. Busta

Injury and Recovery of Yeasts and Mold
K. E. Stevenson and T. R. Graumlich

Injury and Repair of Gram-Negative Bacteria, with Special Consideration of the Involvement of the Cytoplasmic Membrane
L. R. Beuchat

Heat Injury of Bacterial Spores
Daniel M. Adams

The Involvement of Nucleic Acids in Bacterial Injury
M. D. Pierson, R. F. Gomez, and S. E. Martin

SUBJECT INDEX

Volume 24

Preservation of Microorganisms
Robert J. Heckly

Streptococcus mutans Dextransucrase: A Review
Thomas J. Montville, Charles L. Cooney and Anthony J. Sinskey

Microbiology of Activated Sludge Bulking
Wesley O. Pipes

Mixed Cultures in Industrial Fermentation Processes
David E. F. Harrison

Utilization of Methanol by Yeasts
Yoshiki Tani, Nobuo Kato, and Hideaki Yamada